MODERN COMPUTATIONAL METHODS

Series in Computational and Physical Processes in Mechanics and Thermal Sciences

W. J. Minkowycz and E. M. Sparrow, *Editors*

Anderson, Tannehill, and Pletcher, Computational Fluid Mechanics and Heat Transfer
Aziz and Na, Perturbation Methods in Heat Transfer
Baker, Finite Element Computational Fluid Mechanics
Beck, Cole, Haji-Shiekh, and Litkouhi, Heat Conduction Using Green's Functions
Carey, Computational Grids: Generation, Adaptation, and Solution Strategies
Chung, Editor, Numerical Modeling in Combustion
Comini, Del Giudice, and Nonino, Finite Element Analysis in Heat Transfer: Basic Formulation and Linear Problems
Heinrich and Pepper, Intermediate Finite Element Method: Fluid Flow and Heat Transfer Applications
Jaluria, Computer Methods for Engineering
Jaluria and Torrance, Computational Heat Transfer
Koenig, Modern Computational Methods
Patankar, Numerical Heat Transfer and Fluid Flow
Pepper and Heinrich, The Finite Element Method: Basic Concepts and Applications
Shih, Numerical Heat Transfer
Shyy, Udaykumar, Rao, and Smith, Computational Fluid Dynamics with Moving Boundaries
Tannehill, Pletcher, and Anderson, Computational Fluid Mechanics and Heat Transfer, Second Edition
Willmott, Modeling and Dynamics of Regenerative Heat Transfer

PROCEEDINGS

Chung, Editor, Finite Elements in Fluids: Volume 8

Haji-Sheikh, Editor, Integral Methods in Science and Engineering-90

Shih, Editor, Numerical Properties and Methodologies in Heat Transfer: Proceedings of the Second National Symposium

FORTHCOMING

Averill and Reddy, Computational Methods in Damage Mechanics

MODERN COMPUTATIONAL METHODS

Herbert A. Koenig
Department of Mechanical Engineering
University of Connecticut
Storrs, Connecticut

USA	Publishing Office:	Taylor & Francis 325 Chestnut Street Philadelphia, PA 19106 Tel: (215) 625-8900 Fax: (215) 625-2940
	Distribution Center:	Taylor & Francis 47 Runway Road Levittown, PA 19057 Tel: (215) 269-0400 Fax: (215) 269-0363
UK		Taylor & Francis 1 Gunpowder Square London EC4A 3DE Tel: +44 171 583 0490 Fax: +44 171 583 0581

MODERN COMPUTATIONAL METHODS

1 2 3 4 5 6 7 8 9 0

Printed by Hamilton Printing Co., Castleton, NY, 1998.

A CIP catalog record for this book is available from the British Library.
⊗ The paper in this publication meets the requirements of the ANSI Standard Z39.48-1984 (Permanence of Paper).

Library of Congress Cataloging-in-Publication Data

Koenig. Herbert A.
Modern computational mechanics / Herbert A. Koenig.
p. cm. — (Series in computational and physical processes in mechanics and thermal sciences)
Includes bibliographical references and index.
ISBN 1-56032-468-6 (case : alk. paper)
1. Engineering mathematics. I. Title. II. Series.
TA330.K64 1999
620'.001'51—dc21 98-21966
CIP

ISBN 1-56032-468-6

This book is dedicated to my immigrant parents, who did without, so that I may have.

This book is also dedicated to my wife, my friend, my partner, who is truly the wind beneath my wings.

CONTENTS

PREFACE

This textbook is written to introduce students to the powerful world of computational mechanics. It is intended to be used by engineers and scientists who desire to obtain numerical results to difficult problems that cannot be solved by simple computational formulas. This work is aimed at undergraduate senior students in a course on engineering, physics, computer science, or any of the other scientific fields. Novice first-year graduate students could also benefit greatly from the material in this book.

Because of the possible diversity of the readership, this book has been subdivided into the following sections. Part one is written for students who require some computational ability in their work, but do not necessarily desire to solve physical equations. The techniques that are included in this section are linear systems and matrices, nonlinear systems, approximation, interpolation and curve fitting, along with Numerical Integration. These tools will allow even novice students to perform computational analysis on experimental data, linear, and nonlinear simultaneous equation sets, to perform numerical integration of difficult formulas, and to approximate data into desired functional forms. These basic tools will reinforce the methods in calculus, physics, and other basic undergraduate courses in science and engineering.

Part two of this textbook introduces students to the basic computational tools of physics. Students are introduced to techniques that simulate physical processes when their equations have been derived. Once students have been exposed to the engineering fields of structural mechanics, heat transfer, and fluid flow, they will be able to use the methods of Part two to obtain solutions to problems that involve complex geometries and unusual material properties. In this section, finite-difference techniques are completely explained and applied to several problems whose FORTRAN codes are contained in each Appendix found on the accompanying computer disk. Students may therefore write their own codes or use the existing code to solve typical problems. An exploration of the various types of differential equations is made to distinguish the types of engineering processes that occur in nature. These methods are applied to initial-value problems, ordinary differential equations, and certain types of partial differential equations. In each chapter, several examples are solved, and their computer algorithms are included.

Part two continues with the delineation of energy methods that are to be applied to the solution of complex engineering problems that present difficulties with the application of the finite-difference techniques. These energy methods are converted into the development of the robust and powerful finite-element techniques. These methods are fully explored, and several example problems are solved for the consideration of the readers. Once again, the data sets that were used in the solution of the above-mentioned problems are presented in the appendixes.

Part two concludes with a delineation of dynamic methods that are required in problems in which loading histories are present. These techniques supplement the methods that have already been presented.

In all the chapters of Parts one and two there are numerous examples and homework problems for students to consider. In all the examples, the actual computer programs that were used are shown in the appendixes of those chapters. A serious student who has the desire to become computationally competent will find these sections extremely useful.

The textbook concludes with Part three, which contains advanced topic that may be extremely useful to researchers in the areas of wave propagation, deformable-body dynamics, compressible fluid flow, and biomechanics. The method of characteristics that is presented in Chap. 12 can be, and has been, used in compressible fluid flow in which shocks occur and in the study of the wave propagation in engineering materials. Chapter 13 contains finite-volume methods that are unique approaches to the solution of problems in fluid flow and structures. These techniques avoid the use of differential equations entirely since they apply the physical laws directly to the geometric entities. These techniques are mathematically less complex than either the finite-difference or the finite-element techniques. Finally, Chap. 14 deals with some of the biomechanical applications that are readily amenable to the computational techniques of this book. It is here, in this chapter, that readers are exposed to the synergistic nature of engineering solution techniques with those applications that occur in nontraditional areas of concern (e.g., orthodontics). Readers can use the ideas in this chapter to contribute greatly to areas that are not considered strictly of engineering endeavor.

This textbook is a compendium of the author's 30 years of computational experiences as a consultant to major industrial firms and as principal investigator in biomechanical research for 20 years. The author has found all the tools in this book to be extremely useful and robust and urges readers of this work to immerse themselves in the area of modern computational mechanics.

Herbert A. Koenig

ACKNOWLEDGMENT

The author wishes to thank the MARC Corporation for providing him with the software that was used to solve the Finite Element Problems in this textbook.

PART ONE

BASIC COMPUTATIONAL TOOLS

CHAPTER

ONE

LINEAR SYSTEMS AND MATRICES

NOMENCLATURE

a_{ij}	elements of matrix **A**
x_i	component of the solution vector
b_i, h_i	right-hand-side vectors
N	number of equations
U_{ij}	component of matrix **U**
V_{ij}	component of matrix **V**
T_{ij}	component of matrix **T**
n_i	component of vector **n**
t_i	component of vector **t**
$\|\mathbf{A}\|$	norm of matrix **A**
$\mathbf{C}(\mathbf{A})$	condition number of matrix **A**
$\mathbf{I}$	identity matrix
$\mathbf{E}$	error vector
$\mathbf{A}^{-1}$	inverse of matrix **A**
A_i	component of normalized tridiagonal matrix
B_i	component of normalized tridiagonal matrix
A'_i	component of manipulated tridiagonal matrix
B'_i	component of manipulated tridiagonal matrix
$\mathbf{L}$	lower-diagonal matrix in Cholesky decomposition
$\mathbf{U}$	upper-diagonal matrix in Cholesky decomposition
$\mathbf{y}$	intermediate column vector in Cholesky decomposition
ε	convergence tolerance
$\|a_{ij}\|$	absolute value of the component a_{ij}

1-1 INTRODUCTION

The emphasis in this textbook is on computational mechanics. In the following chapters, such diverse topics as curve fitting, numerical integration, finite-difference approximations, and other numerical tools that are required by the engineer for modeling physical processes are discussed.

The eventual outcome from each of the above-mentioned numerical strategies is the need to solve systems of equations that represent numerical approximations to the physical laws that are being modeled. These systems, whether linear or nonlinear, require the inversion (or at least solution) of a matrix set for calculating the physical variables of the problem.

Therefore in this chapter the approaches that can be used to solve a linear system of equations are discussed, while Chap. 2 attempts to approach nonlinear matrix systems.

Linear matrices result from those physical systems that obey laws that are path independent. This means that the solution to the problem does not depend on the path that was taken to arrive at the final result. The theory of elasticity, which is a subset of structural mechanics, is not only linear in nature, but also yields symmetric matrices in the solution process. Alternatively, the fields of fluid mechanics and fluid dynamics generally contain physical laws that are path dependent and yield nonlinear matrix approximations. Path-dependent systems require the careful documentation of each step in the solution process, since the final result depends on the manner in which that solution was attained. It is, of course, easier and more accurate to solve a system of linear equations than a nonlinear set.

In this chapter matrix formalities, condition number, error effects, and solution techniques for linear systems are explored. The nonlinear presentation is reserved for Chap. 2.

1-2 MATRIX FORMALITIES

The discussion of matrices begins with the introduction of some basic notation that will aid in the efficient delineation of sets of matrices. The notational protocol that is used throughout this textbook is either index notation or matrix notation. Both of these methods result in shorthand ways of describing systems of equations. An efficient way of describing the methods of computational mechanics in this textbook is then achieved. Without this notation, large systems of equations would have to be written individually, also known as the scalar method of expressing large systems.

1-2.1 Index Notation

We begin by considering a set of N linear simultaneous equations, i.e.,

$$\begin{aligned} a_{11}x_1 + a_{12}x_2 + a_{13}x_3 + \cdots + a_{1N} &= b_1, \\ a_{21}x_1 + a_{22}x_2 + a_{23}x_3 + \cdots + a_{2N} &= b_2, \\ &\cdots \\ a_{N1}x_1 + a_{N2}x_2 + a_{N3}x_3 + \cdots + a_{NN} &= b_N, \end{aligned} \tag{1-1}$$

where N is a known but unspecified number. We denote the coefficients $\mathbf{a}$ of this set of equations in the following way. The first subscript denotes the equation number while the second subscript indicates the column.

This set of equations may be shown in the traditional matrix form:

$$\begin{bmatrix} a_{11} & a_{12} & \cdots & a_{1N} \\ a_{21} & a_{22} & \cdots & a_{2N} \\ \cdots & \cdots & \cdots & \cdots \\ a_{N1} & a_{N2} & \cdots & a_{NN} \end{bmatrix} \begin{bmatrix} x_1 \\ x_2 \\ \cdots \\ x_N \end{bmatrix} = \begin{bmatrix} b_1 \\ b_2 \\ \cdots \\ b_N \end{bmatrix} \tag{1-2}$$

In order to reduce the amount of writing (especially for large systems), index notation is introduced. The following two rules apply.

Rule of the free index A free index is an index that occurs only *once* in each term. A free index may take on *any* of its admissible values.

Examples of the use of the free index are

$$1. \quad a_k + b_k = c_k \qquad (k = 1, 2, 3),$$

which is index notation for

$$\begin{aligned} a_1 + b_1 &= c_1, \\ a_2 + b_2 &= c_2, \\ a_3 + b_3 &= c_3. \end{aligned}$$

The symmetry of a matrix $\mathbf{U}$ may be expressed by

$$2. \quad U_{ij} = U_{ji} \qquad (i, j = 1, 2, 3),$$

which is the index notation equivalent of

$$\begin{aligned} U_{12} &= U_{21}, \\ U_{13} &= U_{31}, \\ U_{23} &= U_{32}, \end{aligned}$$

i.e., off-diagonal terms are equal.

The skew symmetry of a matrix $\mathbf{V}$ may be expressed as

$$3. \quad V_{ij} = -V_{ji} \qquad (i, j = 1, 2).$$

Here the scalar equations are

$$\begin{aligned} V_{11} &= 0, \\ V_{22} &= 0, \\ V_{12} &= -V_{21}. \end{aligned}$$

These show the conditions for a skew-symmetric matrix, i.e.,

1. The off-diagonal terms have opposite signs.
2. The diagonal terms are zero.

When considering the index notation of the last example, we might be tempted to conclude that only the off-diagonal terms are expressed. The vanishing diagonal of the matrix is also contained within the index notation formulation. Therefore, while index notation creates an efficient means of writing a set of conditions, it may also lead to erroneous conclusions if we are not careful when interpreting the equation.

With the rule of the free index, the matrix equations [which are described by Eqs. (1-1)] may now be written as

$$a_{i1}x_1 + a_{i2}x_2 + \cdots + a_{iN}x_N = b_i \qquad (i = 1,\ 2, \ldots,\ N). \tag{1-3}$$

We may now define the rank of a variable (or its matrix representation) by referring to the number of free indices for the symbol. Thus, in the above example, **a** is a second-rank variable while **x** and **b** are first-rank variables.

Rule of the summation index If an index occurs twice in a term, in which case it is called a summation index, then the term is to be summed with respect to this index over the entire range of its admissible values.

Some examples of the use of the summation (or dummy) index are

$$1. \quad a_i b_i = 0 \qquad (i = 1, 2, 3),$$

which is the index notation expression for the dot product of vectors **a** and **b**, i.e.,

$$a_1 b_1 + a_2 b_2 + a_3 b_3 = 0.$$

$$2. \quad A_{ii} = A_{kk} = 0 \qquad (i, k = 1, 2, 3),$$

which states that the sum of the diagonal terms in matrix **A** vanishes, i.e.,

$$A_{11} + A_{22} + A_{33} = 0.$$

With the rule of the summation index, the original set of simultaneous equations [Eqs. (1-1)] may be written as

$$A_{ij}x_j = b_i \qquad (i, j = 1, 2, \ldots, N). \tag{1-4}$$

1-2.2 Matrix Notation

While index notation simplifies the writing of large systems of equations, it is natural to inquire whether we have exhausted all possibilities for abbreviation. Indeed we have not, for it is possible to devise a matrix notation that represents the ultimate in simplicity. In this notation, Eq. (1-4) becomes

$$\mathbf{Ax} = \mathbf{b}, \tag{1-5}$$

where the boldfaced letters **A**, **x**, and **b** signify that Eq. (1-5) does not relate scalar quantities but rather sets of vectors or matrices.

In order to erase the subscripts from Eq. (1-4) and transform it to matrix notation, we must establish an orderly procedure so that we may be able to reestablish the subscripts and, ultimately, the scalar equations. In other words, we must be able to reverse the

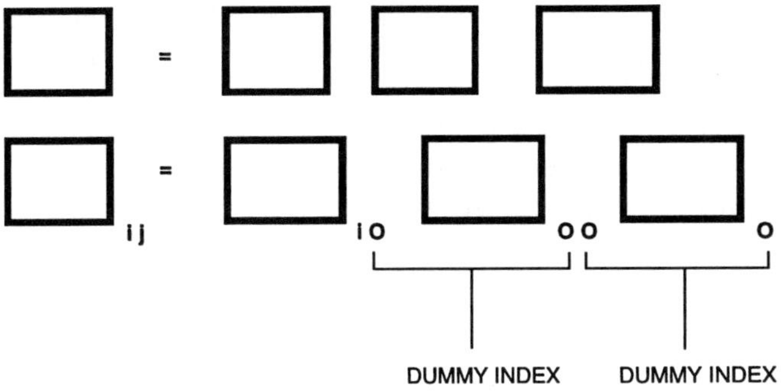

Figure 1-1 Symbolic formula for traslating index notation into direct notation.

process that allows us to express a system of scalar equation in matrix notation. We begin by representing the symbols **A**, **x**, and **b** by ordered arrays, viz.,

$$\|\mathbf{A}\| = \left\| \begin{matrix} A_{11} & A_{12} & A_{13} \\ A_{21} & A_{22} & A_{23} \\ A_{31} & A_{32} & A_{33} \end{matrix} \right\|, \qquad \|\mathbf{x}\| = \left\| \begin{matrix} x_1 \\ x_2 \\ x_3 \end{matrix} \right\|, \qquad \|\mathbf{b}\| = \left\| \begin{matrix} b_1 \\ b_2 \\ b_3 \end{matrix} \right\|.$$

The *first index labels the rows; the second index labels the columns.*

In the case of second-rank variables such as **A**, the methodology that translates index notation into direct (or matrix) notation (and vice versa) is best illustrated by the symbolic formula shown in Fig. 1-1.

Here the blocks stand for any second-rank variable, and the joined circles indicate pairs of summation indices that connect the adjacent members. Except for the choice of the summation index, which is arbitrary, the arrangement of the indices in Fig. 1-1 is completely determined. As soon as the free indices are selected on the left-hand side, the first free index goes into the first place on the right and the second (i.e., last) free index goes into the last place on the right. The interior indices are summation indices connecting the adjacent members. Figure 1-2 shows the method by which the indices may

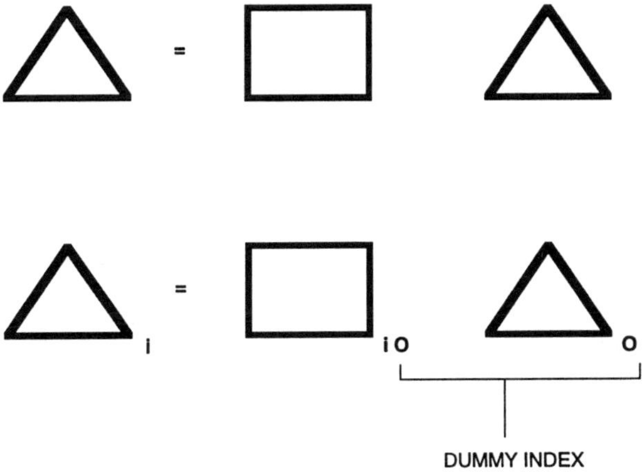

Figure 1-2 Method for erasing indices from an equation.

be erased from an equation that has a second-rank matrix and a vector. For example, the index notation formula $T_{ij}n_j = t_i$ may be transformed into the matrix notation equivalent $\mathbf{t} = \mathbf{Tn}$.

We may also reverse the above process when the index notation abbreviation is desired. Index notation also shows how to carry out the operation of matrix multiplication and is directly transferable into a computer code. As an example, the matrix equation

$$\mathbf{AB} = \mathbf{C}$$

becomes the index notation

$$A_{ij}B_{jk} = C_{ik}.$$

In this equation there are two free indices and one summation index.

1-3 CONDITION OF MATRIX SYSTEMS

If, in a system of equations, a small change in a coefficient makes only a slight change in the solution, then the system is said to be well conditioned. Otherwise the system is ill-conditioned. This aspect of matrix systems is always an important consideration in evaluating the probable success in determining the solution vector. Specifically, the consideration of the condition of a matrix becomes particularly important in Chap. 10, when the construction and the choice of finite elements are discussed.

The discussion of a matrix condition begins with the definition of a quantity called a condition number. This is a measure of how sensitive the matrix will be during computations when some error is introduced into the system. If we define the norm of a matrix as

$$|\mathbf{A}| = \max\left(\max_j \sum_{i=1}^{n} |a_{ij}|, \max_i \sum_{j=1}^{n} |a_{ij}|\right),$$

(i.e., the maximum value of the sum of the absolute values of the components of a particular row or the sum of the absolute values of the components of a particular column), then the condition number for matrix $\mathbf{A}$ is defined as

$$\mathbf{C}(\mathbf{A}) = |\mathbf{A}| \cdot |\mathbf{A}^{-1}|. \tag{1-6}$$

Since the condition number of the identity matrix $\mathbf{I}$ is

$$\mathbf{C}(\mathbf{I}) = 1,$$

it can be shown that

$$\mathbf{C}(\mathbf{A}) = |\mathbf{A}| \cdot |\mathbf{A}^{-1}| \geq |\mathbf{I}| = 1.$$

In other words, the identity matrix has the lowest condition number.

We can now estimate the effect that input errors in matrix $\mathbf{A}$ will have on solution vector $\mathbf{x}$ in the matrix equation $\mathbf{Ax} = \mathbf{b}$. If we allow matrix $\mathbf{A}$ to have an error $\mathbf{E}$, then the system becomes

$$(\mathbf{A} + \mathbf{E})x_e = \mathbf{b},$$

where x_e is the solution to the system that contains the error. If we now combine the error system with $\mathbf{Ax} = \mathbf{b}$, we obtain

$$\mathbf{A}(x_e - \mathbf{x}) = -\mathbf{E}x_e.$$

This then leads to an estimation of the error in the solution, viz.,

$$\frac{|x_e - \mathbf{x}|}{|x_e|} = \mathbf{C}(\mathbf{A})\frac{|\mathbf{E}|}{|\mathbf{A}|}. \tag{1-7}$$

As an example, consider the system

$$\begin{aligned} x_1 + \frac{1}{2}x_2 + \frac{1}{3}x_3 &= 1, \\ \frac{1}{2}x_1 + \frac{1}{3}x_2 + \frac{1}{4}x_3 &= 0, \\ \frac{1}{3}x_1 + \frac{1}{4}x_2 + \frac{1}{5}x_3 &= 0. \end{aligned} \tag{1-8}$$

The solution to Eqs. (1-8) is $\mathbf{x} = [9 \quad -36 \quad 30]$.

If we now approximate the system as

$$\begin{bmatrix} 1.0 & 0.50 & 0.33 \\ 0.50 & 0.33 & 0.25 \\ 0.33 & 0.25 & 0.20 \end{bmatrix} \begin{bmatrix} x_1 \\ x_2 \\ x_3 \end{bmatrix} = \begin{bmatrix} 1.0 \\ 0.0 \\ 0.0 \end{bmatrix}, \tag{1-9}$$

so that an error is introduced into some of the terms of the coefficient matrix, then the solution to Eq. (1.9) is $\mathbf{x} = [-0.6 \quad 17 \quad -21]$. The solutions to systems (1-8) and (1-9), which are supposed to be identical, are greatly different. An examination of the condition number of the original matrix **A** will reveal the reason for this gross error in the solution. The inverse of the matrix in Eqs. (1-8) can be shown to be

$$\begin{bmatrix} 9 & -36 & 30 \\ -36 & 192 & -180 \\ 30 & -180 & 180 \end{bmatrix} = \mathbf{A}^{-1},$$

whose norm is 408. The norm of matrix **A** is $11/6$. By Eq. (1-6), the condition number $\mathbf{C}(\mathbf{A}) = 748$. The error estimate in the solution **x** due to an error norm of 0.005 is

$$\frac{|x_e - \mathbf{x}|}{|x_e|} = 748\left(\frac{0.005}{\frac{11}{6}}\right) = 2.04,$$

which indicates a relative error in excess of 200%. This means that a 0.5% error in the coefficients of matrix set (1-8) could lead to an error of 204% in the calculated solution.

It has been demonstrated, with a simple matrix example, that a large condition number in the matrix of the governing system leads to erroneous results. Therefore it is extremely important in the work of numerical modeling to ensure that the constructed matrix has a low condition number. This idea is explored again in Chap. 10 with reference to finite-element analysis.

1-4 TECHNIQUES FOR MATRIX SOLUTION

We now turn our attention to the manner in which systems of equations are solved for unknown variables, say **x**. In all the systems that are discussed in this chapter, we assume that the coefficient matrix **A** is linear (i.e., the components of **A** are constant) and that all components of the right-hand-side vector **b** (*or* **h**) are known and constant. (In the following chapters of this book, the nonlinearity of the coefficient matrix **A** is addressed. Also, the cases in which some components of **x** are known and some components of **b** are known are discussed. This latter condition is often encountered in physical problems).

There are many wonderful matrix solvers available to students that address a variety of concerns such as sparcity and size. Therefore it is the task in this chapter to describe merely the underlying strategies that are used in a variety of matrix inversion algorithms. Several examples of these techniques are shown in the Appendixes of this chapter (found on the accompanying computer disk), but we will not pretend to construct a robust algorithm; rather, the emphasis will be on methodology rather than creativity.

Matrix solution techniques are divided in general into direct methods and iterative methods. Each of these techniques is now discussed. At the end of this chapter, an attempt is made to guide students toward the choice of one of these solution strategies.

1-4.1 Direct Methods

Direct methods for the solution of matrix equations involve the manipulation of the original matrix into a form that will lead to a convenient calculation of the solution **x**. Direct methods always provide the correct solution, although the number of calculations can be large.

1-4.1.1 Tridiagonal matrices. We will see in later chapters, namely those that are concerned with finite-difference techniques and finite-element techniques, that the nature of the matrix that describes the physically modeled systems depends, in part, on the manner in which the points at which the unknown quantities are desired are numbered. For example, if the points in the body are consecutive, then the resulting system matrix tends to be tridiagonal. This term is given to a matrix in which all terms are zero except for those along the main diagonal and on either side of it. For example, consider the tridiagonal system:

$$\begin{bmatrix} a_{11} & a_{12} & 0 & 0 & \cdots & 0 \\ a_{21} & a_{22} & a_{23} & 0 & \cdots & 0 \\ 0 & a_{32} & a_{33} & a_{34} & \cdots & 0 \\ \cdots & \cdots & \cdots & \cdots & \cdots & \cdots \\ 0 & 0 & 0 & \cdots & a_{n,n-1} & a_{n,n} \end{bmatrix} \begin{bmatrix} x_1 \\ x_2 \\ x_3 \\ \cdots \\ x_n \end{bmatrix} = \begin{bmatrix} b_1 \\ b_2 \\ b_3 \\ \cdots \\ b_n \end{bmatrix}. \tag{1-10}$$

We begin the solution process by normalizing each equation with respect to its diagonal coefficient. This yields the system

$$\begin{bmatrix} 1.0 & -A_1 & 0 & 0 & \cdots & 0 \\ -B_2 & 1.0 & -A_2 & 0 & \cdots & 0 \\ 0 & -B_3 & 1.0 & -A_3 & \cdots & 0 \\ \cdots & \cdots & \cdots & \cdots & \cdots & \cdots \\ & & & & -B_n & 1.0 \end{bmatrix} \begin{bmatrix} x_1 \\ x_2 \\ x_3 \\ \cdots \\ x_n \end{bmatrix} = \begin{bmatrix} C_1 \\ C_2 \\ C_3 \\ \cdots \\ C_n \end{bmatrix}. \tag{1-11}$$

The ith equation has the form

$$-B_i x_{i-1} + x_i - A_i x_{i+1} = C_i. \tag{1-12}$$

We can solve for the value of $x_i = C_i + A_i x_{i+1} + B_i x_{i-1}$.

We would now like to develop a formula by which the ith value of $\mathbf{x}$ is expressible in terms of the $(i+1)$th value of $\mathbf{x}$, namely

$$x_i = A'_i x_{i+1} + B'_i. \tag{1-13}$$

From the first of the equation system (1-11), we have $x_1 = C_1 + A_1 x_2$, from which we conclude that $A'_1 = A_1$ and $B'_1 = C_1$. The second equation in system (1-11) yields $x_2 = C_2 + A_2 x_3 + B_2 x_1$. If we substitute the value of x_1, then

$$x_2 = \frac{A_2}{1 - B_2 A'_1} x_3 + \frac{C_2 + B_2 B'_1}{1 - B_2 A'_1},$$

from which we can infer the general recursive formula, viz.,

$$A'_i = \frac{A_i}{1 - B_i A'_{i-1}}, \qquad B'_i = \frac{B_i B'_{i-1} + C_i}{1 - B_i A'_{i-1}}. \tag{1-14}$$

From Eqs. (1-14), all the N values of A'_i and the N values of B'_i can be computed from the original coefficient matrix before the solution process begins. We begin the solution of x_n by noting that $A'_n = 0$, making $x_n = B'_n$. Then the remaining values of x_i may be calculated from Eq. (1-13).

As an example of the solution process for a tridiagonal system of equations, let us consider the set

$$\begin{bmatrix} 1 & 1 & 0 & 0 \\ 2 & -1 & 5 & 0 \\ 0 & 3 & -4 & 2 \\ 0 & 0 & 2 & 6 \end{bmatrix} \begin{bmatrix} x_1 \\ x_2 \\ x_3 \\ x_4 \end{bmatrix} = \begin{bmatrix} 5 \\ -9 \\ 19 \\ 2 \end{bmatrix}.$$

We begin the solution process by normalizing the equations with respect to the diagonal coefficients, i.e.,

$$\begin{bmatrix} 1 & 1 & 0 & -0 \\ -2 & 1 & -5 & 0 \\ 0 & -3/4 & 1 & -1/2 \\ 0 & 0 & 1/3 & -1 \end{bmatrix} \begin{bmatrix} x_1 \\ x_2 \\ x_3 \\ x_4 \end{bmatrix} = \begin{bmatrix} 5 \\ 9 \\ -19/4 \\ 1/3 \end{bmatrix}.$$

We can then calculate the coefficients A' and B':

	A'	B'
1	-1.0	5.0
2	1.6667	6.3333
3	-2.0	0.0
4	0.0	1.0

so that the solution for the values of **x** may now proceed by backsubstitution as

$$\begin{aligned} x_4 &= B_4' = 1.0, \\ x_3 &= A_3' x_4 + B_3' = -2.0, \\ x_2 &= A_2' x_3 + B_2' = 3.0, \\ x_1 &= A_1' x_2 + B_1' = 2.0. \end{aligned}$$

Because of the special configuration of a tridiagonal matrix, this special procedure may be used with a great savings of time in the solution process. Appendix 1.1 has an example of a code that may be used to solve a tridiagonal system (**tri.f**).

1-4.1.2 Triangular form. A square matrix is triangular if all the elements above or below the main diagonal vanish. When this happens, the solution process is simplified. The demonstration of how any square matrix can be transformed into a triangular matrix is now given.

Consider the system $\mathbf{Ax} = \mathbf{h}$, i.e.,

$$\begin{aligned} a_{11}x_1 + a_{12}x_2 + \cdots + a_{1N}x_N &= h_1, \\ a_{21}x_1 + a_{22}x_2 + \cdots + a_{2N}x_N &= h_2, \\ &\cdots \\ a_{N1}x_1 + a_{N2}x_2 + \cdots + a_{NN}x_N &= h_N. \end{aligned} \tag{1-15}$$

We wish to reduce this system to the form

$$\begin{bmatrix} b_{11} & b_{12} & \cdots & \cdots & b_{1N} \\ 0 & b_{22} & b_{23} & \cdots & b_{2N} \\ \cdots & \cdots & \cdots & \cdots & \cdots \\ 0 & 0 & \cdots & b_{N-1,N-1} & b_{N-1,N} \\ 0 & 0 & \cdots & \cdots & b_{NN} \end{bmatrix} \begin{bmatrix} x_1 \\ x_2 \\ \cdots \\ x_{N-1} \\ x_N \end{bmatrix} = \begin{bmatrix} c_1 \\ c_2 \\ \cdots \\ c_{N-1} \\ c_N \end{bmatrix}. \tag{1-16}$$

In order to accomplish this reduction, we eliminate one variable at a time from Eqs. (1-15) to transform them into the equation system (1-16). We indicate the coefficients of the transformed set $b_{ij}x_j = c_i$ as

$$\begin{aligned} b_{1k}^{(1)} &= a_{1k} \qquad (k = 1, 2, \ldots, n), \\ c_1^{(1)} &= h_1, \end{aligned}$$

where the superscript indicates the number of the current pass through the system of equations. The number (1) indicates the first pass through the equations, etc.

In order to eliminate the coefficients in the first column of Eqs. (1-15), we attempt to make $b_{i1}^{(1)} = 0$ $(i = 2, \ldots, n)$. First we solve for x_1 in the first equation of Eqs. (1-15), viz.,

$$x_1 = \frac{h_1}{a_{11}} - \frac{a_{1i}}{a_{11}} x_i \qquad (i = 2, \ldots, n).$$

From the second equation of Eqs. (1-15), we get, on substitution of x_1,

$$a_{21}\left\{\frac{h_1}{a_{11}} - \frac{a_{1i}}{a_{11}} x_i\right\} + a_{2j}x_j = h_2 \qquad (i, j, 2, \ldots, n)$$

or

$$\left\{a_{2j} - \frac{a_{21}}{a_{11}} a_{1j}\right\} x_j = h_2 - \frac{a_{21}}{a_{11}} h_1 \qquad (j = 2, \ldots, n).$$

We then conclude that

$$b_{2j}^{(1)} = a_{2j} - \frac{a_{21}}{a_{11}} a_{1j},$$

$$c_2^{(1)} = h_2 - \frac{a_{21}}{a_{11}} h_1.$$

In general, to eliminate the x_1 term from all equations in set (1-15) except the first equation, we must have

$$b_{kj}^{(1)} = a_{kj} - \frac{a_{k1}}{a_{11}} a_{1j},$$

$$c_k^{(1)} = h_k - \frac{a_{k1}}{a_{11}} h_1.$$

After pass one, the matrix system has become

$$\begin{bmatrix} b_{11}^{(1)} & b_{12}^{(1)} & b_{13}^{(1)} & \cdots & b_{1N}^{(1)} \\ 0 & b_{22}^{(1)} & b_{23}^{(1)} & \cdots & b_{2N}^{(1)} \\ 0 & b_{32}^{(1)} & b_{33}^{(1)} & \cdots & b_{3N}^{(1)} \\ 0 & b_{42}^{(1)} & \cdots & \cdots & b_{4N}^{(1)} \\ 0 & \cdots & \cdots & \cdots & \cdots \\ 0 & \cdots & \cdots & \cdots & \cdots \\ 0 & b_{N2}^{(1)} & \cdots & b_{N,N-1}^{(1)} & b_{NN}^{(1)} \end{bmatrix} \begin{bmatrix} x_1 \\ x_2 \\ \cdots \\ \cdots \\ \cdots \\ x_{N-1} \\ x_N \end{bmatrix} = \begin{bmatrix} c_1^{(1)} \\ c_2^{(1)} \\ \cdots \\ \cdots \\ \cdots \\ c_{N-1}^{(1)} \\ c_N^{(1)} \end{bmatrix}. \qquad (1\text{-}17)$$

For the next pass, we ignore the first equation of set (1-15) and use the second equation to eliminate the x_2 term from the third equation through the N equation. If we now solve the second equation for x_2,

$$b_{22}^{(1)} x_2 + b_{2j}^{(1)} x_j = c_2^{(1)},$$

$$x_2 = \frac{c_2^{(1)}}{b_{22}^{(1)}} - \frac{b_{2j}^{(1)} x_j}{b_{22}^{(1)}} \qquad (j = 3, \ldots, n).$$

From the third equation,

$$b_{32}^{(1)}\left\{\frac{c_2^{(1)}}{b_{22}^{(1)}}-\frac{b_{2j}^{(1)}x_j}{b_{22}^{(1)}}\right\}+b_{3j}^{(1)}x_j=c_3^{(1)}\qquad (j=3,\ldots,n),$$

$$x_j\left\{b_{3j}^{(1)}-\frac{b_{32}^{(1)}}{b_{22}^{(1)}}b_{2j}^{(1)}\right\}=c_3^{(1)}-\frac{b_{32}^{(1)}}{b_{22}^{(1)}}c_2^{(1)}\qquad (j=3,\ldots,n).$$

We conclude that

$$b_{3j}^{(2)}=b_{3j}^{(1)}-\frac{b_{32}^{(1)}}{b_{22}^{(1)}}b_{2j}^{(1)},$$

$$c_3^{(2)}=c_3^{(1)}-\frac{b_{32}^{(1)}}{b_{22}^{(1)}}c_2^{(1)}.$$

In general,

$$b_{kj}^{(2)}=b_{kj}^{(1)}-\frac{b_{k2}^{(1)}}{b_{22}^{(1)}}b_{2j}^{(1)}\qquad (j=3,\ldots,n),$$

$$c_k^{(2)}=c_k^{(1)}-\frac{b_{32}^{(1)}}{b_{22}^{(1)}}c_2^{(1)}.$$

After the second pass, we now arrive at

$$\begin{bmatrix} a_{11} & a_{12} & a_{13} & \cdots & a_{1N}\\ 0 & b_{22}^{(1)} & b_{23}^{(1)} & \cdots & b_{2N}^{(1)}\\ 0 & 0 & b_{33}^{(2)} & \cdots & b_{3N}^{(2)}\\ \cdots & \cdots & \cdots & \cdots & \cdots\\ 0 & 0 & b_{N3}^{(2)} & \cdots & b_{NN}^{(2)} \end{bmatrix}\begin{bmatrix} x_1\\ x_2\\ x_3\\ \cdots\\ x_N\end{bmatrix}=\begin{bmatrix} h_1\\ c_2^{(1)}\\ c_3^{(2)}\\ \cdots\\ c_N^{(2)}\end{bmatrix}$$

We continue this elimination process for the N equations of the original set (1-15). Once the triangular factorization of the original system is completed, the solution now proceeds as

$$x_N=\frac{c_N}{b_{NN}},$$

$$x_{N-1}=\frac{c_{N-1}-b_{N-1,N}x_N}{b_{N-1,N-1}},$$

$$x_k=\frac{c_k-x_Nb_{kN}-x_{N-1}b_{k,N-1}-\cdots-x_{k+1}b_{k,k+1}}{b_{kk}}\qquad (k=1,2,\ldots,N-2).$$

Please note that in the last equation, index notation has been suspended.

In order to illustrate the process of the triangularization of a matrix, let us look at the system

$$\begin{bmatrix} 1 & 3 & 4 & 1\\ 2 & 5 & 3 & -2\\ 1 & 4 & 6 & -1\\ 2 & 2 & -1 & 2\end{bmatrix}\begin{bmatrix} x_1\\ x_2\\ x_3\\ x_4\end{bmatrix}=\begin{bmatrix} 7\\ 10\\ 8\\ 3\end{bmatrix}.$$

We now apply the above methodology in three successive passes.
After pass 1,

$$\begin{bmatrix} 1 & 3 & 4 & 1 \\ 0 & -1 & -5 & -4 \\ 0 & 1 & 2 & -2 \\ 0 & -4 & -9 & 0 \end{bmatrix} \begin{bmatrix} x_1 \\ x_2 \\ x_3 \\ x_4 \end{bmatrix} = \begin{bmatrix} 7 \\ -4 \\ 1 \\ -11 \end{bmatrix}.$$

After pass 2,

$$\begin{bmatrix} 1 & 3 & 4 & 1 \\ 0 & -1 & -5 & -4 \\ 0 & 0 & -3 & -6 \\ 0 & 0 & 11 & 16 \end{bmatrix} \begin{bmatrix} x_1 \\ x_2 \\ x_3 \\ x_4 \end{bmatrix} = \begin{bmatrix} 7 \\ -4 \\ -3 \\ 5 \end{bmatrix}.$$

After pass 3,

$$\begin{bmatrix} 1 & 3 & 4 & 1 \\ 0 & -1 & -5 & -4 \\ 0 & 0 & -3 & -6 \\ 0 & 0 & 0 & -6 \end{bmatrix} \begin{bmatrix} x_1 \\ x_2 \\ x_3 \\ x_4 \end{bmatrix} = \begin{bmatrix} 7 \\ -4 \\ -3 \\ -6 \end{bmatrix}.$$

The solution of this system can now be achieved through backsubstitution to yield

$$\mathbf{x} = [-5 \quad 5 \quad -1 \quad 1].$$

An example of the matrix triangularization approach is shown in a code that is delineated in Appendix 1.2 (**factor.f**).

1-4.1.3 Matrix factorization. The discussion of direct matrix solution methods ends with the introduction of an important technique that is used in many practical applications.

We can accomplish the process of converting a matrix system into a factored form by using matrix operations. Starting with $\mathbf{Ax} = \mathbf{h}$, we can transform $\mathbf{A}$ into the product of two factors, one, a matrix with all elements above the diagonal zero, and the other, a matrix with all elements below the diagonal zero.

We begin by choosing $\mathbf{LU} = \mathbf{A}$, where $\mathbf{L}$ is the lower-diagonal matrix and $\mathbf{U}$ is the upper diagonal matrix, i.e.,

$$\begin{bmatrix} L_{11} & 0 & 0 \\ L_{21} & L_{22} & 0 \\ L_{31} & L_{32} & L_{33} \end{bmatrix} \begin{bmatrix} 1 & U_{12} & U_{13} \\ 0 & 1 & U_{23} \\ 0 & 0 & 1 \end{bmatrix} = \begin{bmatrix} a_{11} & a_{12} & a_{13} \\ a_{21} & a_{22} & a_{23} \\ a_{31} & a_{32} & a_{33} \end{bmatrix}. \tag{1-18}$$

If we use the rows of the first factor and the first column of the second, we have

$$L_{11} = a_{11},$$
$$L_{21} = a_{21},$$
$$L_{31} = a_{31}.$$

Then,

$$L_{11}U_{12} = a_{12} \quad \text{or} \quad U_{12} = \frac{a_{12}}{a_{11}},$$

$$L_{11}U_{13} = a_{13} \quad \text{or} \quad U_{13} = \frac{a_{13}}{a_{11}},$$

$$L_{21}U_{12} + L_{22} = a_{22},$$

from which we calculate

$$L_{22} = a_{22} - \frac{a_{21}a_{12}}{a_{11}}$$

$$L_{32} = a_{32} - \frac{a_{31}a_{12}}{a_{11}}.$$

Finally we note that

$$L_{21}U_{13} + L_{22}U_{23} = a_{23},$$

from which we calculate

$$U_{23} = \frac{a_{23} - L_{21}U_{13}}{L_{22}},$$

$$L_{33} = a_{33} - L_{32}U_{23} - L_{31}U_{13}.$$

We have thus decomposed matrix $\mathbf{A}$ into an upper-diagonal matrix factor $\mathbf{U}$ and a lower-diagonal matrix factor $\mathbf{L}$. The solution process is now straightforward. Consider the matrix equation

$$\mathbf{Ax} = (\mathbf{LU})\mathbf{x} = \mathbf{h}.$$

If we call $\mathbf{y} = \mathbf{Ux}$, we can solve the auxiliary equation $\mathbf{Ly} = \mathbf{h}$ for the intermediate variable $\mathbf{y}$ and then solve

$$\mathbf{Ux} = \mathbf{y}$$

for the desired unknown vector $\mathbf{x}$, i.e.,

$$\begin{bmatrix} L_{11} & 0 & 0 \\ L_{21} & L_{22} & 0 \\ L_{31} & L_{32} & L_{33} \end{bmatrix} \begin{bmatrix} y_1 \\ y_2 \\ y_3 \end{bmatrix} = \begin{bmatrix} h_1 \\ h_2 \\ h_3 \end{bmatrix}$$

yields the following:

$$y_1 = \frac{h_1}{L_{11}},$$

$$L_{21}y_1 + L_{22}y_2 = h_2,$$

$$y_2 = \frac{1}{L_{22}}\left[h_2 - L_{21}\left(\frac{h_1}{L_{11}}\right)\right],$$

$$L_{31}y_1 + L_{32}y_2 + L_{33}y_3 = h_3,$$

so that

$$y_3 = \frac{1}{L_{33}}\{h_3 - L_{31}y_1 - L_{32}y_2\}$$

If we now use the **y** vector, we can solve for **x**, viz.,

$$\begin{bmatrix} 1 & U_{12} & U_{13} \\ 0 & 1 & U_{23} \\ 0 & 0 & 1 \end{bmatrix} \begin{bmatrix} x_1 \\ x_2 \\ x_3 \end{bmatrix} = \begin{bmatrix} y_1 \\ y_2 \\ y3 \end{bmatrix},$$

$$x_3 = y_3,$$

$$x_2 + U_{23}x_3 = y_2, \qquad x_2 = y_2 - U_{23}x_3,$$

$$x_1 + U_{12}x_2 + U_{13}x_3 = y_1,$$

$$x_1 = y_1 - U_{12}x_2 - U_{13}x_3.$$

For large systems, then, decompose **A** as

$$\mathbf{A} = \mathbf{LU}.$$

Then, using $\mathbf{Ly} = \mathbf{h}$, solve for **y** by using forward substitution. Finally using $\mathbf{Ux} = \mathbf{y}$, solve for **x** by backsubstitution.

The above solution technique is also known as the Cholesky decomposition. To illustrate the matrix factorization process, we consider the linear system

$$\begin{bmatrix} 1 & 3 & 4 \\ 2 & 5 & 3 \\ 1 & 4 & 6 \end{bmatrix} \begin{bmatrix} x_1 \\ x_2 \\ x_3 \end{bmatrix} = \begin{bmatrix} 6 \\ 12 \\ 9 \end{bmatrix}.$$

With reference to the basic forms of Eq. (1-18) and the subsequent calculations for L_{11}, L_{21}, and L_{31}, etc., we can immediately form parts of the **L** matrix and parts of the **U** matrix, viz.,

$$\mathbf{L} = \begin{bmatrix} 1 & 0 & 0 \\ 2 & L_{22} & 0 \\ 1 & L_{32} & L_{33} \end{bmatrix}, \qquad \mathbf{U} = \begin{bmatrix} 1 & 3 & 4 \\ 0 & 1 & U_{23} \\ 0 & 0 & 1 \end{bmatrix}.$$

Then we can calculate the missing elements of **L** and **U** by

$$L_{22} = 5 - \frac{2(3)}{1} = -1,$$

$$L_{32} = 4 - \frac{(1)(3)}{1} = 1,$$

$$U_{23} = \frac{3 - 2(4)}{-1} = 5,$$

$$L_{33} = 6 - (1)(5) - (1)(4) = -3.$$

We have now decomposed the original matrix **A** so that

$$\mathbf{A} = \begin{bmatrix} 1 & 0 & 0 \\ 2 & -1 & 0 \\ 1 & 1 & -3 \end{bmatrix} \begin{bmatrix} 1 & 3 & 4 \\ 0 & 1 & 5 \\ 0 & 0 & 1 \end{bmatrix}.$$

Solving **LY** = **H**, we find

$$\begin{bmatrix} 1 & 0 & 0 \\ 2 & -1 & 0 \\ 1 & 1 & -3 \end{bmatrix} \begin{bmatrix} y_1 \\ y_2 \\ y_3 \end{bmatrix} = \begin{bmatrix} 6 \\ 12 \\ 9 \end{bmatrix},$$

$$\mathbf{y} = [6 \quad 0 \quad -1].$$

Then,

$$\begin{bmatrix} 1 & 3 & 4 \\ 0 & 1 & 5 \\ 0 & 1 & 1 \end{bmatrix} \begin{bmatrix} x_1 \\ x_2 \\ x_3 \end{bmatrix} = \begin{bmatrix} 6 \\ 0 \\ -1 \end{bmatrix}$$

$$\mathbf{x} = [-5 \quad 5 \quad -1].$$

Appendix 1.3 has a formulation that uses a modified version of the **LU** decomposition technique (**mainlud.f**, **decomp.f**, and **back.f**).

The techniques that have been described to this point always yield a solution to a given set of equations. However, while direct techniques produce a solution, the number of operations that are required for completing the solution process is directly related to N. For example, for the **LU** decomposition technique, the number of multiplications and divisions will be N^2 while there will be $N^2 - N$ subtractions. When a large system of equations is encountered (as in *any* physical problem), the large number of operations will result in long solution times and possibly computer round-off errors.

For large systems, which are mostly sparse, direct methods are time consuming and costly so that iterative methods should be considered.

1-4.2 Iterative Methods

Iterative methods are not guaranteed to converge to the correct solution. These techniques, while more efficient for large systems, may not converge, or worse, may converge to a physically unrealistic solution. Therefore we must be careful when using iterative techniques. At the end of this section some ideas that may guide readers toward these iterative techniques are discussed.

1-4.2.1 Jacobi iteration. Consider the system

$$\begin{aligned} a_{11}x_1 + a_{12}x_2 + \cdots + a_{1N}x_N &= b_1, \\ a_{21}x_1 + a_{22}x_2 + \cdots + a_{2N}x_N &= b_2, \\ &\cdots \\ a_{N1}x_1 + a_{N2}x_2 + \cdots + a_{NN}x_N &= b_N. \end{aligned}$$

We begin the solution process by making an initial guess for the unknown vector **x**. It is somewhat obvious that a good estimate of the initial iterate will greatly enhance the chances of a successful convergence. In physical applications, therefore, the use of an approximate solution is generally used for the initial iterate. We use this guess (the zeroth iteration) to calculate the next set of the vector **x**. This process continues until the iteration converges to a given accuracy level. To demonstrate this approach, we calculate the $k+1$st iteration of **x** from the kth iteration by

$$x_1^{k+1} = \{b_1 - a_{12}x_2^k - a_{13}x_3^k - \cdots - a_{1N}x_N^k\}/a_{11},$$

$$x_2^{k+1} = \{b_2 - a_{21}x_1^k - a_{23}x_3^k - \cdots - a_{2N}x_N^k\}/a_{22},$$

where the superscript $(k+1)$ refers to the $k+1$st iterate. In this process, the values of the entire kth iteration are used to calculate the $k+1$st guess.

The iteration process is started with an initial guess $(x_1^0, x_2^0, \ldots, x_N^0)$ and continues until convergence occurs in the form

$$\frac{|x_i^{k+1} - x_i^k|}{x_i^k} < \varepsilon \qquad (i = 1, \ldots, N)$$

where ε is a small convergence criterion (say 0.001). The question arises: Will this process converge?

1-4.2.2 Gauss–Seidel iteration. In this technique, which is a modification of the Jacobi method, the updated values of the variables are used immediately after they are calculated, i.e.,

$$\begin{aligned}
x_1^{k+1} &= \{b_1 - a_{12}x_2^k - a_{13}x_3^k - \cdots - a_{1N}x_N^k\}/a_{11},\\
x_2^{k+1} &= \{b_2 - a_{21}x_1^{k+1} - a_{23}x_3^k - \cdots - a_{2N}x_N^k\}/a_{22},\\
x_3^{k+1} &= \{b_3 - a_{31}x_1^{k+1} - a_{32}x_2^{k+1} - a_{34}x_4^k - \cdots - a_{3N}x_N^k\}/a_{33},\\
&\cdots\\
x_N^{k+1} &= \{b_N - a_{N1}x_1^{k+1} - a_{N2}x_2^{k+1} - \cdots - a_{N,N-1}x_{N-1}^{k+1}\}/a_{NN}.
\end{aligned}$$

The convergence from the Gauss–Seidel technique is generally more rapid than that from the Jacobi method. However, the question still arises: Will this process converge?

In the discussion of direct matrix solution techniques, the question of convergence was not addressed since those methods always yield a solution to the matrix system. For iterative techniques, however, there is considerable concern about the accuracy and the convergence of the solution process. We now discuss some ideas about convergence.

A matrix **A** of dimensions $N \times N$ is said to be strictly diagonally dominant if

$$|a_{k,k}| > |a_{k,1}| + |a_{k,2}| + \cdots + |a_{k,k-1}| + |a_{k,k+1}| + \cdots + |a_{k,N}| \qquad (k = 1, 2, \ldots, N).$$

Basically, if the magnitude of the diagonal term in the matrix is greater than the sum of the absolute values of the other members in that equation, the matrix is said to be strictly diagonally dominant.

A matrix that possesses this characteristic will converge and yield a solution to the system of equations within a required accuracy. Therefore it is important that a given set of equations be reordered, if possible, in order to achieve diagonal dominance. Additionally, if the iteration process is started with a trial solution that is near to the final solution, convergence generally will occur.

In general, iterative techniques have been shown to be more efficient than direct techniques for large matrix systems such as those that are developed in a physical application. Appendix 1.4 presents an algorithm that solves a symmetric system of equations by the Gauss–Seidel method (**snbt.f**). We always generate symmetric matrices when modeling structural applications.

1-4.2.3 Sparse matrix systems. Before we leave the topic of matrix solution techniques, we consider sparse matrix systems. In most applications in which a physical system is modeled, a sparse matrix is developed. By this, we mean that in an $N \times N$ matrix system, the number of coefficients that do not vanish is very small (say, less than 10%) compared with N^2. The greater the sparcity of the system, the more effective the sparse matrix solution techniques. Sparse techniques are always iterative and make use of the fact that most of the coefficient matrix **A** has vanished.

Basically, the ideas that are used in sparse matrix algorithms are

1. A vector that contains the row location of the nonzero terms in the **A** matrix is constructed.
2. A vector that contains the column locations of the nonzero terms in the **A** matrix is constructed.
3. A vector that contains the nonzero terms in the matrix **A** is constructed.

To illustrate the above ideas, let us consider the sparse matrix

$$\begin{bmatrix} 1 & 2 & 0 & 0 & 0 & 0 & 0 & 0 & 0 \\ 2 & 3 & 4 & 5 & 0 & 0 & 0 & 0 & 0 \\ 0 & 4 & 6 & 0 & 0 & 0 & 0 & 0 & 0 \\ 0 & 0 & 0 & 7 & 0 & 0 & 0 & 0 & 0 \\ 0 & 0 & 0 & 0 & 8 & 9 & 0 & 0 & 0 \\ 0 & 0 & 0 & 0 & 0 & 10 & 0 & 0 & 0 \\ 0 & 0 & 0 & 0 & 0 & 0 & 11 & 12 & 0 \\ 1 & 0 & 0 & 0 & 0 & 0 & 0 & 13 & 0 \\ 0 & 0 & 0 & 0 & 0 & 0 & 0 & 0 & 14 \end{bmatrix}.$$

The vector that contains the row locations of the nonzero matrix entries is of length 17, viz.,

$$\begin{bmatrix} 1 & 1 & 2 & 2 & 2 & 2 & 3 & 3 & 4 & 5 & 5 & 6 & 7 & 7 & 8 & 8 & 9 \end{bmatrix}.$$

The vector that contains the column locations of the nonzero matrix entries is of length 17, viz.,

$$\begin{bmatrix} 1 & 2 & 1 & 2 & 3 & 4 & 2 & 3 & 4 & 5 & 6 & 6 & 7 & 8 & 1 & 8 & 9 \end{bmatrix}.$$

Finally, the vector that contains the nonzero matrix entries is of length 17, viz.,

$$\begin{bmatrix} 1 & 2 & 2 & 3 & 4 & 5 & 4 & 6 & 7 & 8 & 9 & 10 & 11 & 12 & 1 & 13 & 14 \end{bmatrix}.$$

In this way, a smaller storage requirement for the coefficients is needed than N^2. Once the three vectors described above are constructed, an iterative solution process is used to achieve the system unknowns. Appendix 1.5 shows a series of computer algorithms that order the original matrix system and then use sparse matrix techniques to obtain the solutions.

1-5 MIXED BOUNDARY CONDITIONS

In many applications, there is a mixed set of knowns and unknowns, namely, some of the **x**'s and some of the **b**'s are known. This condition often occurs in structural problems in which the displacements (the **x** values) are given along some portion of the surface area of the body and the tractions (the **b** values) are known along the remainder of the surface of the structure.

In normal matrix operations, the known values of the **x** vector would be subtracted from the set of simultaneous equations and the size of the matrix system would be reduced. This is, however, impractical and extremely time consuming in a large system of matrix equations. Instead, the procedure that is described below is used for more efficient computer usage. Consider the system

$$\begin{bmatrix} a_{11} & a_{12} & a_{13} & a_{14} \\ a_{21} & a_{22} & a_{23} & a_{24} \\ a_{31} & a_{32} & a_{33} & a_{34} \\ a_{41} & a_{42} & a_{43} & a_{44} \end{bmatrix} \begin{bmatrix} x_1 \\ x_2 \\ x_3 \\ x_4 \end{bmatrix} = \begin{bmatrix} b_1 \\ b_2 \\ b_3 \\ b_4 \end{bmatrix}. \tag{1-19}$$

In this system, b_1^0, b_3^0, b_4^0, and x_2^0 are specified. The unknown variables are therefore x_1, b_2, x_3, and x_4.

To reduce this system without reducing the matrix size, we adopt the following procedure. First, we store the coefficients in row two for later use in calculating the actual value of b_2. Next, we place a 1 in the row and the column where the specified value of **x** is given (e.g., row 2 and column 2 in this example) and zero the remaining elements in that row and in that column. Additionally, we subtract the product of the given variable (x_2) and its column coefficient from the right-hand side of each equation in the matrix set, viz.,

$$\begin{bmatrix} a_{11} & 0 & a_{13} & a_{14} \\ 0 & 1 & 0 & 0 \\ a_{31} & 0 & a_{33} & a_{34} \\ a_{41} & 0 & a_{43} & a_{44} \end{bmatrix} \begin{bmatrix} x_1 \\ x_2 \\ x_3 \\ x_4 \end{bmatrix} = \begin{bmatrix} b_1^0 & -a_{12} & x_2^0 \\ & x_2^0 & \\ b_3^0 & -a_{32} & x_2^0 \\ b_4^0 & -a_{42} & x_2^0 \end{bmatrix}. \tag{1-20}$$

Instead of expending the cost of a matrix reduction, the above technique has transformed Eqs. (1-19) into Eqs. (1-20) at a nominal cost. For large matrices, this technique is far superior to the usual matrix size reduction. After the remaining unknown **x** values are calculated, we can calculate the values of the unknown **b** quantities (b_2, in this case) by using the stored values of the coefficients of row two and by performing a direct calculation, i.e.,

$$b_2 = a_{21}x_1 + a_{22}x_2 + a_{23}x_3 + a_{24}x_4.$$

1-6 CONCLUSION

In this chapter several matrix solution techniques have been presented that may be applied when students are modeling physical problems by the techniques that are described in this textbook. As stated above, the delineation of these techniques in this chapter is by no means intended to explore all the numerous matrix inversion schemes that are currently being used in the field of computational mechanics. Instead, the main purpose of this chapter has been to expose students to the ideas that encompass the creation of matrix inverters and to explore the decisions that need to be made when one desires to choose a particular solution strategy.

The question of whether to use a direct solution technique or an iterative solution technique has been addressed. The effect that errors in the coefficient matrix will have on the ultimate solution was addressed. The condition number of a matrix and its effect on the solution accuracy have been discussed. The consideration of the condition of a matrix is presented again in Chapter 10, in which a discussion of the construction of a finite element is presented.

Most of the matrices that are constructed with the computational methods in this textbook will be symmetric matrices. However, when numerical modeling is applied to the field of fluid mechanics, one generally encounters nonsymmetric matrices. The area of computational fluid dynamics is not a major concern of this textbook. A brief presentation of this field is included in Chaps. 11 and 12. Therefore, for the sake of completeness, Appendix 1.6 presents an algorithm that will solve a system of unsymmetric equations (**unsyms.f**).

Finally, a full understanding of the techniques that are presented in this chapter is of enormous value in the successful application of the rest of the computational strategies that are explained in the remainder of this book.

FURTHER READING

Arden, B. W. and Astill, K. N. (1970). *Numerical Algorithms—Origins and Applications*, Addison-Wesley, Reading, Massachusetts.

Fuller, L. E. (1962). *Basic Matrix Theory*, Prentice-Hall, Englewood Cliffs, New Jersey.

Kardestuncer, H. (1974). *Elementary Matrix Analysis of Structures*, McGraw-Hill, New York.

Mathews, J. H. (1987). *Numerical Methods*, Prentice-Hall, Englewood Cliffs, New Jersey.

Press, W. H., Teukolsky, S. A., Vetterling, W. T., and Flannery, B. P. (1986). *Numerical Recipes in FORTRAN*, Cambridge University Press, New York.

Salvadori, M. G. and Baron, M.L. (1952). Numerical Methods in Engineering, Prentice-Hall, Englewood Cliffs, New Jersey.

Zlatev, Z., Barker, V.A., and Thomsen, P.G. (1978). *SSLEST*, Numerisk Institut, Danmarks Tekniske Hojskole, 2800 Lyngby, Denmark.

PROBLEMS

Solve the following problems by writing a computer code for the solution technique that is requested.

In most cases, there is a computer code in the Appendixes of this chapter against which students can verify their work.

Problem Set 1: Tridiagonal Matrices

The problems in this set are to be solved by providing an algorithm that is applicable to tridiagonal matrices. Appendix 1.1 contains a code, called **tri.f**, that can be used by students as a guide to their assignments.

1-1.1 Write a program to solve the tridiagonal system

$$\begin{bmatrix} 4 & -1 & 0 & 0 & 0 & 0 \\ -1 & 4 & -1 & 0 & 0 & 0 \\ 0 & -1 & 4 & -1 & 0 & 0 \\ 0 & 0 & -1 & 4 & -1 & 0 \\ 0 & 0 & 0 & -1 & 4 & -1 \\ 0 & 0 & 0 & 0 & -1 & 4 \end{bmatrix} \begin{bmatrix} x_1 \\ x_2 \\ x_3 \\ x_4 \\ x_5 \\ x_6 \end{bmatrix} = \begin{bmatrix} 0 \\ 1 \\ 1 \\ 1 \\ 1 \\ 0 \end{bmatrix}.$$

SOLUTION $\mathbf{x} = [0.09756 \quad 0.3902 \quad 0.4634 \quad 0.4634 \quad 0.3902 \quad 0.09756]$

1-1.2 Write a program to solve the tridiagonal system

$$\begin{bmatrix} 3 & -1 & 0 \\ -1 & 2 & -1 \\ 0 & -1 & 1 \end{bmatrix} \begin{bmatrix} x_1 \\ x_2 \\ x_3 \end{bmatrix} = \begin{bmatrix} 1 \\ 4 \\ -3 \end{bmatrix}.$$

SOLUTION $\mathbf{x} = [1 \quad 2 \quad -1]$

1-1.3 Write a program to solve the tridiagonal system

$$\begin{bmatrix} 3 & -1 & 0 \\ -1 & 2 & -1 \\ 0 & -1 & 1 \end{bmatrix} \begin{bmatrix} x_1 \\ x_2 \\ x_3 \end{bmatrix} = \begin{bmatrix} 1 \\ -3 \\ 2 \end{bmatrix}.$$

SOLUTION $\mathbf{x} = [0 \quad -1 \quad 1]$.

1-1.4 Write a program to solve the tridiagonal system

$$\begin{bmatrix} 2 & -1 & 0 \\ -1 & 5 & 2 \\ 0 & 2 & 1 \end{bmatrix} \begin{bmatrix} x_1 \\ x_2 \\ x_3 \end{bmatrix} = \begin{bmatrix} 0 \\ -3 \\ 8 \end{bmatrix}.$$

SOLUTION $\mathbf{x} = [-19 \quad -38 \quad 84]$

1-1.5 Write a program to solve the tridiagonal system

$$\begin{bmatrix} -1 & 3 & 0 \\ 3 & 4 & 1 \\ 0 & 1 & 2 \end{bmatrix} \begin{bmatrix} x_1 \\ x_2 \\ x_3 \end{bmatrix} = \begin{bmatrix} -3 \\ -3 \\ 1 \end{bmatrix}.$$

SOLUTION $\mathbf{x} = [0 \quad 1 \quad 1]$.

1-1.6 Write a program to solve the tridiagonal system

$$\begin{bmatrix} 5 & -4 & 0 \\ -4 & -1 & 2 \\ 0 & 2 & 4 \end{bmatrix} \begin{bmatrix} x_1 \\ x_2 \\ x_3 \end{bmatrix} = \begin{bmatrix} -7 \\ -12 \\ -8 \end{bmatrix}.$$

SOLUTION $\mathbf{x} = [-3 \quad -2 \quad -1]$.

1-1.7 Write a program to solve the tridiagonal system

$$\begin{bmatrix} 7 & 8 & 0 \\ 8 & -3 & 2 \\ 0 & 2 & -1 \end{bmatrix} \begin{bmatrix} x_1 \\ x_2 \\ x_3 \end{bmatrix} = \begin{bmatrix} 8 \\ 1 \\ 0 \end{bmatrix}.$$

SOLUTION $\mathbf{x} = [0 \quad 1 \quad 2]$.

Problem Set 2: Triangularization Method

The problems in this set are to be solved by providing an algorithm that applies the triangularization method to the solution process. Appendix 1.2 contains a code, called **factor.f**, that can be used by students as a guide to their assignments.

1-2.1 Solve the following system of equations by the triangularization method:

$$\begin{aligned} 2x_1 + 4x_2 + 6x_3 &= 2, \\ 6x_1 + 6x_2 + 4x_3 &= 8, \\ 7x_1 + 8x_2 + 10x_3 &= 15. \end{aligned}$$

SOLUTION $\mathbf{x} = [5 \quad -5 \quad 2]$.

1-2.2 Solve the following set of equations by the triangularization method:

$$\begin{aligned} 2x_1 + 4x_2 + 6x_3 &= 2, \\ 6x_1 + 12x_2 + 4x_3 &= 8, \\ 7x_1 + 8x_2 + 10x_3 &= 15. \end{aligned}$$

SOLUTION $\mathbf{x} = \left[\frac{50}{14} \quad -\frac{15}{14} \quad -\frac{2}{14}\right]$.

1-2.3 Solve the following system of simultaneous equations by the triangularization method:

$$\begin{aligned} 0.0001x_1 + 1.00x_2 &= 1.00, \\ 2.00x_1 + 1.00x_2 &= 3.00. \end{aligned}$$

SOLUTION $\mathbf{x} = [1.0005 \quad 0.999]$.

1-2.4 Solve the matrix system in Problem 1-1.1 by the triangularization method.

1-2.5 Solve the matrix system in Problem 1-1.2 by the triangularization method.

1-2.6 Solve the matrix system in Problem 1-1.3 by the triangularization method.

1-2.7 Solve the matrix system in Problem 1-1.4 by the triangularization method.

1-2.8 Solve the matrix system in Problem 1-1.5 by the triangularization method.

1-2.9 Solve the matrix system in Problem 1-1.6 by the triangularization method.

1-2.10 Solve the matrix system in Problem 1-1.7 by the triangularization method.

1-2.11 Solve the matrix system in Problem 1-3.1 by the triangularization method.

1-2.12 Solve the matrix system in Problem 1-4.1 by the triangularization method.

1-2.13 Solve the matrix system in Problem 1-4.2 by the triangularization method.

1-2.14 Solve the matrix system in Problem 1-4.3 by the triangularization method.

Problem Set 3: LU Decomposition

The problems in this set are to be solved by providing an algorithm that applies the **LU** decomposition method to the solution process. Appendix 1.3 contains three codes, called **mainlud.f**, **decomp.f**, and **back.f**, that can be used by students as a guide to their assignments.

1-3.1 Solve the following system by the **LU** decomposition method:

$$\begin{bmatrix} 2 & -1 & 0 \\ 3 & 4 & 1 \\ 5 & 6 & 2 \end{bmatrix} \begin{bmatrix} x_1 \\ x_2 \\ x_3 \end{bmatrix} = \begin{bmatrix} 4 \\ -2 \\ -1 \end{bmatrix}.$$

SOLUTION: $\mathbf{x} = [1 \quad -2 \quad 3]$.

1-3.2 Solve the matrix system of Problem 1-1.1 by the **LU** decomposition method.

1-3.3 Solve the matrix system in Problem 1-1.2 by the **LU** decomposition method.

1-3.4 Solve the matrix system in Problem 1-1.3 by the **LU** decomposition method.

1-3.5 Solve the matrix system in Problem 1-1.4 by the **LU** decomposition method.

1-3.6 Solve the matrix system in Problem 1-1.5 by the **LU** decomposition method.

1-3.7 Solve the matrix system in Problem 1-1.6 by the **LU** decomposition method.

1-3.8 Solve the matrix system in Problem 1-1.7 by the **LU** decomposition method.

1-3.9 Solve the matrix system of Problem 1-2.1 by the **LU** decomposition method.

1-3.10 Solve the matrix system of Problem 1-2.2 by the **LU** decomposition method.

1-3.11 Solve the matrix system of Problem 1-2.3 by the **LU** decomposition method.

1-3.12 Solve the matrix system of Problem 1-4.1 by the **LU** decomposition method.

1-3.13 Solve the matrix system of Problem 1-4.2 by the **LU** decomposition method.

1-3.14 Solve the matrix system of Problem 1-4.3 by the **LU** decomposition method.

Problem Set 4: Gauss–Seidel Iteration Method

The problems in this set are to be solved by providing an algorithm that applies the Gauss–Seidel iteration method to the solution process. Appendix 1.4 contains a code, called **snbt.f**, that can be used by students as a guide to their assignments.

1-4.1 Solve the following matrix set of equations by the Gauss–Seidel iteration method.

$$\begin{bmatrix} 2 & -1 & 0 \\ -1 & -5 & 2 \\ 0 & 2 & 1 \end{bmatrix} \begin{bmatrix} x_1 \\ x_2 \\ x_3 \end{bmatrix} = \begin{bmatrix} 0 \\ -3 \\ 8 \end{bmatrix}.$$

SOLUTION $\mathbf{x} = [1 \quad 2 \quad 4]$.

1-4.2 Solve the following set of equations by the Gauss–Seidel iteration method:

$$\begin{bmatrix} 7 & 8 & 0 \\ 8 & -3 & 2 \\ 0 & 2 & -1 \end{bmatrix} \begin{bmatrix} x_1 \\ x_2 \\ x_3 \end{bmatrix} = \begin{bmatrix} 8 \\ 1 \\ 0 \end{bmatrix}.$$

SOLUTION $\mathbf{x} = [0 \quad 1 \quad 2]$.

1-4.3 Solve the following equation set by the Gauss–Seidel iteration method:

$$\begin{bmatrix} 0 & 2 & -1 \\ 7 & 8 & 0 \\ 8 & -3 & 2 \end{bmatrix} \begin{bmatrix} x_1 \\ x_2 \\ x_3 \end{bmatrix} = \begin{bmatrix} 0 \\ 8 \\ 1 \end{bmatrix}.$$

Compare your results of this problem with the solution to Problem 1-4.2. Discuss convergence. Explain the difficulties in this problem.

1-4.4 Solve the matrix set of Problem 1-1.1 by the Gauss–Seidel iteration method.

1-4.5 Solve the matrix system in Problem 1-1.2 by the Gauss–Seidel iteration method.

1-4.6 Solve the matrix system in Problem 1-1.3 by the Gauss–Seidel iteration method.

1-4.7 Solve the matrix system in Problem 1-1.4 by the Gauss–Seidel iteration method.

1-4.8 Solve the matrix system in Problem 1-1.5 by the Gauss–Seidel iteration method.

1-4.9 Solve the matrix system in Problem 1-1.6 by the Gauss–Seidel iteration method.

1-4.10 Solve the matrix system in Problem 1-1.7 by the Gauss–Seidel iteration method.

1-4.11 Solve the matrix set of Problem 1-2.1 by the Gauss–Seidel iteration method.

1-4.12 Solve the matrix set of Problem 1-2.2 by the Gauss–Seidel iteration method.

1-4.13 Solve the matrix set of Problem 1-2.3 by the Gauss–Seidel iteration method.

1-4.14 Solve the matrix set of Problem 1-3.1 by the Gauss–Seidel iteration method.

CHAPTER

TWO

NONLINEAR SYSTEMS

NOMENCLATURE

x_n	nth iterative for the solution of $\mathbf{x} = f(x)$
α	root of a nonlinear scalar equation $x = f(x)$
$f(x)$	nonlinear functional relationship
$f'(x)$	first derivative of $f(x)$
$f(a)$	value of $f(x)$ at $\mathbf{x} = \mathbf{a}$
$f(b)$	value of $f(x)$ at $\mathbf{x} = \mathbf{b}$
E_n	$\alpha - x_n$, which is the error in the nth iteration
a	beginning of an interval for the bisection method and the false-position method
b	terminus of an interval for the bisection method and the false-position method
c	estimate of root location within the interval $[a, b]$
m	slope of a straight line
ε	convergence criterion
δ	convergence criterion
r_i	residual error of the ith equation in a nonlinear set of equations
δ_{jk}	Kronecker delta (identity matrix)
J_{ik}	Jacobian of the matrix A_{ik}
J_{ik}^{-1}	inverse of the Jacobian
$[\mathbf{A}]$	coefficient matrix (A_{ij} in subscript notation)
$[\mathbf{b}]$	right-hand side of matrix equations (b_i in index notation)
$[\mathbf{x}]$	solution vector (x_i in subscript notation)
∂	partial derivative with respect to a coordinate.

2-1 INTRODUCTION

Chapter 1 dealt with linear systems that result from the numerical simulation of physical systems. In this chapter, we direct our attention to nonlinear systems, both scalar and vectoral, which arise in the modeling of such physical processes as computational fluid dynamics.

It would seem, at first glance, that we should consider nonlinear matrix systems before addressing the area of an individual scalar equation. This would seem to be a logical extension to the material in Chap. 1, i.e., linear matrix systems. This would not, however, be the most efficient approach to the subject material in this chapter. There are many fine solution techniques that are available for scalar systems (one dependent variable) that, once derived, can be directly applied to nonlinear matrix systems (e.g., the Newton–Raphson solution technique). We, therefore derive computational algorithms for scalar applications and apply some of the same techniques to matrix systems.

2-2 SCALAR EQUATIONS

When the modeling of a physical process results in the derivation of nonlinear equations, the solution process becomes extremely difficult. The reasons for this are

1. There are no general procedures that are applicable to all types of nonlinear systems.
2. There is no guarantee that solutions exist for these systems.
3. Nonlinear equations can contain multiple roots that may be mathematically correct, but do not satisfy the physics of the system that they are modeling. It is therefore important to achieve the root of the nonlinear set that will satisfy both the equations and all the physical conditions of the process. In this regard, the selection of a good initial guess (or iterate) becomes extremely important since most nonlinear processes are generally solved by trial and error.

The formal definition of a scalar nonlinear equation is

> Given an equation of the form $f(x) = \mathbf{x}$, a solution is a value of $\mathbf{x} = \alpha$ for which $\alpha = f(\alpha)$.

Since this textbook is concerned with those numerical techniques that are applied to systems in physics and engineering, we desire only the real roots of the nonlinear equations.

2-2.1 Method of Successive Approximations

The solution technique that we explore initially is the seemingly simple method of successive approximations. This procedure would seem to be a natural first choice for the successive iterations that may lead to one or several of the multiple roots of the scalar equation. Starting with an initial guess x_0 that is reasonably close to the root that is desired (say within 10%), the iterative procedure $x_n = f(x_{n-1})$ produces a new iterate $\{x_k\}$.

Mathematically then it can be shown that the following inequality applies, viz.,

$$|\alpha - x_k| = |f(\alpha) - f(x_{k-1})| \leq |\alpha - x_{k-1}|\,|f'(z_{k-1})|, \tag{2-1}$$

where z_k is a value between α and x_{k-1}. This process will converge only if $|f'(x)| < 1$.

To apply this technique, we must express the variable $\mathbf{x}$ as a function $f(x)$. An example of the solution procedure is now delineated to explain the strategies that are required.

Consider the equation

$$f(x) = x^3 + 2x^2 + 10x - 20 = 0. \tag{2-2}$$

We wish to find the real root of Eq. (2-2) by the method of successive approximations. If we factor an $\mathbf{x}$ from Eq. (2-2) and transpose the constant to the right-hand side of the equation, we can then set up the iteration process as

$$x_n = \frac{20}{x_{n-1}^2 + 2x_{n-1} + 10} = f(x_{n-1}). \tag{2-3}$$

Since we do not know where the root lies, we cannot at this point check whether the magnitude of the derivative of Eq. (2-3),

$$f'(x) = \frac{20(2x+2)}{\{x^2 + 2x + 10\}^2},$$

is less than 1 near the root. We must therefore choose an initial guess and monitor the progress (or lack thereof) of the convergence. If we choose $x_0 = 1$ and use the formula shown in Eq. (2-3), we arrive at the next iteration with $x_1 = (20/13) \approx 1.5385$. We may now calculate the derivative near $\mathbf{x} = 1$, i.e, $f'(1) = -0.47$, whose magnitude is less than 1. Thus we are confident that the iterative procedure will converge. The value of 0.47 is large, however, and will lead to a slow rate of convergence of the solution. Continuing the above process of successive iterations, the solution continues as

n	x_n
0	1.0000
1	1.5385
2	1.2950
3	1.4018
4	1.3542
5	1.3753
6	1.3659
7	1.3701
8	1.3682
9	1.3691
10	1.3687
11	1.3689
12	1.3688
13	1.3688

The process converges to four decimal places in 13 iterations. If an initial iterate of $\mathbf{x} = 5$ were chosen, then the method of successive approximations would achieve a converged solution in 14 iterations. Both of the above results were obtained with the code called **success_1.f**, which is given in Appendix 2.1.

We can speed the process by noting that the error in the process may be estimated by Eq. (2-1), i.e.,

$$E_n = \alpha - x_n = f(\alpha) - f(x_{n-1}) = f'(\eta)(\alpha - x_{n-1}). \tag{2-4}$$

Then, by applying Eq. (2.4) successively to the $n + 1$ and the $n + 2$ iterations, we obtain

$$\alpha - x_{n+1} \approx f'(\eta)(\alpha - x_n),$$
$$\alpha - x_{n+2} \approx f'(\eta)(\alpha - x_{n+1}),$$

or

$$\frac{\alpha - x_{n+1}}{\alpha - x_{n+2}} \approx \frac{\alpha - x_n}{\alpha - x_{n+1}}.$$

Solving for α, we get

$$\alpha = x_{n+2} - \frac{(x_{n+2} - x_{n+1})^2}{x_{n+2} - 2x_{n+1} + x_n}. \tag{2-5}$$

Equation (2-5) can be applied after the second iteration and can greatly accelerate the convergence.

It is important to point out here that the iterative process that was established in Eq. (2-3) is not the only formulation that can be established from Eq. (2-2). Alternatively, we could have manipulated Eq. (2-2) as

$$x_n = \frac{20 - 2x_{n-1}^2 - x_{n-1}^3}{10} = f(x_{n-1}). \tag{2-6}$$

Here, if we begin our iteration once again with $x_0 = 1$, we calculate $x_1 = 1.70$. If we calculate the derivative $f'(x)$ at both $\mathbf{x} = 1$ and at $\mathbf{x} = 1.70$, we get

$$f'(1) = -\frac{7}{10},$$
$$f'(1.7) = -1.547.$$

Since the magnitude of the derivative is less than 1 at $\mathbf{x} = 1$ and is greater than 1 at $\mathbf{x} = 1.70$, we would expect convergence near $\mathbf{x} = 1$ and divergence near $\mathbf{x} = 1.7$. This is exactly what happens in the iteration. The solution process hunts for convergence but never achieves this goal. The results from this formulation are

n	x_n
1	1.0
2	1.7
3	0.9307
4	1.7461
19	0.5669
20	1.9175
29	0.5502
30	1.9228
39	0.5490
40	1.9232
49	0.54895
50	1.9232

The above result was obtained with the above-described code, called **success_2.f**, which is given in Appendix 2.1.

This exercise was chosen to demonstrate that for a nonlinear equation, the solution strategy is crucial. The wrong choice of a solution technique will result in a divergent (or a nonconvergent) formulation. While the method of successive approximations is simple to apply, it may not lead to the desired results. We turn, therefore, to a more aggressive solution approach.

2-2.2 Bisection Method

It has just been shown that the method of successive approximations can lead to a divergent, rather than a convergent, iteration process. Equally important, the need for a good initial guess to begin the iteration cannot be overemphasized. In this subsection, a method is introduced that can be used either to locate a root of a nonlinear equation or, in some cases, be used to find a proper initial iterate for other solution techniques. We begin by introducing the following theorem.

> If $f(x)$ is continuous in the interval [**a**, **b**], and if $f(a)$ and $f(b)$ have opposite signs, then there is at least one real root of $f(x) = 0$ between **a** and **b**.

Figure 2-1 shows the results of the above theorem. The following rules apply to the iteration and to the strategy that is used to calculate the next iterate in the process. If we desire to choose an intermediate point **c** (between **a** and **b**) as the next iterate, then

1. If $f(a)$ and $f(c)$ have opposite signs, a root lies in [**a**, **c**].
2. If $f(c)$ and $f(b)$ have opposite signs, a root lies in [**c**, **b**].
3. If $f(c) = 0$, we have found a root at $\mathbf{x} = \mathbf{c}$.

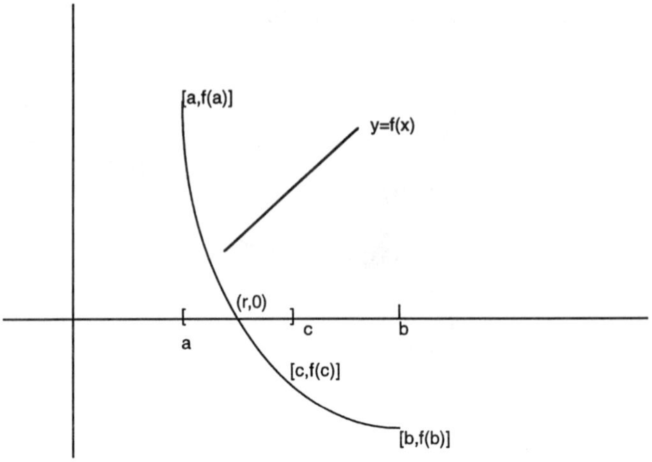

Figure 2-1 Bracketing root r in interval [**a, b**].

To apply these rules, we now adopt the following methodology:

A. To find a root in the interval [**a, b**], first choose the interval bisection point

$$\mathbf{c} = (\mathbf{a} + \mathbf{b})/2.$$

B. If rule 1 applies, redefine the search interval from [**a, b**] to $[a_1, b_1]$ by choosing $a_1 = \mathbf{a}$ and $b_1 = \mathbf{c}$ such that now .

$$c_1 = (a_1 + b_1)/2.$$

C. If rule 2 applies, choose $a_1 = \mathbf{c}$ and $b_1 = \mathbf{b}$ such that now

$$c_1 = (a_1 + b_1)/2.$$

This bisection process continues until convergence is achieved. Thus, for the $n + 1$ iteration,

$$[a_{n+1}, b_{n+1}] = [a_n, c_n] \qquad \text{or} \qquad [a_{n+1}, b_{n+1}] = [c_n, b_n].$$

To illustrate the bisection method for finding roots of nonlinear equations, we return to the above-mentioned problem. We desire to find the root of the cubic polynomial equation (2-2). If we start with the same initial guess, $\mathbf{x} = 1$, we bracket the initial interval $[\mathbf{a}, \mathbf{b}] = [1, 2]$. Using the above methodology, we find that the first value of **c** that is calculated by the bisection method is 1.5. We then use the above rules of the technique to judge whether the root lies in the interval [**a, c**] or in the interval [**c, b**]. This technique is continued until convergence is achieved. We get

n	**a**	**b**	**c**
1	1.0	2.0	1.5
2	1.0	1.5	1.25
3	1.25	1.5	1.375
4	1.25	1.375	1.3125
5	1.3125	1.375	1.34375
6	1.34375	1.375	1.359375
7	1.359375	1.375	1.367188
8	1.367188	1.375	1.371094
9	1.367188	1.371094	1.369141
10	1.367188	1.369141	1.368164

Convergence is achieved in 10 iterations within a 0.1% degree of accuracy. The bisection method is seen here to be 23% more robust than the method of successive approximations and, more importantly, is guaranteed to converge. The above result was obtained from the code **Bisect/False/Newt.f**, which is shown in Appendix 2.2.

2-2.3 Method of False Position (*regula falsi* Method)

The bisection method converges at a fairly slow speed. We would like to use a method that would accelerate the process of convergence. The false-position technique starts with the bracketing of the root, as in subsection 2-2.2. In this technique, however, we find the next iterate on the interval [**a**, **b**] by drawing a straight line between points [**a**, $f(a)$] and [**b**, $f(b)$] intersecting the x axis at a point **c**, which becomes the next iterate. The slope of the line that connects **a** and **b** is

$$m = \frac{f(b) - f(a)}{\mathbf{b} - \mathbf{a}} = \frac{0 - f(b)}{\mathbf{c} - \mathbf{b}},$$

from which we can solve for c, viz.,

$$c_n = b_n - \frac{f(b_n)(b_n - a_n)}{f(b_n) - f(a_n)}. \tag{2-7}$$

Figures 2-2 and 2-3 show the two possibilities that can occur when the intersection point $\mathbf{x} = \mathbf{c}$ is determined.

Case 1 (Fig. 2-2): $f(a)$ and $f(c)$ have opposite signs. The next interval of choice is [**a**, **c**].

Case 2 (Fig. 2-3): $f(c)$ and $f(b)$ have opposite signs. The next interval of choice is [**c**, **b**].

This process continues until a root of $f(x) = 0$ is attained to a specific degree of accuracy. The false-position technique will converge to a root much more rapidly than

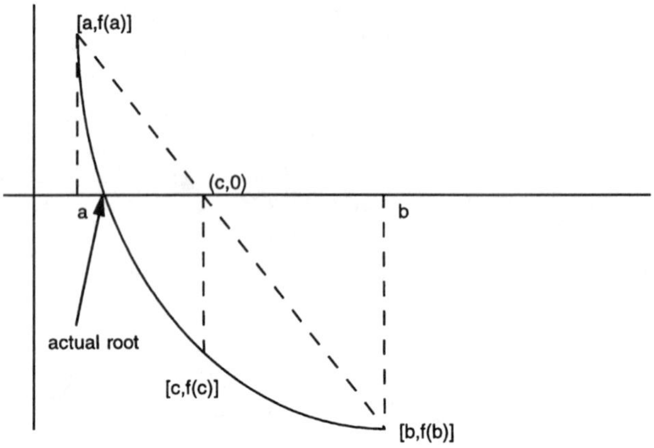

Figure 2-2 $f(a)$ and $f(c)$ have opposite signs.

the bisection method. A caution should be noted here. If $f(x) = 0$ has several roots in the search interval [**a**, **b**], then a different starting iteration point must be used for each root. Otherwise the same root will always be achieved.

To illustrate the false-position method, we once again choose Eq. (2-2). Following the rules that have been delineated above, the false position process yields

n	**a**	**b**	**c**
1	1.0	2.0	1.304348
2	1.304348	2.0	1.357912
3	1.357912	2.0	1.366978
4	1.366978	2.0	1.368501
5	1.368501	2.0	1.368757

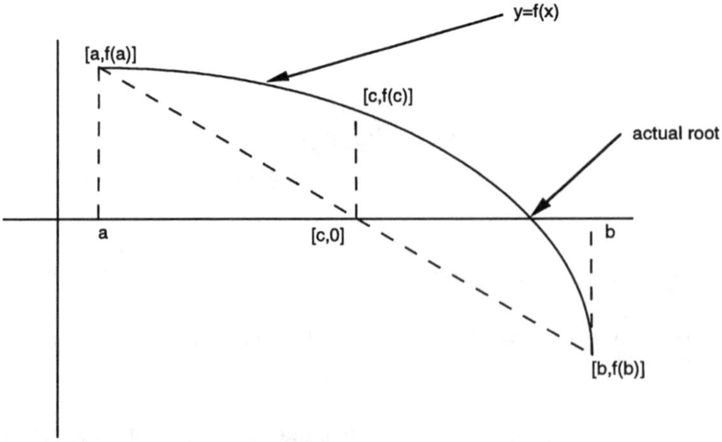

Figure 2-3 $f(c)$ and $f(b)$ have opposite sign.

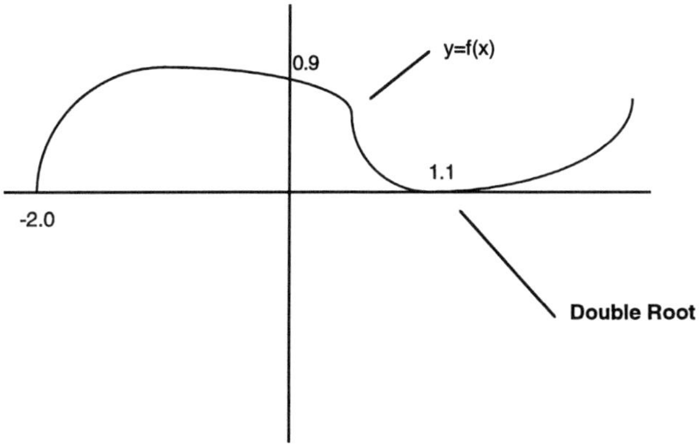

Figure 2-4 Double root.

In this case, convergence to within 0.1% was achieved in only five iterations when we started from the same initial guess as in the above-mentioned methods. It has therefore been demonstrated that the false-position method is more robust than the bisection technique. The above result was obtained from the code **Bisect/False/Newt.f**, which is shown in Appendix 2.2.

The above techniques are called *globally convergent techniques*. These methods generally produce convergent results. However, as in any numerical technique, certain special cases can occur that would make the convergence difficult to achieve:

1. Many equations possess a double root. Figure 2-4 shows this possibility. In this case, both the bisection method and the false-position method will fail to discover the duality of the root. These techniques will, at best, indicate that only a single root is found. In the worst case, these techniques may cause the initial search process to pass over the root completely, since no change in the sign of the function will be achieved between $f(a)$ and $f(b)$.
2. During the iteration process, especially when the root is near the current guess, the system may exhibit either a steep rate of convergence or a very shallow rate of convergence. In these events, it will become difficult to obtain an accurate root with either the bisection or the false-position methods. These phenomena are illustrated in Fig. 2-5.

In the cases in which either a steep- or a shallow-convergence band is encountered, then a combined convergence criterion is required, viz.,

$$|x_n - \alpha| < \delta, \qquad |f(x)| < \varepsilon. \tag{2-8}$$

This constitutes a rectangle of convergence and is shown in Figs. 2-6 and 2-7.

Except for the few possibilities that have just been described, globally convergent methods always converge to a solution of $\mathbf{x} = f(x)$.

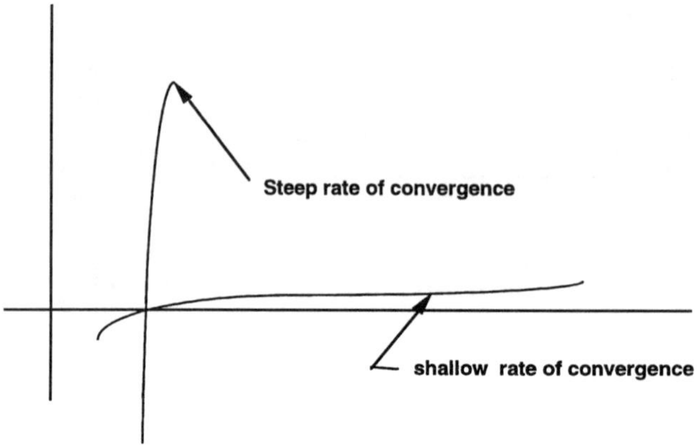

Figure 2-5 Steep- and rapid-convergence configurations.

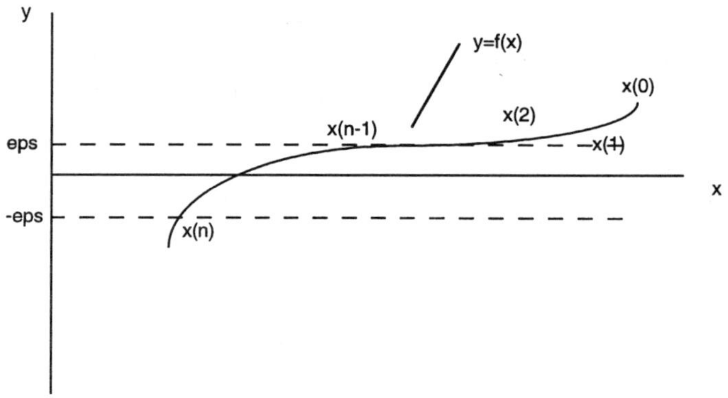

Figure 2-6 Horizontal-convergence band.

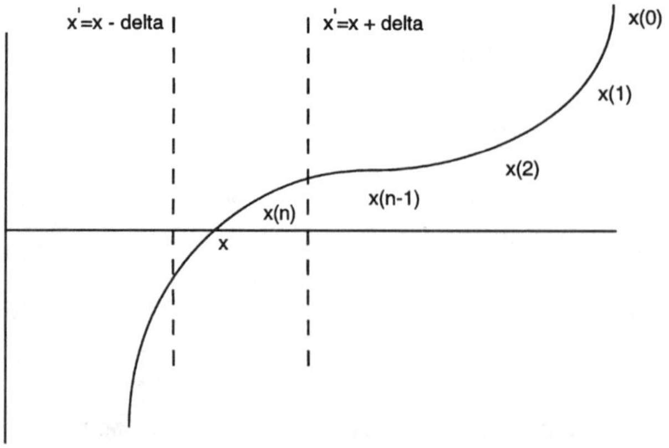

Figure 2-7 Vertical-convergence band.

2-2.4 Locally Convergent Methods

Scalar Newton–Raphson Method. There is a set of convergent techniques for nonlinear problems that are called locally convergent methods. The one that is used in most applications is the Newton–Raphson method. This technique calculates the slope of the function at a point that has been already determined for the present iterate in order to calculate where the next iterate is to be chosen. It uses this slope and a small interval perturbation of the variable to estimate the location of the desired root.

As with the above-mentioned techniques, we start the iteration process with an initial estimate of the root x_0. The slope of the function is then $m = f'(x_0)$. The next iterate, x_1, is then estimated by

$$x_1 = x_0 - \frac{f(x_0)}{f'(x_0)}, \tag{2-9}$$

or, in general,

$$x_k = g(x_{k-1}) = x_{k-1} - \frac{f(x_{k-1})}{f'(x_{k-1})} \qquad k = 1, 2, \ldots. \tag{2-10}$$

It must be stated, however, that the original guess x_0 must be close to the actual root α or this technique may not converge. Figure 2-8 shows how the process is applied in a successive manner and how the convergence may proceed. The iterate step that chooses x_1 appears to be converging toward the root α. However, when x_1 and the slope of the function at $\mathbf{x} = x_1$ are used in the Newton–Raphson method to estimate x_2, the process appears to diverge.

We return again to Eq. (2-2) to illustrate how the Newton–Raphson technique may be applied to nonlinear scalar equations. For this method, we also need to calculate the derivative of the function, which is $f'(x) = 3x^2 + 4x + 10$. The iteration now proceeds

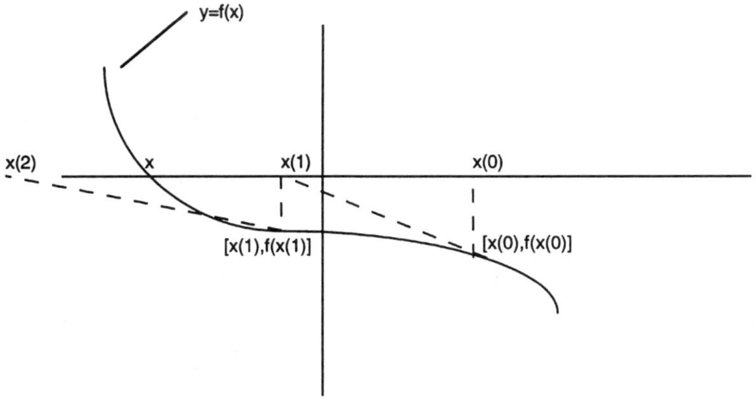

Figure 2-8 Newton–Raphson method.

as described above, namely,

n	**x**
1	1.0
2	1.411765
3	1.369336
4	1.368808

Convergence is now achieved, to the same accuracy as that of the above techniques in only four iterations. When the process does converge, the Newton–Raphson iterative techniques are probably the most robust of all the solution techniques that have been described in this chapter. However, once again, the reader must be cautioned that locally convergent methods tend to converge only when the procedure is begun with a value near the final solution. The above result was obtained from the code **Bisect/False/Newt.f**, which is shown in Appendix 2.2.

2-3 MATRIX EQUATIONS

Nonlinear matrix equations require iteration procedures for obtaining the solution vector. In these cases the iterative methodologies must begin with an estimate of the entire solution vector. The two techniques that are described below are generally used for this purpose.

2-3.1 Matrix Newton–Raphson Method

The Newton–Raphson technique may also be applied to nonlinear matrix systems of equations. Consider the system of equations

$$[\mathbf{A}][\mathbf{x}] = [\mathbf{b}] \qquad \text{or} \qquad A_{ij}x_j = b_i. \tag{2-11}$$

(Note that all the derivations that are given in this section use the subscript notation defined in Chapter 1. Care should be used when trying to interpret the formulas in this section.)

We begin, once again, by choosing an initial guess for the solution **x**. In this case, however, the solution is a vector, not a scalar. The initial guess must include all components of the vector. In actual physical applications in which nonlinear matrix systems occur (such as fluid mechanics), a reasonable estimate for the starting iteration vector is usually the solution to the corresponding linear problem. The corresponding linear problem consists of the physical geometry, the boundary conditions, and the initial conditions of the original problem solved by a system of equations in which the nonlinear terms have been removed.

Forming the error residual due to the current value of the iteration for the vector **x**, we obtain

$$A_{ij}x_j - b_i = r_i. \tag{2-12}$$

Since x_j is not the correct solution to Eqs. (2-11), an error in each component of the solution vector occurs. We call this vector the residual r_i. Differentiating Eq. (2-12) with respect to x_k, we get

$$A_{ij}\frac{\partial x_j}{\partial x_k} + x_j\frac{\partial A_{ij}}{\partial x_k} - \frac{\partial b_i}{\partial x_k} = \frac{\partial r_i}{\partial x_k}. \tag{2-13}$$

But

$$\frac{\partial x_j}{\partial x_k} = \delta_{jk}. \tag{2-14}$$

Then Eq. (2-13) becomes

$$A_{ik} + x_j\frac{\partial A_{ij}}{\partial x_k} - \frac{\partial b_i}{\partial x_k} = \frac{\partial r_i}{\partial x_k}. \tag{2-15}$$

If we now define a nonlinear Jacobian J_{ik} as

$$J_{ik} = A_{ik} + x_j\frac{\partial A_{ij}}{\partial x_k} - \frac{\partial b_i}{\partial x_k}, \tag{2-16}$$

we arrive at the equation that will govern the iteration, i.e.,

$$J_{ik} = \frac{\Delta r_i}{\Delta x_k}, \tag{2-17}$$

where $\Delta r_i = -r_i$. Since the goal is to achieve zero residual error, we arrive at the relationship that will select the next iteration of the vector, viz.,

$$\Delta x_k = J_{ik}^{-1}\Delta r_i, \tag{2-18}$$

and therefore the next iterate for the vector **x** may be calculated from

$$x_k^{i+1} = x_k^i + \Delta x_k. \tag{2-19}$$

Here the superscripts denote the iteration number. We can easily verify that Eq. (2-16) reduces to the standard formula for the Jacobian in the linear case, i.e.,

$$J_{ik} = A_{ik}.$$

It must be pointed out that the solution to Eq. (2-18) is carried out by the same matrix solution techniques that were described in Chap. 1. Now, however, since the Jacobian matrix was formed by components of the solution vector, it must be reformed after each new iterate is calculated from Eq. (2-19). Furthermore, if the solution to Eq. (2-18) is done by one of the iterative techniques that were described in Chap. 1, then the Newton–Raphson techniques will produce an iteration within an iteration. This means that there will be an iteration to achieve the solution to Eq. (2-18) and another iteration to update matrix **A** with the new iterate. This is a time-consuming task. The results of this method, however, are usually worth the effort that is required for implementing this formulation since the achievement of a robust solution algorithm will result.

As an example of the use of this powerful technique, let us consider a small example of a 3×3 system of nonlinear equations, viz.,

$$\begin{bmatrix} u_1^2 + 1.0 & u_1u_2 + 2.0 & u_1u_3 + 3.0 \\ u_2u_1 + 4.0 & u_2^2 + 5.0 & u_2u_3 + 6.0 \\ u_3u_1 + 7.0 & u_3u_2 + 8.0 & u_3^2 + 9.0 \end{bmatrix} \begin{bmatrix} u_1 \\ u_2 \\ u_3 \end{bmatrix} = \begin{bmatrix} u_3^3 + 1.0 \\ 10.0 * u_2 * u_3 \\ 92.0 \end{bmatrix}.$$

It is immediately apparent that this set of equations is nonlinear, not only in the coefficient matrix, but also on the right-hand-side vector. Because this system is nonlinear, multiple solutions may exist. The algorithm, **jacobian.f**, which is shown in Appendix 2.3, was written to solve this system of equations. If one chooses an initial iterate,

$$\mathbf{u} = [0.1 \quad 0.1 \quad 0.1]$$

then the code, when the Newton–Raphson matrix iteration technique is used, converges in 221 iterations with a solution and a residual as

$$\begin{aligned}\mathbf{u} &= [0.939 \quad -4.51 \quad 3.07],\\ \mathbf{r} &= [-0.343e{-}04 \quad 0.952e{-}02 \quad -0.420e{-}04].\end{aligned}$$

Note: e is computer shorthand for power notation, e.g., $-0.343e{-}04 = -0.343 \times 10^{-4}$.

If we now attempt to solve the above system by choosing an initial iterate,

$$\mathbf{u} = [100 \quad 100 \quad 100],$$

then the solution process converges in 267 iterations to a solution and a residual; thus

$$\begin{aligned}\mathbf{u} &= [0.381 \quad 2.9 \quad 2.68],\\ \mathbf{r} &= [0.191e{-}05 \quad 0.994e{-}02 \quad -0.114e{-}04].\end{aligned}$$

It is somewhat disturbing that the two solutions are very different. Additionally, a solution of

$$\mathbf{u} = [1 \quad 2 \quad 3]$$

is also shown to be a solution of the above system, with a null residual vector.

This simple example demonstrates how the choice of an initial iterate will influence the resulting solution. The importance of obtaining a good initial guess (as explained above) is clearly demonstrated. Additionally, if this system represented a physical system, then there would be concern as to the correctness of a particular solution and the satisfaction of the other attributes of the problem (e.g., the boundary conditions).

Appendix 2.3 contains a sample program, called **jacomp.f**, written in FORTRAN, that uses the Newton–Raphson Jacobian method to solve a system of nonlinear matrix equations.

The formulation of the Jacobian in Eq. (2-16) contains a series of terms that are difficult to calculate unless a simple analytical expression is available for matrix **A** (which is not the usual case). The terms $x_j[(\partial A_{ij})/(\partial x_k)]$ are difficult to calculate. Additionally if the number of equations that are represented by **A** is N, then the formula requires N^3 operations. It would therefore be reasonable to investigate whether the Jacobian may be calculated by an alternative means, even if this pursuit results in a Jacobian that is an approximation to the exact expression that is delineated in Eq. (2-16).

2-3.2 Approximate Jacobian Formulation

We now derive a formulation that calculates an approximation to the Jacobian, but one that is easy to calculate in successive iterations. If we begin by denoting the approximate

Jacobian by [**B**], then the iterative equivalent to Eq. (2-17) for the ith iteration is

$$B_{ij}^i \delta x_j^i = -r_i^i, \tag{2-20}$$

where $\delta x_j^i = x_{j+1}^i - x_j^i$ and r_i^i is the ith iteration of the residual vector. In Eq. (2-20), the superscript refers to the iteration number while the subscripts are to be interpreted as normal indices. Then the condition on the next iterate B_{ij}^{i+1} is

$$B_{ij}^{i+1} \delta x_j^{i+1} = \delta r_i^{i+1}, \tag{2-21}$$

where $\delta r_i^i = r_i^{i+1} - r_i^i$. We now seek to obtain an estimate for the update to the approximate Jacobian, B_{ij}^{i+1}. Using the method of Broyden[1], we can show that

$$B_{ij}^{i+1} = B_{ij}^i + \frac{\left(\delta r_i^i - B_{ik}^i \delta x_k^i\right)}{\delta x_i^i \delta x_i^i} \delta x_j^i. \tag{2-22}$$

Equation (2-22) shows that the updated Jacobian may be calculated from the previous matrix and knowledge of the incremental changes in the solution vector δx_i and the residual vector δr_i.

It therefore follows that the updated expression for the inverse Jacobian may be written as

$$\left(B_{ij}^{i+1}\right)^{-1} = \left(B_{ij}^i\right)^{-1} + \frac{\left(\delta x_i^i - \left[B_{ik}^i\right]^{-1} \delta r_k^i\right)\left(\delta x_m^i \left[B_{mj}^i\right]^{-1}\right)}{\delta x_k^i \left(B_{km}^i\right)^{-1} \delta x_m^i}. \tag{2-23}$$

Equation (2-23) may be calculated for the $i+1$st iteration from the results of the ith iteration. Once a new estimate of the Jacobian is achieved, we can use the inverse equation (2-23) in conjunction with Eq. (2-18) to calculate the next vector **x** by using Eq. (2-19). The iteration process can then proceed to the next iteration.

2-4 CONCLUSION

This chapter has illustrated the techniques, both scalar and matrix, that can be used to solve a single nonlinear equation or a set of nonlinear equations. The common theme among all the techniques is the idea that the solution process may not converge. Globally convergent techniques are more apt to produce a root to the equation(s) than locally convergent methods. However, even here, while this root may be mathematically correct it may not have physical significance in the engineering problem that is being simulated. In other words, it may be the wrong root of the equation.

Nonlinear equations have, in general, multiple roots that exist on different branches of a solution curve. In engineering problems it is essential that not only the equations be satisfied, but that the boundary and the initial conditions also be satisfied. A weak iteration process or a poor initial guess at the initial iterate may result in the convergence of the process toward the wrong root or may result in a divergent iteration technique. In either of the latter cases, the numerical simulation of the physical process has failed to produce a solution to the physical process.

For these reasons it is extremely important that the iterative process produce a convergent solution and that this solution have physical significance. One technique that is used by researchers to achieve the above objectives is the use of the solution to the linear version of the problem (as explained above) as the initial iterate in the solution process. This approach has multiple benefits. First, it begins the iteration process with a solution that satisfies both the linearized equations and the boundary conditions. Second, if the system is not too highly nonlinear, then this approach provides an initial iterate that is close to the actual solution of the nonlinear system.

REFERENCE

1. Press, W. H., Teukolsky, S. A., Vetterling, W. T., and Flannery, B. P. (1986). *Numerical Recipes in FORTRAN*, Cambridge University Press, New York.

FURTHER READING

Arden, B. W. and Astill, K. N. (1970). *Numerical Algorithms - Origins and Applications*, Addison-Wesley, Reading, Massachusetts.

Mathews, J. H. (1987). *Numerical Methods for Computer Science, Engineering, and Mathematics*, Prentice-Hall, Englewood Cliffs, New Jersey.

Pennington, R. H. (1965). *Introductory Computer Methods and Numerical Analysis*, Macmillan, New York.

Scheid, F. (1968). *Schaum's Outline of Theory and Problems of Numerical Analysis*, McGraw-Hill, New York.

PROBLEMS

For each problem in this chapter, write a computer program to solve for the unknown quantities. Sample codes are contained in Appendixes 2.1–2.3, which can be used as checks on the work of the students.

For nonlinear scalar equations

1. **success_1.f** solves problems by the method of successive approximations.
2. **success_2.f** repeats the solution of **success_1.f**, but with a different approach to the successive approximations.
3. **bisect/false/newt.f** uses, alternately, the bisection method, the false-position method, and the Newton–Raphson method to achieve the same solution as codes 1 and 2. This program is found in Appendix 2.2.

Students need merely to change the functional expression within each of these codes to solve any particular problem.

For nonlinear matrix equations Appendix 2.3 contains a robust algorithm, **jacomp.f**, that will solve any size set of nonlinear matrix equations. Students can modify the matrix description, the right-hand side, and the derivative description in this code to use it in solving any individual problem.

2-1 Find the roots of $4 \sin 3x = x^2 + 3$ by using the method of successive approximations. How many roots exist?

2-2 Solve Problem 2-1 by using the bisection method. How many roots exist?

2-3 Solve Problem 2-1 by using the false-position method. How many roots exist?

2-4 Solve Problem 2-1 by using the Newton–Raphson method. How many roots exist?

2-5 For a projectile that is fired at an elevation angle of 45° from the horizontal, the position of the projectile at any time t is given by

$$\mathbf{y} = f(t) = (Cv_y + 32C^2)\left[1 - \exp\left(-\frac{t}{C}\right)\right] - 32Ct,$$

$$\mathbf{x} = r(t) = Cv_x\left[1 - \exp\left(-\frac{t}{C}\right)\right]$$

Given $v_x = v_y = 160$ ft/s and $C = 10$, find the elapsed time until impact with the ground and find the range.

Use the method of successive approximations to solve this problem.

2-6 Use the bisection method to solve Problem 2-5.

2-7 Use the false-position method to solve Problem 2-5.

2-8 Use the Newton–Raphson method to solve Problem 2-5.

2-9 Using the method of successive approximations, find one root of

$$8^x + x^3 - x = 0.$$

ANSWER $\mathbf{x} = -1.05197$.

2-10 Solve Problem 2-9 by the bisection method.

2-11 Solve Problem 2-9 by the false-Position method.

2-12 Solve Problem 2-9 by the Newton–Raphson method.

2-13 Using the method of successive approximations, calculate the two real roots of

$$e^x - 4x = 0.$$

ANSWER $\mathbf{x} = 0.356884$ and $\mathbf{x} = 2.15462$.

2-14 Solve Problem 2-13 by the bisection method.

2-15 Solve Problem 2-13 by the false-position method.

2-16 Solve Problem 2-13 by the Newton–Raphson method.

2-17 Use the method of successive approximations to find all the roots of

$$e^{-x} - \sin\frac{\pi x}{2} = 0$$

in the interval $\mathbf{x} = 0$ to $\mathbf{x} = 10$.

2-18 Solve Problem 2-17 by the bisection method.

2-19 Solve Problem 2-17 by the false-position method.

2-20 Solve Problem 2-17 by the Newton–Raphson method.

2-21 Using the method of successive approximations, write a program that calculates $\sqrt{\mathbf{x}}$. Use this program to solve for $\sqrt{5}$. Specify the number of iterations that are required for converging to an answer within 1%.

2-22 Redo Problem 2-21 using the false-position method. Compare the number of iterations to convergence with that of Problem 2-21.

2-23 Redo Problem 2-21 using the bisection method. Compare the number of iterations to convergence with that of Problems 2-21 and 2-22.

2-24 Redo Problem 2-21 using the Newton–Raphson method. Compare the number of iterations to convergence with that of Problems 2-21, 2-22, and 2-23.

2-25 Using the Newton–Raphson iteration method, solve for the roots of the equation

$$f(x) = x^3 - A,$$

thereby establishing an iteration procedure to calculate the cube root of any number A. Using this algorithm, calculate $\sqrt[3]{-7}$.

2-26 Solve Problem 2-25 by the method of successive approximations.

2-27 Solve Problem 2-25 by the false-position method.

2-28 Solve Problem 2-25 by the bisection method.

2-29 Given the nonlinear coefficient matrix **A**, where

$$\mathbf{A}\begin{bmatrix} U_1U_1 & U_1U_2 & U_1U_3 & 0 & 0 & 0 \\ U_2U_1 & U_2U_2 & U_2U_3 & U_2U_4 & 0 & 0 \\ U_3U_1 & U_3U_2 & U_3U_3 & U_3U_4 & U_3U_5 & 0 \\ 0 & U_4U_2 & U_4U_3 & U_4U_4 & U_4U_5 & U_4U_6 \\ 0 & 0 & U_5U_3 & U_5U_4 & U_5U_5 & U_5U_6 \\ 0 & 0 & 0 & U_6U_4 & U_6U_5 & U_6U_6 \end{bmatrix},$$

and the right-hand-side vector **b**, where

$$\mathbf{b} = \begin{bmatrix} U_2(1+U_6) \\ 2U_5U_6 \\ 20U_2U_4 + U_5 \\ 10U_6^2 \\ 10\left(U_6^2 + U_4 + U_3\right) \\ 10\left(U_6^2 + U_2U_5\right) + 2U_1 \end{bmatrix}$$

such that $[\mathbf{A}][\mathbf{U}] = [\mathbf{b}]$, write a computer code using the Newton–Raphson Jacobian iteration technique to solve for the unknown vector **U**. Try to establish a method for estimating a good initial iterate.

By trying several initial iterates, determine what effect the initial guess has on the final solution.

ANSWER $\mathbf{U} = [1.0 \quad 2.0 \quad 3.0 \quad 4.0 \quad 5.0 \quad 6.0]$.

2-30 We would like to investigate the effects of an interference between two circular cylinders on the stress intensity near the interference. Figure 2-9 shows the arbitrary surfaces $S_1(\mathbf{x}, \mathbf{y})$ and $S_2(\mathbf{x}, \mathbf{y})$ that have a maximum interference **I**. The extremities of these intersections occur at (**a**, **b**) and (**c**, **d**), respectively.

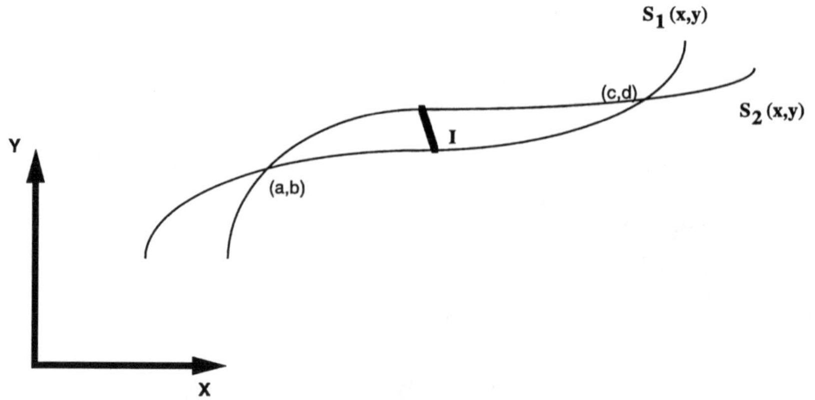

Figure 2-9 Intersection of two arbitrary surfaces.

For a given value of **I**, then, we desire the calculation of the points (**a**, **b**) and (**c**, **d**). Write a FORTRAN code that solves for these two points when the surfaces are two intersecting circles, viz.,

$$S_1(\mathbf{x}, \mathbf{y}):(\mathbf{x} - e_1)^2 + (\mathbf{y} - e_2)^2 = r_1^2,$$
$$S_2(\mathbf{x}, \mathbf{y}):(\mathbf{x} - f_1)^2 + (\mathbf{y} - f_2)^2 = r_2^2,$$

where

$$e_1 = 0.0,$$
$$e_2 = 0.0,$$
$$r_1 = 1.0,$$
$$f_1 = 4.0,$$
$$f_2 = 4.0,$$
$$r_2 = 1.0.$$

Solve this problem by the method of successive approximations.

2-31 Solve Problem 2-30 by the bisection method.

2-32 Solve Problem 2-30 by the false-position method.

2-33 Solve Problem 2-30 by the Newton–Raphson method.

2-34 Write a FORTRAN code that solves Problem 2-30 for any arbitrarily given functional values, such as involute surfaces that represent gear teeth. Your code should be general enough to accommodate any functional form.

CHAPTER

THREE

APPROXIMATION, INTERPOLATION, AND CURVE FITTING

NOMENCLATURE

m_k	$S_k''(x)$
T_i	Chebyshev polynomial
a_i	coefficient in curve-fitting approximation
b_i	coefficient in curve fitting approximation
θ	independent variable
$[\mathbf{a}, \mathbf{b}]$	interval of data set
$\{S_{i,j}\}$	spline coefficients
$S_k(x)$	spline curve for the kth interval
E	measure of error in least-squares method
F	dependent variable
f	functional relationship
$i$pos	row location in a matrix
$j$pos	column location in a matrix
t	transformation variable
x	independent variable
y	dependent variable

3-1 INTRODUCTION

Curve fitting is used to put data into a convenient mathematical form (e.g., an algebraic polynomial equation), to obtain accurate derivatives and integrals, and to attempt to obtain functional values at points where data are not available. In Chapter 9 the theory and the application of finite-element analyses for physical problems are developed. It is shown that certain physical variables need to be calculated in order to obtain a complete solution to the specific problem that is being considered. In that case, curve fitting can be used to determine the values of the dependent variables at specially designated points that are called the integration points. In geometric modeling, a curve fit of the data can be used to evaluate the local curvatures and twists, as well as the local unit normals of an arbitrary curve in space.

In this section, several techniques are presented to curve fit discrete or continuous data. These include

1. Algebraic polynomial curve fitting
2. Chebyshev polynomial fitting
3. Least-squares polynomial fitting
4. Least-squares Chebyshev polynomial fitting
5. Fourier series
6. Spline functions

Two common functions, $\mathbf{y} = \sin 2x$ and $\mathbf{y} = \sin x$, are used throughout this chapter to illustrate the curve-fitting techniques being presented. We refer to these functions as control functions. In all the cases herein, the following conditions are assumed to apply:

> Given a table of values of $[x_i, f(x_i)]$ at $n + 1$ points, a mathematical function that represents this set of data is required. All the subsequent curve-fitting techniques are designed to satisfy this requirement.

A note of caution must be introduced at this point. The techniques within this chapter are to be used only when a set of discrete data points is available and an approximate functional relationship is desired. If the analytical expression for the data is known, then the ideas in this chapter are not needed. We use the control functions only as benchmarks to determine how well the approximating curves match the data.

3-2 POLYNOMIAL CURVE FIT

This section introduces the simplest curve-fitting approach that can be used in problems for which it is necessary to obtain an analytical expression that will approximate a set of discrete data points. The method, algebraic polynomial curve fitting, is easy to formulate and apply but is limited to low-order polynomials, for reasons that are subsequently explained. Given a set of $n + 1$ data points, viz.,

x	$f(x)$
x_0	$f(x_0)$
x_1	$f(x_1)$
...	...
x_n	$f(x_n)$

we attempt to fit all the data points by using an algebraic polynomial approximation of order n, i.e.,

$$f(x) = a_0 + a_1 x + a_2 x^2 + \cdots + a_n x^n, \tag{3-1}$$

where $a_0, a_1, \ldots, a_n$ are unknown coefficients that remain to be determined. We can evaluate the $n + 1$ coefficients by inserting each data point into Eq. (3-1), viz.,

$$\begin{aligned} f(x_0) &= a_0 + a_1 x_0 + a_2 x_0^2 + \cdots + a_n x_0^n, \\ f(x_1) &= a_0 + a_1 x_1 + a_2 x_1^2 + \cdots + a_n x_1^n, \\ &\cdots \\ f(x_n) &= a_0 + a_1 x_n + a_2 x_n^2 + \cdots + a_n x_n^n. \end{aligned} \tag{3-2}$$

These equations can be expressed in matrix form:

$$\begin{bmatrix} 1 & x_0 & x_0^2 & x_0^3 & \cdots & x_0^n \\ 1 & x_1 & x_1^2 & x_1^3 & \cdots & x_1^n \\ \cdots & \cdots & \cdots & \cdots & \cdots & \cdots \\ 1 & x_n & x_n^2 & x_n^3 & \cdots & x_n^n \end{bmatrix} \begin{bmatrix} a_0 \\ a_1 \\ \cdots \\ a_n \end{bmatrix} = \begin{bmatrix} f(x_0) \\ f(x_1) \\ \cdots \\ f(x_n) \end{bmatrix}. \tag{3-3}$$

Equations (3-3) may now be solved for the unknown coefficients $a_0, a_1, \ldots, a_n$. The functional approximation that is given in Eq. (3-1) is now uniquely determined. While these polynomials pass through each data point, they are usually inaccurate between data points, where it is likely for the user to desire the calculation of a functional value. The reason for this inaccuracy is that an nth-degree polynomial has a total of (n-1) maxima and minima. At these points, the function's derivative must vanish. In order to accomplish the dual tasks of passing through all of the data points and have $n - 1$ extrema, the function can have large derivatives (i.e., the function may wiggle between the given data points) and not produce a good interpolating function. The larger the order of the polynomial, the higher will be the probability that this would happen. Equation (3-1) may be compactly expressed as

$$f(x) = \sum_{k=0}^{n} a_k x^k. \tag{3-4}$$

To illustrate this curve-fitting procedure, let us try to approximate $\mathbf{y} = \sin 2x$ by a quintic polynomial, i.e.,

$$\mathbf{y} = a_0 + a_1 x + a_2 x^2 + a_3 x^3 + a_4 x^4 + a_5 x^5. \tag{3-5}$$

Figure 3-1 shows the curve $\mathbf{y} = \sin 2x$. Since Eq. (3-5) has six unknown coefficients to be evaluated, we arbitrarily choose the equally spaced points,

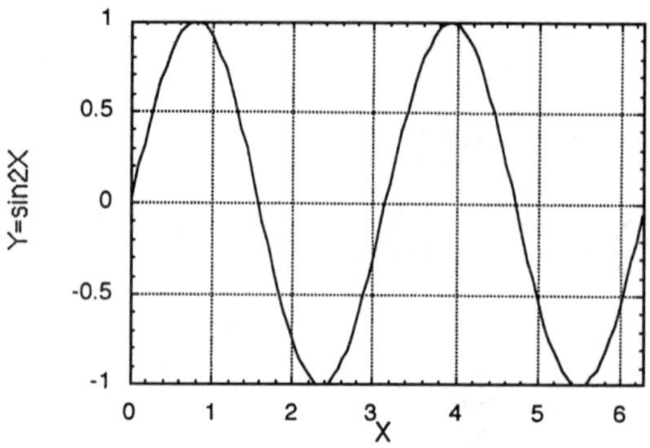

Figure 3-1 Function sin 2x.

X	$f(x) = \sin 2X$
0	0.0
$\frac{2\pi}{5} = 1.2566$	0.5878
$\frac{4\pi}{5} = 2.5133$	−0.9511
$\frac{6\pi}{5} = 3.7699$	0.9511
$\frac{8\pi}{5} = 5.0265$	−0.5878
$2\pi = 6.2832$	0.0

with which to attempt our curve fit.

The application of Eq. (3-5) to the above data yields

$$\begin{bmatrix} 1 & 0 & 0 & 0 & 0 & 0 \\ 1 & 1.26 & 1.58 & 1.98 & 2.49 & 3.13 \\ 1 & 2.51 & 6.31 & 15.9 & 39.8 & 100 \\ 1 & 3.77 & 14.2 & 53.5 & 202 & 760 \\ 1 & 5.02 & 25.2 & 127 & 637 & 3200 \\ 1 & 6.28 & 39.4 & 248 & 1560 & 9770 \end{bmatrix} \begin{bmatrix} a_0 \\ a_1 \\ a_2 \\ a_3 \\ a_4 \\ a_5 \end{bmatrix} = \begin{bmatrix} 0 \\ 0.588 \\ -0.951 \\ 0.951 \\ -0.588 \\ 0 \end{bmatrix}. \tag{3-6}$$

These equations may be solved so that

$$\begin{aligned} a_0 &= 0, \\ a_1 &= 9.23, \\ a_2 &= -12.6, \\ a_3 &= 5.7, \\ a_4 &= -1.04, \\ a_5 &= 0.0663. \end{aligned}$$

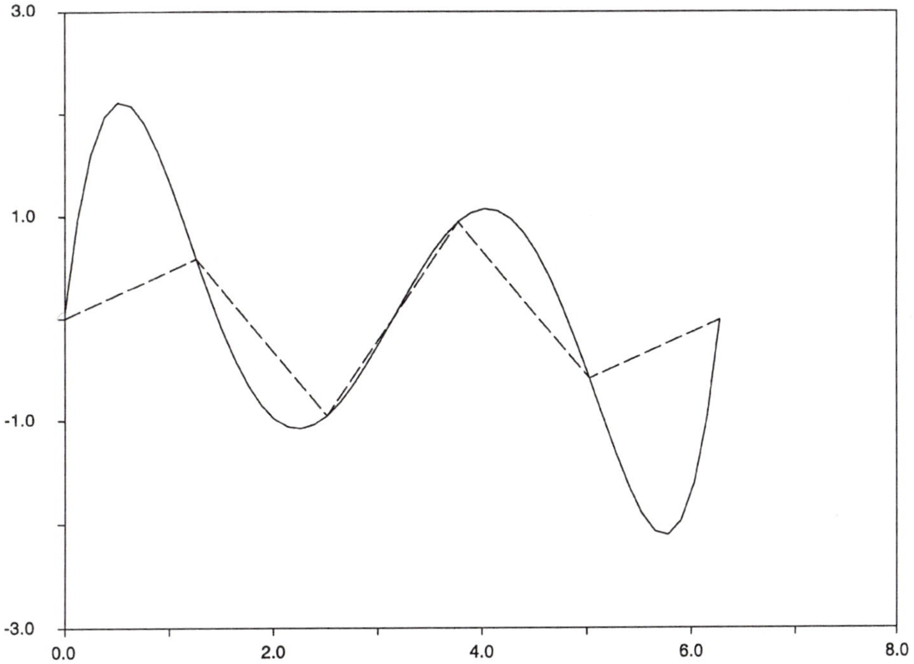

Figure 3-2 Fifth-order polynomial curve fit.

The quintic polynomial that approximates the data then becomes

$$\mathbf{y} = 9.23x - 12.6x^2 + 5.7x^3 - 1.04x^4 + 0.0663x^5. \tag{3-7}$$

In order to evaluate the accuracy of this approximation (with reference to $\mathbf{y} = \sin 2x$), we evaluate Eq. (3-7) at a point that lies within the range of the data set, i.e., $\mathbf{x} = \pi$. We choose this point for evaluation since the function has a null value. Thus

$$\mathbf{y} = 9.23(\pi) - 12.6(\pi)^2 + 5.7(\pi)^3 - 1.04(\pi)^4 + .0663(\pi)^5 = 0.359.$$

The correct value for y is, of course, 0.00.

Figure 3-2 shows a direct comparison between the given data and the polynomial approximation that was derived above. Additionally, if one recalls the Taylor series expansion for $\sin 2x$, i.e.,

$$\begin{aligned}\mathbf{y} = \sin 2x &\approx 2x - \frac{(2x)^3}{3!} + \frac{(2x)^5}{5!} - \cdots \\ &= 2x - 1.333x^3 + 0.267x^5 - \cdots,\end{aligned}$$

a comparison with the polynomial coefficients shows that although Eq. (3-7) is a good fit of the data, it is a poor fit of $\mathbf{y} = \sin 2x$ and will be inaccurate outside the interval that contains the data. Additionally, Fig. 3-2 shows that the quintic algebraic polynomial

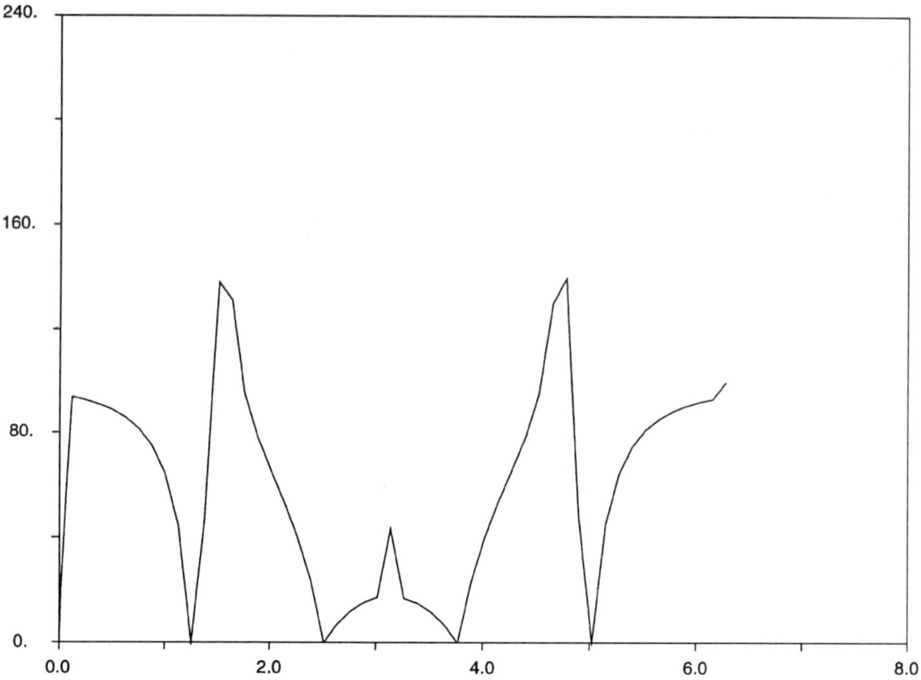

Figure 3-3 Error characteristics of a fifth-order polynomial fit.

oscillates wildly between data points, rendering it useless for the determination of intermediate points or derivatives. The user would not use this polynomial in an interpolative evaluation. Figure 3-3 displays the error between the function $\mathbf{y} = \sin 2x$ and the quintic polynomial that has been derived. It is obvious that unacceptable errors are produced at points that do not coincide with the data.

The results in this section were obtained with the computer code **poly.for**, which is given in Appendix 3.1.

We may also apply the above methodology to a function that has two independent variables. Then,

$$f(x, y) = \sum_{i=0}^{n} \sum_{j=0}^{m} a_{ij} x^i y^j, \tag{3-8}$$

which produces an $(n \times m)$ polynomial approximation of the function $f(x, y)$. The implementation of the technique for a two-dimensional algebraic polynomial fit is left to the readers, but it proceeds directly from the methods that are described above.

3-3 CHEBYSHEV POLYNOMIAL FIT

Some of the difficulties with the polynomial equation approximation that may occur (such as erroneous derivatives) can be reduced or eliminated entirely by use of Chebyshev polynomials instead of the algebraic polynomials of Section 3-2. The definition of a Chebyshev polynomial is contained in the following rules:

1. A Chebyshev polynomial is defined over the interval $[-1, 1]$.
2. The range of the independent variable must then be

$$-1 < x_0 < x_1 < \cdots < x_n < 1.$$

3. The zeroth-order Chebyshev polynomial is $T_0(x) = 1$.
4. The first-order Chebyshev polynomial is $T_1(x) = x$.
5. The second-order Chebyshev polynomial is $T_2(x) = 2x^2 - 1$.

A Chebyshev polynomial has a recursive formula that is used to calculate higher-order polynomials, viz.,

$$T_k(x) = 2xT_{k-1}(x) - T_{k-2}(x) \qquad (k = 2, 3 \ldots).$$

For a function that has only one independent variable, the Chebyshev approximating formulation yields

$$f(x) = \sum_{k=0}^{n} a_k T_k(x). \tag{3-9}$$

As an example of this technique, let us return to the six data points that are available to find an approximation of $y = \sin 2x$. However, since our interval of interest is

$$[\mathbf{a}, \mathbf{b}] = [0, 2\pi] = \left[0, \frac{2\pi}{5}, \frac{4\pi}{5}, \frac{6\pi}{5}, \frac{8\pi}{5}, 2\pi\right]$$

and the Chebyshev polynomials are defined over only the interval $[-1,\ 1]$, we must transform the actual physical interval to the Chebyshev interval by using the transformation

$$\mathbf{t} = 2\left(\frac{x-a}{b-a}\right) - 1. \tag{3-10}$$

The variable $\mathbf{t}$ is the independent variable in the transformed Chebyshev domain and plays the same role as $\mathbf{x}$ in the physical domain.

Applying Eq. (3-10) to the interval $[0, 2\pi]$, we get

$$\mathbf{t} = 2\left(\frac{x-0}{2\pi}\right) - 1 = \frac{x}{\pi} - 1,$$

$$\mathbf{t} = [-1 \quad -0.6 \quad -0.2 \quad +0.2 \quad +0.6 \quad +1] = \mathbf{x}',$$

so that

$$f(x') = a_0T_0(x') + a_1T_1(x') + a_2T_2(x') + a_3T_3(x') + a_4T_4(x') + a_5T_5(x'),$$

where

$$
\begin{aligned}
T_0(x') &= 1, \\
T_1(x') &= x', \\
T_2(x') &= 2x'^2 - 1, \\
T_3(x') &= 2x'T_2(x') - T_1(x') = 4x'^3 - 3x', \\
T_4(x') &= 8x'^4 - 8x'^2 + 1, \\
T_5(x') &= 16x'^5 - 20x'^3 + 5x'.
\end{aligned}
$$

The approximating function becomes

$$
\begin{aligned}
f(x') = a_0 + a_1 x' + a_2\{2x'^2 - 1\} + a_3\{4x'^3 - 3x'\} \\
+ a_4\{8x'^4 - 8x'^2 + 1\} + a_5\{16x'^5 - 20x'^3 + 5x'\}.
\end{aligned}
\tag{3-11}
$$

If we evaluate Eq. (3-11) at the $\mathbf{x}'$ data points, we get

$$
\begin{bmatrix}
1 & -1 & 1 & -1 & 1 & -1 \\
1 & -0.6 & -0.28 & 0.936 & -0.843 & 0.0758 \\
1 & -0.2 & -0.92 & 0.568 & 0.693 & -0.845 \\
1 & 0.2 & -0.92 & -0.568 & 0.693 & 0.845 \\
1 & 0.6 & -0.28 & -0.936 & -0.843 & -0.0758 \\
1 & 1 & 1 & 1 & 1 & 1
\end{bmatrix}
\begin{bmatrix} a_0 \\ a_1 \\ a_2 \\ a_3 \\ a_4 \\ a_5 \end{bmatrix}
=
\begin{bmatrix} 0 \\ 0.588 \\ -0.951 \\ 0.951 \\ -0.588 \\ 0 \end{bmatrix}.
\tag{3-12}
$$

Equations (3-12) may now be solved to yield

$$\mathbf{a} = [0.000164 \quad -1.09 \quad -0.000726 \quad -0.176 \quad -0.00262 \quad 1.27],$$

and the functional approximation now becomes

$$f(x') = 0.000164T_0 - 1.09T_1 - 0.000726T_2 - 0.176T_3 - 0.00262T_4 + 1.27T_5.$$

If we now evaluate this equation at $\mathbf{x} = \pi(\mathbf{x}' = 0)$, we get

$$f(0) = 0.000164 + 0.000726 - 0.00262 = -0.00173,$$

which is only 0.173% from the correct answer.

Figure 3-4 shows a comparison between the Chebyshev approximating polynomial function and the data. Once again, there are excessive fluctuations in the curve. The same detriments that were valid in the algebraic polynomial fit are applicable here. Figure 3-5 displays the errors in the fit of the data versus the control function. Unacceptably large errors also occur in this fit.

In the same manner as with the polynomial approximations, a two-dimensional Chebyshev polynomial fit becomes

$$f(x, y) = \sum_{i=0}^{n} \sum_{j=0}^{m} a_{ij} T_i(x) T_j(y). \tag{3-13}$$

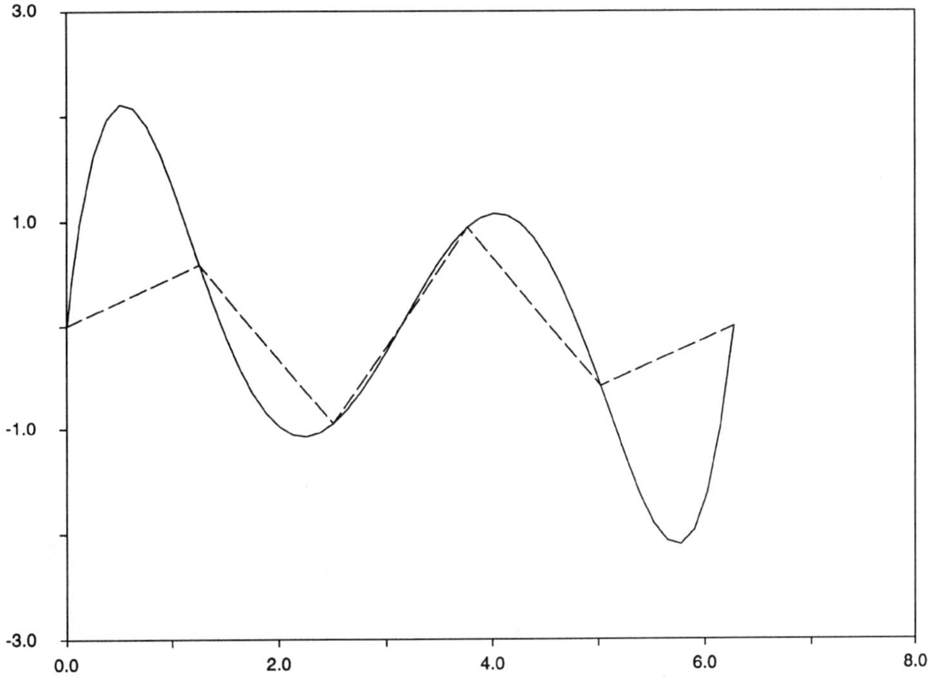

Figure 3-4 Fifth-order Chebyshev polynomial fit.

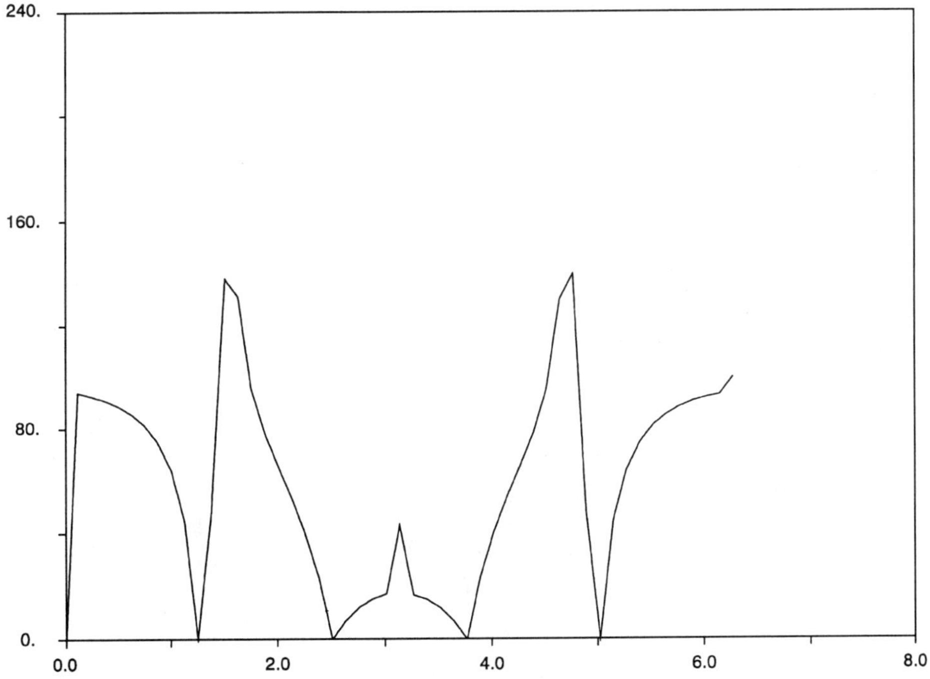

Figure 3-5 Error characteristics of a fifth-order Chebyshev polynomial fit.

In general, the physical interval of the interpolation of **x** is [**a**, **b**] while the physical interval of **y** is [**c**, **d**]. To transform these intervals to the mathematical interval [−1, 1], we must change the variable **x** to **t**:

$$\mathbf{t} = 2\left(\frac{x-a}{b-a}\right) - 1,$$

so that when $\mathbf{x} = \mathbf{a}$, $\mathbf{t} = -1$, and when $\mathbf{x} = \mathbf{b}$, $t = +1$. Similarly, **y** gets transformed to **u** by

$$\mathbf{u} = 2\left(\frac{y-c}{d-c}\right) - 1.$$

The results in this section were achieved with the code **cheby.for**, which is shown in Appendix 3.2.

3-4 LEAST-SQUARES POLYNOMIAL FIT

We have seen the difficulties that arise when we attempt to fit all the available data by an algebraic polynomial fit or by a Chebyshev polynomial fit. If there is a large amount of data, the above methodologies would require a high-order polynomial fit. When a lower-order approximation is desired in order to obtain an analytical expression, it is desirable to approximate the data set by a simple polynomial curve that will not go through all the data points. This would provide a smoother curve and would not oscillate excessively between the given data points.

In this case, the order of the polynomial is preselected, and the coefficients of that curve are calculated so that the deviation of the approximating curve from the given data is a minimum. This procedure is called the least-squares polynomial fit.

If we choose an algebraic polynomial of order p to fit n data points, then

$$\mathbf{y} = \sum_{k=1}^{p+1} a_k x^{k-1}$$

[where $p \leq n$ (data points)] is the expression for this curve. The error that defines the deviation of this polynomial from the actual data is defined as

$$\mathbf{E} = \sum_{i=1}^{n} \left\{ \sum_{k=1}^{p+1} a_k x_i^{k-1} - y_i \right\}^2, \tag{3-14}$$

where $(x_i, y_i; i = 1, n)$ are the data points that are known.

The error in Eq. (3-14) is squared so that the deviation of the approximating function from one data point is always added to the deviation of the previous data point.

Before we develop a general formula for a least-squares algebraic polynomial fit, let us look at an example. If we have n available data points, but choose a straight line as our approximation function, then,

$$\mathbf{y} = a_1 + a_2 x$$

so that the error **E** [from Eq. 3-14] becomes

$$\mathbf{E} = \sum_{i=1}^{n} \{y_i - a_1 - a_2 x_i\}^2. \tag{3-15}$$

To minimize the error in Eq. (3-15), we must choose the coefficients a_1 and a_2 so that **E** is a minimum. This requires that each of the derivatives $[(\partial E)/(\partial a_1)]$ and $[(\partial E)/(\partial a_2)]$ vanish. Thus,

$$\frac{\partial E}{\partial a_1} = \sum_{i=1}^{n} 2\{a_1 + a_2 x_i - y_i\} = 0,$$

or

$$a_1 \sum_{i=1}^{n} 1 + a_2 \sum_{i=1}^{n} x_i = \sum_{i=1}^{n} y_i,$$

$$n a_1 + \left(\sum_{i=1}^{n} x_i \right) a_2 = \sum_{i=1}^{n} y_i. \tag{3-16}$$

Similarly,

$$\frac{\partial E}{\partial a_2} = \sum_{i=1}^{n} 2 x_i \{a_1 + a_2 x_i - y_i\} = 0,$$

or

$$a_1 \sum_{i=1}^{n} x_i + a_2 \sum_{i=1}^{n} x_i^2 = \sum_{i=1}^{n} x_i y_i. \tag{3-17}$$

Equations (3-16) and (3-17) may be solved simultaneously for the unknown coefficients a_1 and a_2.

We now develop a general approach for a least-squares algebraic polynomial fit [Eq. (3-14)]. There will be $(p + 1)$ coefficients to be determined. The minimization process for the rth coefficient requires that

$$\frac{\partial E}{\partial a_r} = 0 = \sum_{i=1}^{n} 2 r x_i^{r-1} \sum_{k=1}^{p+1} \left(a_k x_i^{k-1} - y_i \right) \qquad (r = 1, 2, \ldots, p+1)$$

or

$$0 = \sum_{i=1}^{n} 2 r x_i^{r-1} \left[a_1 + a_2 x_i + a_3 x_i^2 + \cdots + a_k x_i^{k-1} + \cdots - y_i \right],$$

from which we finally get the general formula

$$a_1 \left\{ \sum_{i=1}^{n} x_i^{r-1} \right\} + a_2 \left\{ \sum_{i=1}^{n} x_i^{r} \right\} + \cdots + a_k \left\{ \sum_{i=1}^{n} x_i^{r-1+k-1} \right\} = \sum_{i=1}^{n} y_i x_i^{r-1},$$

$$(r = 1, 2, \ldots, p+1). \tag{3-18}$$

The simultaneous Eq. (3-18) may then be solved for the coefficients $a_1, a_2, \ldots, a_{p+1}$.

As an example of how the least-squares polynomial fit may be used to obtain an approximating curve, let us return to the control function $\mathbf{y} = \sin 2x$. We use the quintic approximation of the above sections. In this case, however, we assume that there are 15 data points that are available, i.e.,

X	Y= sin 2X
0	0
0.449	0.782
0.897	0.975
1.35	0.435
1.79	−0.435
2.24	−0.975
2.69	−0.782
3.14	0
3.59	0.780
4.04	0.976
4.49	0.438
4.93	−0.438
5.38	−0.976
5.83	−0.780
6.28	0

If we apply the least-squares approximation to these data and solve for the unknown coefficients, we get

$$\mathbf{a} = [-0.0772 \quad 5.26 \quad -7.23 \quad 3.28 \quad -0.601 \quad 0.0383], \tag{3-19}$$

so that the quintic algebraic polynomial that fits the 15 data points with the least error becomes

$$\mathbf{y} = -0.0772 + 5.26x - 7.23x^2 + 3.28x^3 - 0.601x^4 + 0.0383x^5. \tag{3-20}$$

It should be noted here that, while the least-squares polynomial fit provides a means of approximation that is shorter than a point-by-point polynomial fit, care must be taken when we are choosing the order of the polynomial. If we choose an order of the polynomial that is too small , then an inaccurate function will be derived, regardless of the number of data points that are available for the fit.

Figure 3-6 shows that this approximation closely represents the control function and can be used as an analytic representation for $\mathbf{y} = \sin 2x$. Figure 3-7 shows that the point-by-point error between this function and the control function is considerably less than with the algebraic polynomial fit of Section 3-2. Appendix 3.3 contains an algorithm, **least.for**, that was used to solve the preceding example. It may also be used to create any order algebraic least-squares fit of an arbitrary data set.

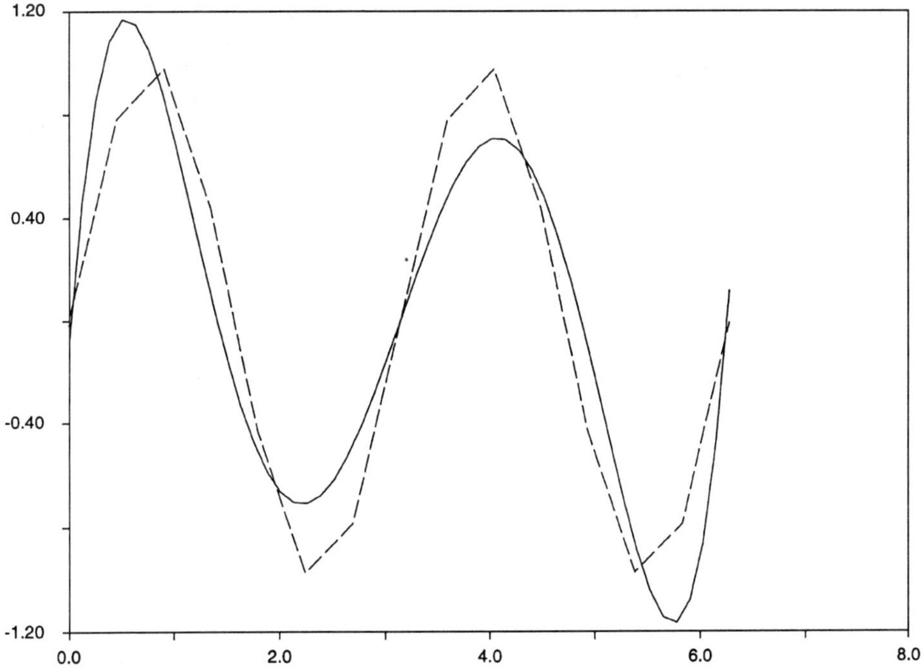

Figure 3-6 Fifth-order least-squares polynomial approximation.

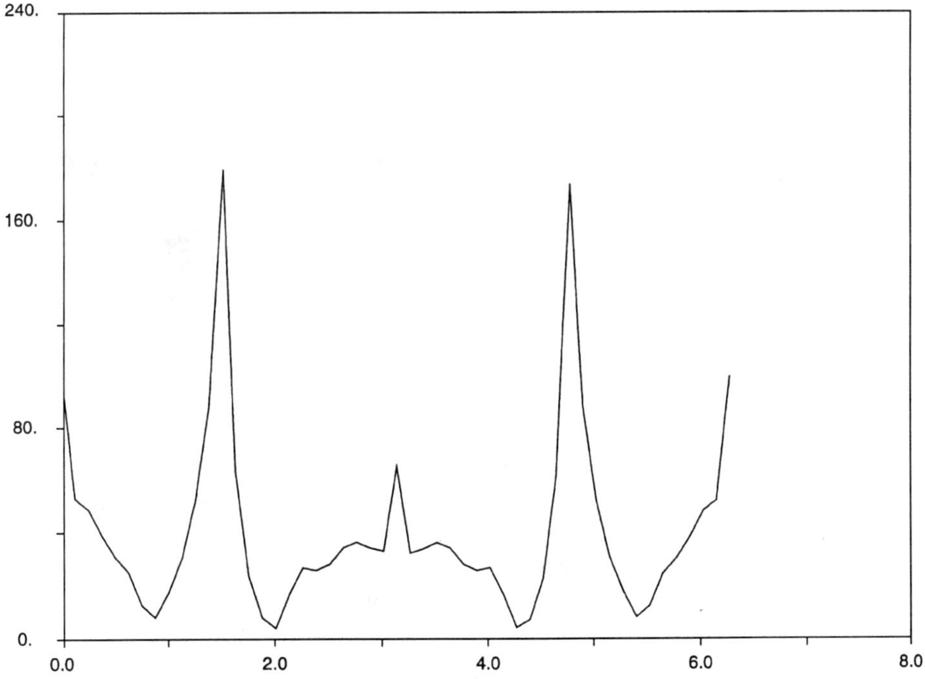

Figure 3-7 Error characteristics of a least-squares fifth-order polynomial.

3-5 LEAST-SQUARES CHEBYSHEV POLYNOMIAL FIT

In the manner that was just described, we can fit the data points with a least-squares Chebyshev polynomial fit instead of a least-squares algebraic polynomial fit. This will lead to a fit that will have fewer oscillations. It will also produce greater accuracy when we are interpolating for the functional values at locations other than at the given data points. We begin by choosing a two-dimensional interpolating function:

$$f(x, y) = \sum_{j=1}^{m}\sum_{k=1}^{n} a_{jk}T_j(x)T_k(y), \tag{3-21}$$

where $T_j(x)$ is the jth-order Chebyshev polynomial function of $\mathbf{x}$. For the least-squares formulation, we form the error

$$\mathbf{E} = \sum_{i=1}^{p}\{F_i - F(x_i, y_i)\}^2,$$

where F_i are the data points at $\{x_i, y_i\}$ and p (in this case) is the number of data points that is available.

The error minimization then proceeds as shown above. We are required to determine all the $(m \times n)$ coefficients. For the coefficient a_{st},

$$\frac{\partial E}{\partial a_{st}} = 0 = \sum_{i=1}^{p} 2\{F_i - F(x_i, y_i)\}\frac{\partial F(x_i, y_i)}{\partial a_{st}} \qquad (s = 1, \ldots, m) \quad (t = 1, \ldots, n)$$

or

$$\sum_{i-1}^{p}\{F_i - F(x_i, y_i)\}T_s(x_i)T_t(y_i) = 0.$$

Then,

$$\sum_{i=1}^{p}\left[\left\{F_i - \sum_{j=1}^{m}\sum_{k=1}^{n} a_{jk}T_j(x_i)T_k(y_i)\right\}T_s(x_i)T_t(y_i)\right] = 0,$$
$$(s = 1, 2, \ldots, m) \qquad (t = 1, 2, \ldots, n),$$

and finally,

$$\sum_{j=1}^{m}\sum_{k=1}^{n} a_{jk}\sum_{i=1}^{p} T_j(x_i)T_k(y_i)T_s(x_i)T_t(y_i) = \sum_{i=1}^{p} F_iT_s(x_i)T_t(y_i) \tag{3-22}$$

yields a set of $(m \times n) = N$ simultaneous equations for the unknown coefficients a_{jk}.

To form the solution matrix for the coefficients a_{jk} in Eq. (3-22), the following strategy must be used in order to identify which equation represents a particular coefficient.

1. The number of coefficients is $(m \times n) = N$. Establish a counter that will locate the position of the coefficient a_{jk} in the resulting set of matrix equations. To locate each coefficient, choose

$$j\text{pos} = (j - 1)^*n + k \qquad (j = 1, \ldots, m) \quad (k = 1, \ldots, n).$$

2. The number of rows in the matrix set of equations is

$$(s \times t) = N,$$

such that

$$\mathbf{b}(ipos) = \text{right-hand side}$$

with

$$ipos = (s - 1)^* m + t.$$

3. The matrix coefficient $\mathbf{A}(ipos, jpos)$ as well as the right-hand-side vector $\mathbf{b}(ipos)$ may then be calculated from Eq. (3-22).

3-6 FOURIER SERIES OF DISCRETE SYSTEMS

Fourier series approximations are usually reserved to represent periodic functions whose analytical expressions are known. The conventional development of a Fourier series formulation is widely covered in other textbooks and is not examined herein. Instead, we are concerned with the Fourier series approximation when the function is known at a discrete number of equally spaced points rather than as an algebraic equation. The data do not have to represent a periodic function. However, the choice of this approximating technique will yield better results if the data closely approximate a periodic function. By performing a variable transformation, we can transform the physical interval by using a new independent variable θ that has the range $[0, 2\pi]$ from some given interval $[\mathbf{a}, \mathbf{b}]$. We then subdivide this interval into $2N$ equally spaced parts by using $\Delta\theta = (\pi/N)$. The function y_k is then known at the points $\theta_i = (i - 1)^* \Delta\theta$. There are $2N$ known values of the function through which the series will be fitted. Then we have

$$\begin{aligned} y_0 &= f(0), \\ y_1 &= f(\Delta\theta), \\ &\vdots \\ y_k &= f(k\Delta\theta), \\ &\vdots \\ y_{2N-1} &= f[2(N-1)\Delta\theta]. \end{aligned}$$

We define our discrete Fourier series as

$$f(\theta) = \frac{a_0}{2} + \sum_{j=1}^{m} (a_j \cos j\theta + b_j \sin j\theta), \tag{3-23}$$

where m is the number of terms in the series.

We can, of course, put the series through all the $2N$ points. However, if we wish to obtain a smooth curve that will fit the data, we may want to revert to a least-squares approach, as delineated above. In this case, we wish to minimize

$$\mathbf{E} = \sum_{k=0}^{2N-1} [f(\theta) - y_k]^2.$$

Following the procedures of Section 3-5 we can show that the Fourier coefficients are

$$a_j = \frac{1}{N} \sum_{k=0}^{2N-1} f(\theta_k) \cos j\theta_k \qquad j = 0, 1, 2, \ldots, m,$$
$$b_j = \frac{1}{N} \sum_{k=0}^{2N-1} f(\theta_k) \sin j\theta_k \qquad j = 1, 2, \ldots, m, \tag{3-24}$$

where $\theta_k = k\Delta\theta = k(\pi/N)$.

To illustrate the application of the above technique, let us consider four data points that are available to us by using the control function $\mathbf{y} = \sin x$. Here, let us once again consider the data in the interval $[0, 2\pi]$, i.e.,

θ	0	$\frac{2\pi}{3}$	$\frac{4\pi}{3}$	2π
$f(\theta)$	0	$\frac{\sqrt{3}}{2}$	$-\frac{\sqrt{3}}{2}$	0

If we apply the discrete Fourier series methodology to the above data with four points and use a two-term Fourier series (as the least-squares approximating curve), then the series becomes

$$f(\theta) = 0.75 \sin\theta - 0.75 \sin 2\theta. \tag{3-25}$$

Equation (3-25) is a poor approximation of $\mathbf{y} = \sin\theta$. The reason for this can be seen if we plot the four points graphically (Fig. 3-8). If we connect these four points by straight lines, the function looks more like an alternating sawtooth than a sine function. In this case, the poor approximation is due to the insufficient number of data points and is not due to the order of the least-squares approximating curve. If the number of data points that is available in the region $[0, 2\pi]$ is increased to eight (with the same two-term

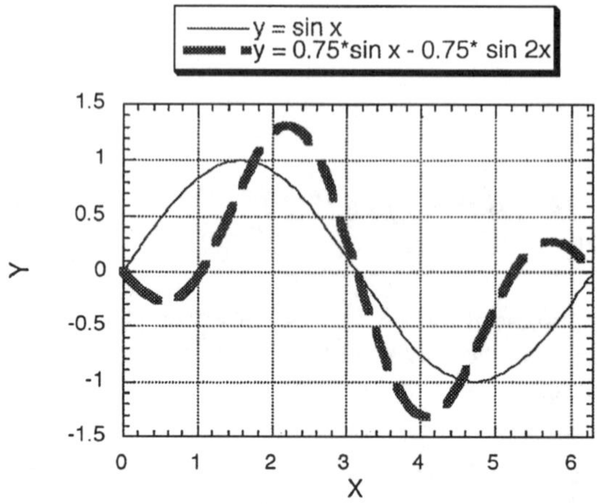

Figure 3-8 Four-data-point discrete Fourier fit.

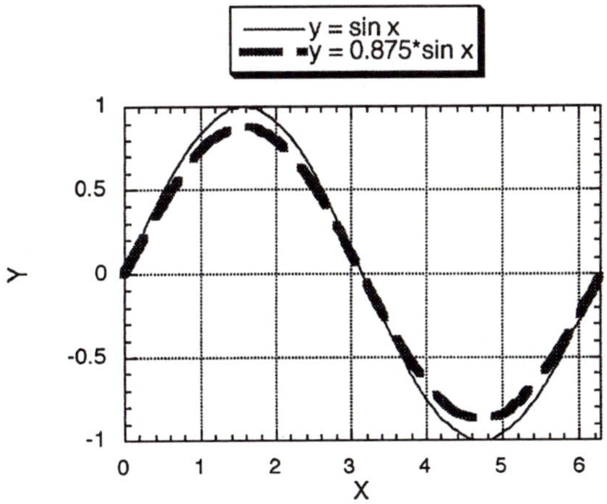

Figure 3-9 Eight-data-point discrete Fourier fit.

Fourier series), then the approximation becomes

$$f(\theta) = 0.875 \sin \theta. \tag{3-26}$$

Equation (3-26) is a better approximation, but still falls short of an acceptable approximation (Fig. 3-9). For 16 data points,

$$f(\theta) = 0.938 \sin \theta, \tag{3-27}$$

and for 32 data points,

$$f(\theta) = 0.98 \sin \theta. \tag{3-28}$$

We then see that the approximation approaches the actual function as the number of data points increases, although the number of terms in the Fourier series still remains at two. Figures 3-10 and 3-11 show the effect of Eqs. (3-27) and (3-28), respectively, on the approximation process.

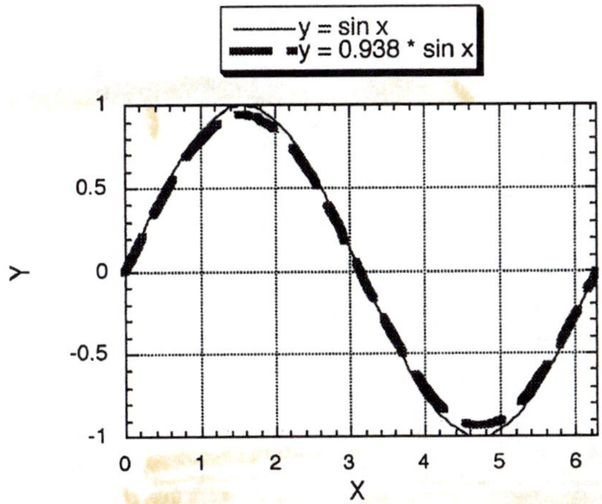

Figure 3-10 16-point discrete Fourier fit.

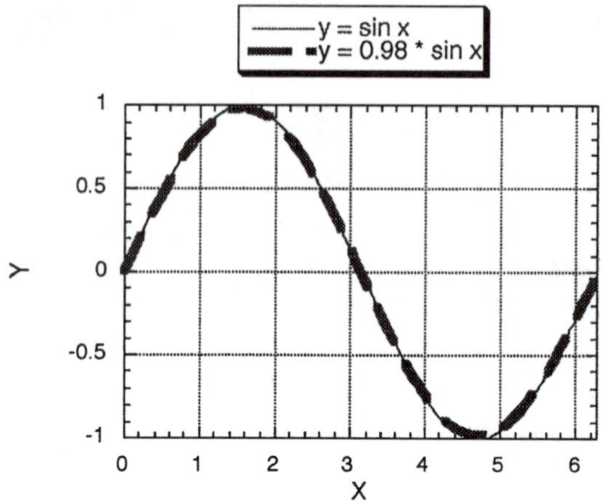

Figure 3-11 32-point discrete Fourier fit.

Thus we see that a least-squares Fourier series approximation can yield a good fit of the data, especially if the function is periodic. If the range of the given data set is not $[0, 2\pi]$, then a transformation of the independent variable must be performed.

Appendix 3.4 contains a FORTRAN; algorithm, called **fourier.for**, that was used to solve the just-described problem in this section.

3-7 SPLINE FUNCTIONS

3-7.1 Piecewise Linear Splines

Polynomial interpolations for a set of $N + 1$ data points $\{x_k, y_k\}$ are frequently unsatisfactory. As described above, polynomials of degree N have $N - 1$ relative maxima and minima, and the approximating polynomials can wiggle in order to pass through the points. An alternative method for curve fitting the data is to piece together the algebraic representations of lower-degree polynomials $S_k(x)$ and interpolate between successive data points. We call these points nodes (or knots). Basically, we form a low-degree polynomial that will approximate the data between any two adjacent nodal points $\{x_k, y_k\}$ and $\{x_{k+1}, y_{k+1}\}$ within the data set $\{x_0, y_0\} - \{x_n, y_n\}$. Since these representations will be valid only within a specific region, they will be more accurate than if we attempted to use them over the entire range of data. Figure 3-12 demonstrates the concept of this piecing together strategy. The two adjacent portions of the curve $\mathbf{y} = S_k(x)$ and $\mathbf{y} = S_{k+1}(x)$ pass through a common knot at (x_{k+1}, y_{k+1}) and are tied together there.

For another example of this concept, let us look at Fig. 3-13. The simplest curve fit is a polynomial of degree 1 (i.e. a straight line), i.e.,

$$S_k(x) = y_k \frac{x - x_{k+1}}{x_k - x_{k+1}} + y_{k+1} \frac{x - x_k}{x_{k+1} - x_k} \qquad \text{for} \quad x_k \le x \le x_{k+1} \tag{3-29}$$

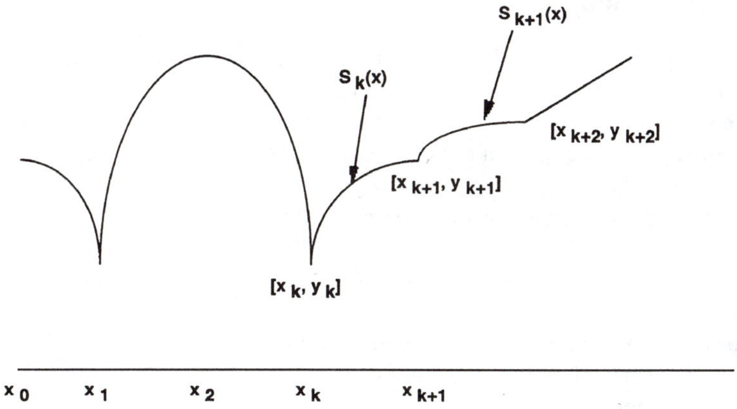

Figure 3-12 Basic configuration for spline methodologies.

or

$$S_k(x) = y_k + d_k(x - x_k), \tag{3-30}$$

where

$$d_k = \frac{y_{k+1} - y_k}{x_{k+1} - x_k}.$$

Thus, for the entire field, a series of straight lines $S(x)$ can be pieced together as

$$S(x) = \begin{cases} y_0 + d_0(x - x_0) & \text{for } x \text{ in } [x_0, x_1] \\ y_1 + d_1(x - x_1) & \text{for } x \text{ in } [x_1, x_2] \\ \quad \cdots & \\ y_{n-1} + d_{n-1}(x - x_{n-1}) & \text{for } x \text{ in } [x_{n-1}, x_n] \end{cases} . \tag{3-31}$$

It must be reemphasized here that each of the above pieces are valid only in the interval that is specified and must not be used outside that interval.

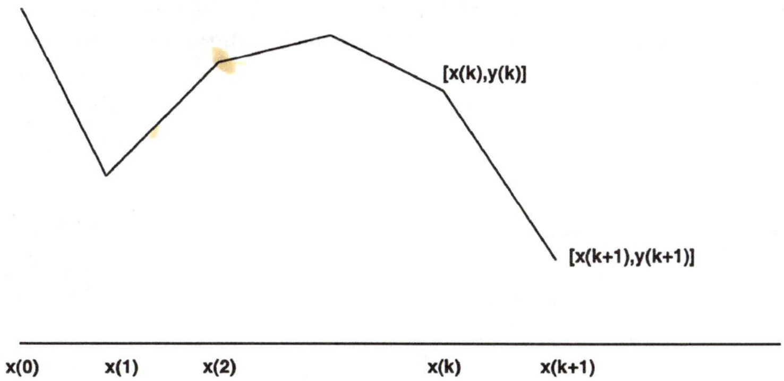

Figure 3-13 Piecewise linear interpolation.

3-7.2 Piecewise Cubic Splines

One of the more useful splines in mechanics is the cubic spline, which has both the first and the second derivatives continuous on the entire interval $[x_0, x_n]$. The continuity of $S'(x)$ means that the graph $\mathbf{y} = S(x)$ will have no sharp corners. The continuity of $S''(x)$ means that the curvature is defined at each point. We therefore derive the methodology for cubic-spline approximations.

These are the requirements for the construction of a cubic spline:

1. $S(x) = S_k(x) = a_{k,0} + a_{k,1}(x - x_k) + a_{k,2}(x - x_k)^2 + a_{k,3}(x - x_k)^3$, $(k = 0, 1, \ldots, N - 1)$.
2. $S_k(x_k) = y_k$ (the spline passes through each given data point).
3. $S_k(x_{k+1}) = S_{k+1}(x_{k+1})$ (the spline is a continuous function).
4. $S'_k(x_{k+1}) = S'_{k+1}(x_{k+1})$ (the spline forms a smooth function).
5. $S''_k(x_{k+1}) = S''_{k+1}(x_{k+1})$ (the second derivative is continuous).

Before proceeding to the task of determining the coefficients $a_{k,j}$, we must determine whether we have enough information to construct these splines. There are n splines (over the range $[x_0, x_n]$) with four unknown coefficients each. Thus we have $4n$ degrees of freedom. To satisfy these unknowns, we have

1. $n + 1$ data points to satisfy property 2 above.
2. Each of properties 3–5 supplies $(n - 1)$ conditions (since we have $n - 1$ common points for the n splines).

Hence there are $n + 1 + 3(n - 1) = 4n - 2$ conditions that can be used to determine the $4n$ unknown coefficients. We therefore need two additional conditions, called end-point constraints, that will have to be added for us to determine all the coefficients completely. These will involve either the slope $S'(x)$ or the second derivative $S''(x)$ at the end points x_0 and x_n. We examine these constraints below.

Before we begin the derivation process below, it may be useful to discuss the theory that drives the mathematics. Since we are attempting to derive a cubic representation, we know that the second derivative will be linear. We begin with this premise. We then carefully integrate this line by using the above conditions 1–5 to evaluate the constants of integration and to produce a recursive formula that can be used to evaluate the coefficients from each cubic-spline segment.

To begin the derivation process, consider that since $S(x)$ is piecewise cubic, $S''(x)$ is piecewise linear on $[x_0, x_n]$. Thus, for the kth spline (which is valid only within $x_k \leq x \leq x_{k+1}$),

$$S''_k(x) = S''(x_k)\frac{x - x_{k+1}}{x_k - x_{k+1}} + S''(x_{k+1})\frac{x - x_k}{x_{k+1} - x_k}. \tag{3-32}$$

If we denote $m_k = S''(x_k)$, $m_{k+1} = S''(x_{k+1})$, and $h_k = x_{k+1} - x_k$, then the second derivative becomes

$$S''_k(x) = \frac{m_k}{h_k}(x_{k+1} - x) + \frac{m_{k+1}}{h_k}(x - x_k) \tag{3-33}$$

for $x_k \leq x \leq x_{k+1}$ and $k = 0, 1, \ldots, N - 1$.

Integrating Eq. (3-33) between the limits x_k and x_{k+1}, we get

$$S'_k(x) = -\frac{m_k}{2h_k}(x_{k+1} - x)^2 + \frac{m_{k+1}}{2h_k}(x - x_k)^2 + p_k + q_k, \tag{3-34}$$

where both p_k and q_k are constants of integration.

Integrating Eq. (3-34) between the above limits yields

$$S_k(x) = \frac{m_k}{6h_k}(x_{k+1} - x)^3 + \frac{m_{k+1}}{6h_k}(x - x_k)^3 + p_k(x_{k+1} - x) + q_k(x - x_k). \tag{3-35}$$

If we substitute $y_k = S_k(x_k)$ at $x = x_k$ and $y_{k+1} = S_k(x_{k+1})$ at $x = x_{k+1}$ into Eq. (3-35), we get

$$y_k = \frac{m_k}{6}h_k^2 + p_k h_k, \tag{3-36}$$

$$y_{k+1} = \frac{m_{k+1}}{6}h_k^2 + q_k h_k. \tag{3-37}$$

We can solve Eqs. (3-36) and (3-37) for p_k and q_k and then substitute them back into Eq. (3-35) yielding,

$$S_k(x) = \frac{m_k}{6h_k}(x_{k+1} - x)^3 + \frac{m_{k+1}}{6h_k}(x - x_k)^3$$
$$+ \left(\frac{y_k}{h_k} - \frac{m_k h_k}{6}\right)(x_{k+1} - x) + \left(\frac{y_{k+1}}{h_k} - \frac{m_{k+1} h_k}{6}\right)(x - x_k). \tag{3-38}$$

At this point, the only quantities that are unknown in Eq. (3-38) are the m_k values. Differentiating Eq. (3-38), we obtain

$$S'_k(x) = -\frac{m_k}{2h_k}(x_{k+1} - x)^2 + \frac{m_{k+1}}{2h_k}(x - x_k)^2$$
$$-\left(\frac{y_k}{h_k} - \frac{m_k h_k}{6}\right) + \left(\frac{y_{k+1}}{h_k} - \frac{m_{k+1} h_k}{6}\right). \tag{3-39}$$

If we evaluate Eq. (3-39) at x_k, we get

$$S'_k(x_k) = -\frac{m_k}{3}h_k - \frac{m_{k+1}}{6}h_k + d_k, \tag{3-40}$$

where

$$d_k = \frac{y_{k+1} - y_k}{h_k}.$$

If we replace k with $k-1$ in Eq. (3-39), then

$$S'_{k-1}(x_k) = \frac{m_k}{3}h_{k-1} + \frac{m_{k-1}}{6}h_{k-1} + d_{k-1}. \tag{3-41}$$

Using Eqs. (3-40) and (3-41), we can satisfy property 4 (i.e, the spline forms a smooth function) by

$$h_{k-1}m_{k-1} + 2(h_{k-1} + h_k)m_k + h_k m_{k+1} = u_k, \tag{3-42}$$

where $u_k = 6(d_k - d_{k-1})$ for $k = 1, 2, \ldots, N-1$.

We use Eq. (3-42) to calculate the values of m and thus completely define all the splines. However, as explained above, we have a deficit of two equations (i.e., the end-point constraints). We therefore need conditions to help us calculate m_0 and m_n at the endpoints.

In order to generalize the end conditions, let us transform Eq. (3-42) into a more convenient form. If we define

$$\begin{aligned} \alpha_i &= \frac{h_{i+1}}{h_{i-1}+h_i}, \\ \beta_i &= 1-\alpha_i = \frac{h_i}{h_{i-1}+h_i}, \\ \gamma_i &= \frac{6}{h_{i-1}+h_i}\left\{\frac{y_{i+1}-y_i}{h_{i+1}} - \frac{y_i-y_{i-1}}{h_i}\right\}, \end{aligned} \tag{3-42a}$$

then Eq. (3.42) becomes

$$\beta_i m_{i-1} + 2m_i + \alpha_i m_{i+1} = \gamma_i. \tag{3-43}$$

If we recall that the m values represent the second derivatives of the curves, then we see that m_{-1} and m_{n+1} are not defined. Thus, from Eq. (3-43), the end conditions may now be conveniently written as

$$\begin{aligned} 2m_0 + \alpha_0 m_1 &= \gamma_0, \\ \beta_N m_{N-1} + 2m_N &= \gamma_N. \end{aligned} \tag{3-44}$$

This formulation allows us to form the following set of linear simultaneous equations that may be used to solve for the unknown values of m:

$$\begin{bmatrix} 2 & \alpha_0 & 0 & & & & \\ \beta_1 & 2 & \alpha_1 & & & & \\ 0 & \beta_2 & 2 & & & & \\ \cdots & \cdots & \cdots & \cdots & \cdots & & \\ & & & 2 & \alpha_{N-2} & 0 \\ & & & \beta_{N-1} & 2 & \alpha_{N-1} \\ & & & 0 & \beta_N & 2 \end{bmatrix} \begin{bmatrix} m_0 \\ m_1 \\ m_2 \\ \vdots \\ m_{N-2} \\ m_{N-1} \\ m_N \end{bmatrix} = \begin{bmatrix} \gamma_0 \\ \gamma_1 \\ \gamma_2 \\ \vdots \\ \gamma_{N-2} \\ \gamma_{N-1} \\ \gamma_N \end{bmatrix}. \tag{3-45}$$

Equations (3-45) form a tridiagonal set that can be solved by the techniques that were described in Chapter 1.

Table 3-1 summarizes the possible ways of stating the end-point constraints. It is obvious that case 3 (extrapolation of the second derivative) in the above table cannot be conveniently put into the α, β, γ formulation. Instead, for case 3 only, the first equation in the set of Eqs. (3-45) must be modified to

$$m_0 - \left\{1 + \frac{h_0}{h_1}\right\} m_1 + \frac{h_0}{h_1} m_2 = 0,$$

Table 3-1 End-point constraints for a cubic spline

Strategy	Coefficients for Eq. (3-44)
1. Clamped Specify $S'(x_0) = y'_0$ and $S'(x_N) = y'_N$. (this is best if derivatives are known at the end points).	$\alpha_0 = \beta_N = 1$ $\gamma_0 = \frac{6}{h_0}\left\{\frac{y_1 - y_0}{h_0} - y'_0\right\}$ $\gamma_N = \frac{6}{h_{N-1}}\left\{y'_N - \frac{y_N - y_{N-1}}{h_{N-1}}\right\}$
2. Natural (if no information is available regarding the end conditions of the spline).	$\alpha_0 = \gamma_0 = \beta_N = \gamma_N = 0$
3. Extrapolate $S''(x)$ to the end points.	See discussion below this table for this case.
4. $S''(x)$ is constant near the end points	$\alpha_0 = -2 = \beta_N$ $\gamma_0 = 0 = \gamma_N$
5. Specify $S''(x) = y''(x)$ at each end point	$\alpha_0 = 0 = \beta_N$ $\gamma_0 = 2y''_0$ $\gamma_N = 2y''_N$

and the last equation in the set of Eqs. (3-45) must be changed to

$$\frac{h_{N-1}}{h_{N-2}} m_{N-2} - m_{N-1}\left\{1 + \frac{h_{N-1}}{h_{N-2}}\right\} + m_N = 0.$$

By choosing one of the end-point strategies from Table 3-1 and adding it to Eq. (3-43), we can solve the matrix set of equations for $[m_k]$. The spline coefficients $\{S_{kj}\}$ for $S_k(x)$ can then be computed by

$$\begin{aligned} \{S_{k,0}\} &= y_k, \\ \{S_{k,1}\} &= d_k - \frac{h_k(2m_k + m_{k+1})}{6}, \\ \{S_{k,2}\} &= \frac{m_k}{2}, \\ \{S_{k,3}\} &= \frac{m_{k+1} - m_k}{6h_k}. \end{aligned} \tag{3-46}$$

Equations (3-46) then allow us to describe the spline for region k, viz.,

$$S_k(x) = [(\{S_{k,3}\}w + \{S_{k,2}\})w + \{S_{k,1}\}]w + y_k, \tag{3-47}$$

where $w = x - x_k$ and Eq. (3-47) is valid only in the interval $x_k \le x \le x_{k+1}$.

To illustrate the use of spline function approximations and to demonstrate the use of end-condition strategies, we return to the approximation of $\mathbf{y} = \sin\, x$ (with four data points). We attempt to approximate this function with two separate end-point strategies. In the first example, we assume that there is no information available concerning the end points of the splines. Therefore we adapt strategy 2 from Table 3-1. It is common to call this choice a natural spline. In the second example, we assume that we know the slopes of the splines at the end points of the overall region and we use end-point strategy 1 from Table 3.1. This choice is known as a clamped spline.

Example I: natural spline Since we are fitting four points, we have three splines to determine. Since the points are equally spaced at $h_i = (\pi/3)$, then from Eqs. (3-42a),

$$\alpha_1 = \alpha_2 = \beta_1 = \beta_2 = \frac{1}{2}.$$

We may also calculate

$$\gamma_1 = \gamma_2 = -\frac{27\sqrt{3}}{2\pi^2},$$

which leads to the equations

$$\begin{aligned} \frac{1}{2}m_0 + 2m_1 + \frac{1}{2}m_2 &= -\frac{27\sqrt{3}}{2\pi^2}, \\ \frac{1}{2}m_1 + 2m_2 + \frac{1}{2}m_3 &= -\frac{27\sqrt{3}}{2\pi^2}. \end{aligned} \tag{3-48}$$

Since we are using a natural spline, Eqs. (3-43) yield $m_0 = m_3 = 0$, so that we may calculate

$$m_1 = m_2 = -\frac{27\sqrt{3}}{5\pi^2}.$$

Equations (3-46) yield the coefficients of the natural cubic spline, viz.,

$S_k(x)$	$S_{k,0}$	$S_{k,1}$	$S_{k,2}$	$S_{k,3}$
$S_0(x)$	0.0	0.992	0.0	−0.151
$S_1(x)$	0.866	0.496	−0.474	0.0
$S_2(x)$	0.866	−0.496	−0.474	0.151

Equation (3-47) may now be used to construct the cubic splines and to evaluate the dependent variable for any selection of independent values that are in the range $[0, \pi]$.

Figure 3-14 shows the comparison of this methodology with the given function and the data points. The fit is excellent.

Example II: derivative end points For this approach to the approximation of the cubic spline to $\mathbf{y} = \sin x$, we want the end-point first derivatives to be equal to the sine derivatives, i.e.,

$$S_1'(0) = 1, \qquad S_3'(\pi) = -1.$$

The end coefficients then become

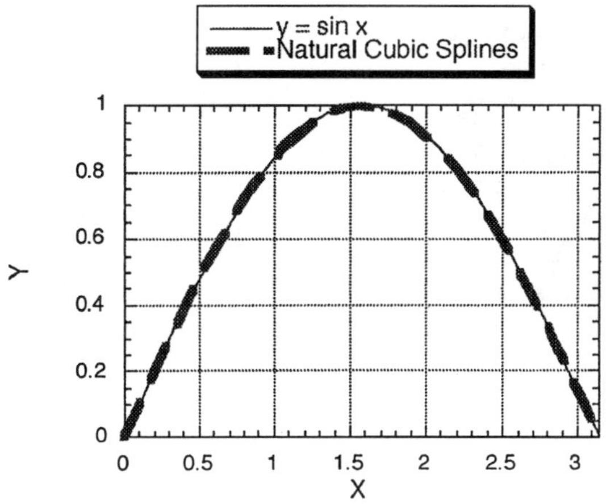

Figure 3-14 Natural cubic-spline fit.

$$\alpha_0 = \beta_3 = 1 \qquad \text{(Table 3.1, strategy 1)},$$

$$\gamma_0 = \gamma_3 = \left(\frac{18}{\pi}\right)\left(\frac{3\sqrt{3}}{2\pi} - 1\right),$$

so that the system of equations for the unknown $\{m\}$ values becomes

$$\begin{aligned} 2m_0 + m_1 &= \left(\frac{18}{\pi}\right)\left(\frac{3\sqrt{3}}{2\pi} - 1\right), \\ \frac{1}{2}m_0 + 2m_1 + \frac{1}{2}m_2 &= \frac{-27\sqrt{3}}{2\pi^2}, \\ \frac{1}{2}m_1 + 2m_2 + \frac{1}{2}m_3 &= \frac{-27\sqrt{3}}{2\pi^2}, \\ m_2 + 2m_3 &= \left(\frac{18}{\pi}\right)\left(\frac{3\sqrt{3}}{2\pi} - 1\right), \end{aligned} \tag{3-49}$$

from which we determine that

$$m_0 = m_3 = \frac{18\sqrt{3}}{\pi^2} - \frac{10}{\pi}, \tag{3-50}$$

$$m_1 = m_2 = \frac{2}{\pi} - \frac{9\sqrt{3}}{\pi^2}. \tag{3-51}$$

The coefficients of the cubic splines that form the approximation of the given data when end-point strategy 1 is used

$S_k(x)$	$S_{k,0}$	$S_{k,1}$	$S_{k,2}$	$S_{k,3}$
$S_0(x)$	0.0	1.0	−0.0121	−0.146
$S_1(x)$	0.866	0.494	−0.471	0.0
$S_2(x)$	0.866	−0.494	−0.471	0.146

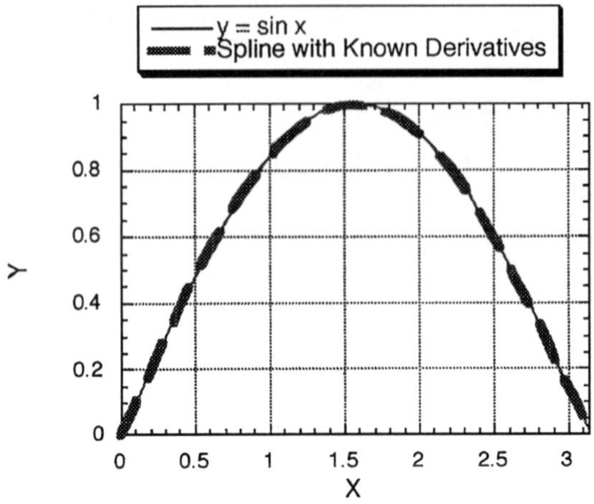

Figure 3-15 Spline with known derivatives.

Figure 3-15 shows the result of this strategy.

A comparison of Figs. 3-14 and 3-15 show that for this function (i.e., $\mathbf{y} = \sin x$), either of the above end-point strategies will yield excellent fits of the data when cubic-spline functions are used as the approximating curves.

Appendix 3-5 has an algorithm, called **spline.for**, that will create the cubic splines and evaluate these splines at a desired value of the independent variable. This code was used in the examples of this section.

3-8 CONCLUSION

In this chapter, several important techniques were presented that can aid students in obtaining smooth algebraic representations of discrete data.

We have explored the algebraic polynomial curve-fitting techniques, the Chebyshev polynomial technique, the least-squares versions of the two above-mentioned techniques, the Fourier series approximation for discrete data, and finally the cubic-spline methodology. The choice of which technique to choose in a particular application is, of course, left to the readers. Criteria such as number of data points, cyclic nature of the data, and requirements for obtaining derivatives, should lead students to one of the robust procedures that have been presented in this chapter. A brief discussion of the advantages and the disadvantages of the methods is made here to aid novice readers in the choice of the appropriate interpolation scheme.

The algebraic polynomial technique is the easiest to implement from the existing data. If the data are relatively smooth or if the amount of data is relatively small, this approach will yield a satisfactory fit. As for all the techniques in this chapter, one should plot the algebraic representation in the same region as the given data. If the resulting comparison is satisfactory to the user, then the process has been successful. On the other hand, if large changes occur in the slopes of the curve and it wiggles greatly, then the determination of the values of the dependent variable between the given data points will be inaccurate. In this event, the algebraic polynomial approach must be abandoned.

The Chebyshev polynomial fitting attempts to rectify some (the wiggling) of the problems that may be encountered in the above-mentioned fitting strategy. Here too, if the amount of data is excessive, the approximating representation will be considered unsatisfactory. Further, one must remember to transform the physical interval to the Chebyshev domain $[-1, +1]$. Once the fit is complete, a transformation back to the physical domain [**a**, **b**] is usually required so that the expression may be used with a physical data set.

Additionally, as in the algebraic polynomial technique, a large data set will yield large matrices that will have to be inverted in order to solve for the coefficients. This can result in huge computer costs and ultimately result in inaccurate values for the coefficients because of a computer round-off error in the matrix inversion.

The least-squares techniques for both algebraic and Chebyshev polynomials are trade-offs. On the one hand, there is less computational effort required because of the lower-order expressions that are assumed. Since lower-order curves are used, there is a possibility of less wiggling. On the other hand, we obtain approximations that will not fit all the data. Further, the success of the least-squares strategies depends almost entirely on the choice of the preselected orders of the polynomials. Thus the reader has to have some unusual prescience as to the nature of the given data points.

The Fourier series fit of discrete data is useful mainly when the data are suspected of being periodic within (and without) the range of interest. If this is not the case, then a choice of one of the other interpolation techniques will probably produce better results.

Finally, we consider the advantages and the disadvantages of cubic-spline methodologies. These are perhaps the most accurate of the fitting methods since they allow for the fitting of curves within a given domain whose size may be selected by the user. One does not attempt to require that the same expression be valid over the entire range of the data. The only adverse aspect of these techniques is the storage of many sets of cubic splines and the subsequent search for the correct spline that is applicable to the evaluation at a specific point (i.e., one must locate which spline controls the region in which the desired evaluative point resides).

The use of the techniques in this chapter can aid materially in the ease of application of the methods that are presented in Chapter 4 on numerical integration.

FURTHER READING

Arden, B. W. and Astill, K. N. (1970). *Numerical Algorithms—Origins and Applications*, Addison-Wesley, Reading, Massachusetts.

Carnahan, B., Luther, H., and Wilkes, J. (1969). *Applied Numerical Methods*, Wiley, New York.

Mathews, J. H. (1987). *Numerical Methods for Computer Science, Engineering, and Mathematics*, Prentice-Hall, Englewood Cliffs, New Jersey.

Pennington, R. H. (1965). *Introductory Computer Methods and Numerical Analysis*, Macmillan, New York.

Press, W. H., Teukolsky, S., Vetterling, W., and Flannery, B. (1986). *Numerical Recipes in FORTRAN*, Cambridge University Press, Cambridge.

Salvadori, M. G. and Baron, M. L. (1952). *Numerical Methods in Engineering*, Prentice-Hall, Englewood Cliffs, New Jersey.
Scheid, F. (1968). *Schaum's Outline of Theory and Problems of Numerical Analysis*, McGraw-Hill, New York.

PROBLEMS

For each problem in this section, write a computer code to perform the required approximation. Several algorithms (which have been described within this chapter) are given in the Appendixes to act as a model for the students.

3-1 Given the function $\mathbf{y} = [1/(1 + 12x^2)]$, construct an algebraic polynomial approximation, of the order of 4, for this function.

3-2 Repeat problem 3-1 with an eighth degree algebraic polynomial fit. Compare your results with those of Problem 3-1.

3-3 Repeat problem 3-1 with a tenth degree algebraic polynomial fit. Compare your results with those of Problems 3-1 and 3-2.

3-4 For the function of Problem 3-1, construct an eighth order Chebyshev polynomial approximation. Compare your results with those of the algebraic polynomial fit of Problem 3-2.

3-5 For the function of Problem 3-1, construct a tenth-order Chebyshev polynomial approximation. Compare your results with those of the algebraic polynomial fit of Problem 3-3.

3-6 Use algebraic polynomials to find a series that will approximate $\mathbf{y} = \sin(\pi/2)x$ over the interval $0 \le x \le 2\pi$.

3-7 Use Chebyshev polynomials to find a series that will approximate $\mathbf{y} = \sin(\pi/2)x$ over the interval $-1 \le x \le 1$.

3-8 Construct a least-squares algebraic polynomial fit for the function $\mathbf{y} = ae^{bx}$. Hint: Express this function as $\log_e y = \log_e a + bx$.

3-9 For the given set of data, construct a least-squares quadratic polynomial function. Use the constructed approximation to smooth the data. Plot both curves.

x	6	8	10	12	14	16	18	20	22	24
y	3.8	3.7	4.0	3.9	4.3	4.2	4.2	4.4	4.5	4.5

ANSWER $a_0 = 3.73, a_1 = 0.11, a_2 = -0.0023$.

3-10 Repeat Problem 3-9 for a least-squares Chebyshev polynomial fit of the data with the function

$$y(x) = \sum_{k=1}^{3} a_i T_i(x).$$

3-11 During an experiment in which a single mass is undergoing a single-degree-of-freedom vibration, the following data are recorded. $\boldsymbol{x}$ is the displacement of the mass and t is the time when the oscillation was recorded.

x	0.0	0.98	0.15	0.01	−0.16	−0.97	0.0
t	0	2	4	6	8	10	12

Determine the first three coefficients of a Fourier series that is fitted to the data.

3-12 Repeat Problem 3-11 for a third-degree algebraic polynomial approximation. Compare the results.

3-13 Repeat Problem 3-11 for a third-order Chebyshev polynomial approximation. Compare the results.

3-14 Repeat Problem 3-11 for a third degree least-squares algebraic polynomial approximation. Compare the results.

3-15 Repeat Problem 3-11 for a third-order least-squares Chebyshev polynomial approximation. Compare the results.

3-16 Find the coefficients of a clamped cubic spline that passes through the points (0, 0), (1, 0.5), (2, 2.0), and (3, 1.5) with the end conditions $S'(0) = 0.2$ and $S'(3) = -1$.

3-17 Repeat the construction of a spline for the data of Problem 3-16, by using a natural-spline fit.

3-18 Construct a cubic spline to fit the following data:

x	0	2	2.5	3	3.5	4	4.5	5	6
y	0	2.0	3.5	3.8	3.5	3.5	3.5	2.6	0

Compare a plot of the original data with a plot of the spline function.

ANSWER $\{m\} = \{-0.23, -0.72, -4.08, 2.65, 0.69, -5.4, -0.7\}$. Use the natural end condition option.

3-19 Repeat Problem 3-18. Attempt to choose an end condition that will approximate the given data more accurately that in Problem 3-18.

CHAPTER

FOUR

NUMERICAL INTEGRATION

NOMENCLATURE

S_k	kth interval spline
$f_j(x)$	functional value at the jth point
$G[f]$	Gauss quadrature
x_j	integration variable for the jth point
$[\mathbf{a}, \mathbf{b}]$	interval of integration
b'	ordinate intercept of a straight line
$S[f]$	Simpson's rule
$T[f]$	trapezoidal rule
w_j	weight for f_j
$E[f]$	error of the approximation
h	grid spacing
I	value of the definite integral
m	slope of a straight line
N	number of grid points in the integration interval
$Q[f]$	quadrature formula
$\mathbf{t}$	transformation variable

4-1 INTRODUCTION

Numerical integration is a primary tool that is used by engineers and scientists to obtain approximate values of definite integrals that cannot be solved analytically. The goal is to approximate the definite integral of $f(x)$ over the interval $[\mathbf{a}, \mathbf{b}]$ by use of a finite number of data points.

In Chapter 3, the curve fitting of data points was presented so that approximations to a data set may be made. These curve fits may then be used to integrate these functions analytically over the range of the data set. Additionally, even when the approximation function is expressed in a piecewise manner (as spline curves), the integration may be carried out analytically.

In this chapter, however, we seek to establish integration formulas for discrete data such as those that are obtained experimentally. Also, when an analytic function does not have a closed-form integral, such as $\int_a^b e^{x^2}\,\mathrm{d}x$, and the definite integral is desired, the techniques of this chapter will be applicable.

4-2 SIMPLE INTEGRATION FORMULAS

The study of numerical integration techniques begins with a definition of the interval and the type of formula that we wish to derive. Suppose the known data points are defined in the interval [**a**, **b**] such that

$$\mathbf{a} = x_0 < x_1 < x_2 < \cdots < x_{n-1} < x_n = \mathbf{b}.$$

We define a series of weighted terms for each data point such that

$$Q[f] = \sum_{j=0}^{n} w_j f(x_j) = w_0 f(x_0) + w_1 f(x_1) + \cdots + w_n f(x_n). \qquad (4\text{-}1)$$

We approximate the definite integral by

$$\int_a^b f(x)\,\mathrm{d}x = Q[f] + E[f],$$

where $E[f]$ is the truncation error and $\{w_i\}$ are the weight functions. All the integration formulas that are derived in this chapter take the form of Eq. (4-1), i.e., the summation of a weighting function multiplied by its functional value.

To create simple (yet powerful) integration formulas, consider choosing an interval [**a**, **b**] over which the desired integration $\int_a^b f(x)\,dx$ is required. We can then use approximating algebraic polynomials of degrees 1, 2, 3, and 4 (similar to those that were discussed in Chap. 3) to arrive at a set of quadrature formulas called the Newton–Cotes formulas.

If we assume equally spaced nodes, i.e.,

$$x_j = x_0 + h_j,$$

where

$$f_j = f(x_j),$$

then the following quadrature formulas can be derived.

From the choice of a linear approximating polynomial, $\mathbf{y} = mx + b'$, we can form the definite integral over the range $[x_0, x_1]$:

$$\begin{aligned} I &= \int_{x_0}^{x_1} (mx + b')\,\mathrm{d}x \\ &= \frac{1}{2}mx^2 + b'x\Big|_{x_0}^{x_1} \\ &= \frac{1}{2}m\left(x_1^2 - x_0^2\right) + b'(x_1 - x_0), \end{aligned}$$

but the $\mathbf{y}$ intercept of the straight line is given by $b' = f(x_0) - mx_0$. Then,

$$\begin{aligned} I &= \frac{1}{2}m\left\{x_1^2 - x_0^2\right\} + (x_1 - x_0)\{f(x_0) - mx_0\} \\ &= m\left\{\frac{1}{2}\left(x_1^2 - x_0^2\right) + x_0(x_1 - x_0)\right\} + hf(x_0) \\ &= m\{x_1 - x_0\}\left\{\frac{1}{2}(x_1 + x_0) - x_0\right\} + hf(x_0), \end{aligned}$$

but the slope of the straight line is

$$m = \frac{f(x_1) - f(x_0)}{(x_1 - x_0)}$$

Substituting

$$\begin{aligned} I &= \frac{f(x_1) - f(x_0)}{(x_1 - x_0)}(x_1 - x_0)\left\{\frac{x_1 - x_0}{2}\right\} + hf(x_0) \\ &= \{f(x_1) - f(x_0)\}\frac{h}{2} + hf(x_0), \end{aligned}$$

we finally arrive at our first quadrature formula,

$$I = \frac{h}{2}\{f(x_0) + f(x_1)\}. \tag{4-2}$$

Equation (4-2) is known as the trapezoidal rule. Figure 4-1 shows how the trapezoidal rule (with $h = 0.5$) approximates the definite integral. The shaded area represents the approximate area under $f(x)$.

If we divide the entire interval into N subintervals, $[x_k, x_{k+1}]$ of width $h = [(b - a)/N]$, then each node is found at

$$x_k = a + kh \qquad (k = 0, 1, \ldots, N),$$

and we find the area under the curve (the definite integral) by summing the areas of the individual trapezoids, i.e.,

$$T[f, h] = \frac{h}{2}\sum_{k=1}^{N}[f(x_{k+1}) + f(x_k)], \tag{4-3}$$

$$\int_a^b f(x)\,\mathrm{d}x \approx T(f, h). \tag{4-4}$$

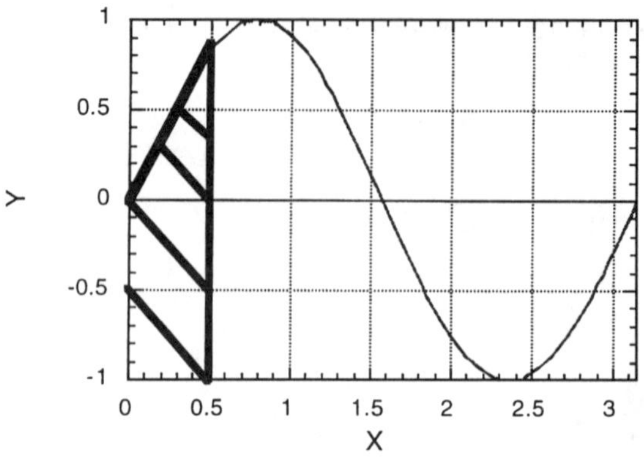

Figure 4-1 The trapezoidal rule is a two-point formula ($h = 0.5$).

Approximation (4-4) is called the composite trapezoidal rule. Figure 4-2 shows how the composite trapezoidal rule approximates the integral.

$T[f, h]$ can also be expressed as

$$T[f, h] = \frac{h}{2}[f_0 + 2f_1 + 2f_2 + \cdots + 2f_{N-1} + f_N]$$

or

$$T[f, h] = \frac{h}{2}[f(a) + f(b)] + h\sum_{k=1}^{N-1} f(x_k). \tag{4-5}$$

A more accurate integration formula, called Simpson's rule, can be derived by assuming that a quadratic polynomial curve of the form $\mathbf{y} = ax^2 + b'x + c$ will approximate $f(x)$ more closely than the linear function of the above discussion. Since there are three unknown coefficients in the quadratic polynomial, we need three data points for their evaluation. The definite integral is then

$$I = \int_{x_0}^{x_2} (ax^2 + b'x + c)\,\mathrm{d}x.$$

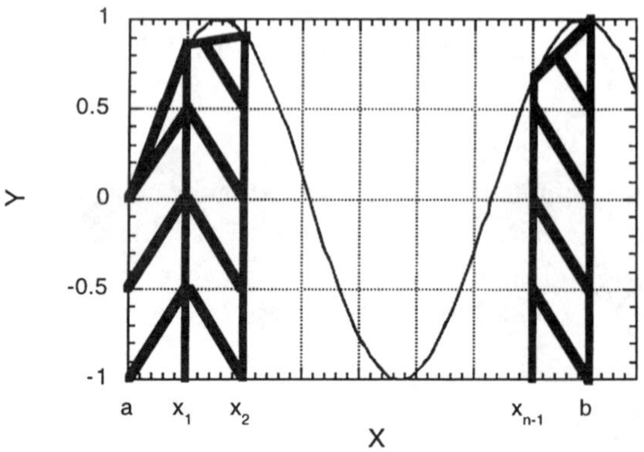

Figure 4-2 Composite trapezoidal rule.

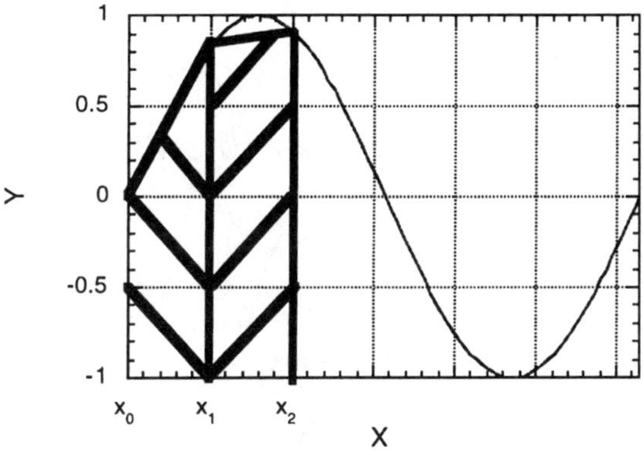

Figure 4-3 Simpson's rule.

Figure 4-3 shows how this quadratic function **y** is used to approximate the integral under the curve $f(x) = ax^2 + b'x + c$.

Since

$$f(x_0) = ax_0^2 + b'x_0 + c,$$
$$f(x_1) = ax_1^2 + b'x_1 + c,$$
$$f(x_2) = ax_2^2 + b'x_2 + c,$$

we can solve for **a**, b', and **c** in terms of $f(x_0)$, $f(x_1)$, and $f(x_2)$, resulting in

$$\int_{x_0}^{x_2} f(x)\,\mathrm{d}x \approx \frac{h}{3}(f_0 + 4f_1 + f_2). \tag{4-6}$$

Approximation (4-6) is called Simpson's rule; it is a three-point formula. From the above discussion for the trapezoidal composite rule, we can infer the composite Simpson's rule.

Suppose that **[a, b]** is subdivided into $2N$ subintervals $[x_k, x_{k+1}]$ of equal width $h = [(b-a)/2N]$ by the relationship

$$x_k = x_0 + kh \qquad (k = 0, 1, \ldots, 2N).$$

Then

$$S[f, h] = \frac{h}{3}\sum_{k=1}^{N}[f(x_{2k-2}) + 4f(x_{2k-1}) + f(x_{2k})] \tag{4-7}$$

such that

$$\int_a^b f(x)\,\mathrm{d}x \approx S[f, h]. \tag{4-8}$$

Also,

$$S[f, h] = \frac{h}{3}[f_0 + 4f_1 + 2f_2 + 4f_3 + \cdots + 2f_{2N-2} + 4f_{2N-1} + f_{2N}]$$

or

$$S[f, h] = \frac{h}{3}[f(a) + f(b)] + \frac{2h}{3}\sum_{k=1}^{N-1} f(x_{2k}) + \frac{4h}{3}\sum_{k=1}^{N} f(x_{2k-1}). \tag{4-9}$$

Of course, a more accurate quadrature formula can be derived if the approximating curve is a third-order algebraic polynomial, viz.,

$$\mathbf{y} = ax^3 + b'x^2 + cx + d. \tag{4-10}$$

Using the techniques of the above discussions and using Eq. (4-10), we can derive

$$\int_{x_0}^{x_3} f(x)\,\mathrm{d}x \approx \frac{3h}{8}(f_0 + 3f_1 + 3f_2 + f_3). \tag{4-11}$$

Approximation (4-11) is a four-point formula and is called Simpson's 3/8 formula. Four data points are required for evaluating the coefficients a, b', c, and d.

Finally, for a fourth-order algebraic polynomial approximation,

$$\mathbf{y} = ax^4 + b'x^3 + cx^2 + dx + e,$$

and using five data points for the evaluation, we can derive the following approximation of the definite integral in the interval $[x_0, x_4]$:

$$\int_{x_0}^{x_4} f(x)\,\mathrm{d}x \approx \frac{2h}{45}(7f_0 + 32f_1 + 12f_2 + 32f_3 + 7f_4). \tag{4-12}$$

Approximation (4-12) is a five-point formula, called Boole's rule.

Each of the above integration formulas were derived by use of an algebraic polynomial of order M and integration of this approximating curve over the desired interval. When the required number of data points was used, the coefficients of the polynomial were established. As discussed in Chapter 3, the higher-order polynomials tend to wiggle between data points so that the resulting value of the definite integral may be adversely affected. The choice of a high-order algebraic polynomial therefore does not guarantee an accurate integration formula. On the other hand, the choice of a lower-order approximating polynomial may lead students to have to consider a small integration interval in order to achieve a desired accuracy. Thus, in order to calculate the definite integral over a large physical region [**a**, **b**], the construction of many subintervals will be required for reasonable accuracy. This condition will result in long computing times and could lead to computer round-off errors.

As in Chapter 3, we seek better approximating strategies in the interval of the data. Once again, we turn to cubic-spline approximations.

4-3 INTEGRATION OF CUBIC SPLINES

Polynomial approximations go through the desired points but oscillate between points, thus making the fit of the curve highly inaccurate. This phenomenon was explained in Chapter 3 and remains valid; a technique of curve fitting, and thus a good candidate for use with numerical integration, was also discussed in the last chapter. This is the **spline-fitting** technique.

Consider the interval $[x_k, x_{k+1}]$. Figure 4-4 shows the way that the cubic-spline approximation is used to simulate the curve that will be integrated in order to provide an estimate of the area under that segment of a functional curve. Recalling (see Chapter 3) that the cubic spline is given by

$$S_k = a_k(x - x_k)^3 + b_k(x - x_k)^2 + c_k(x - x_k) + d_k,$$

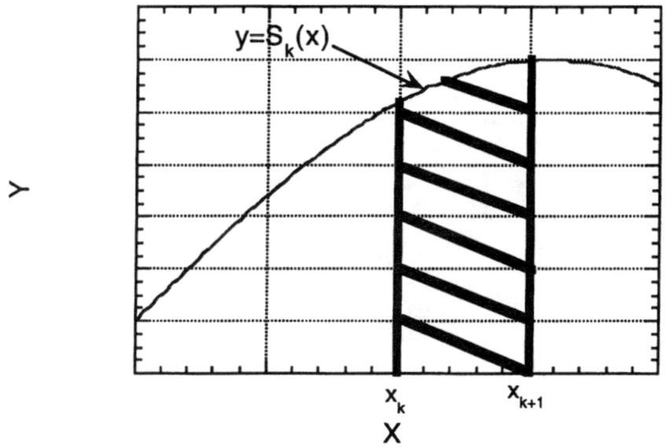

Figure 4-4 Spline approximation for quadrature.

we can write the integral for the kth interval as,

$$I_k = \int_{x_k}^{x_{k+1}} \{a_k(x - x_k)^3 + b_k(x - x_k)^2 + c_k(x - x_k) + d_k\}\,\mathrm{d}x,$$

$$I_k = \left\{\frac{1}{4}a_k(x - x_k)^4 + \frac{1}{3}b_k(x - x_k)^3 + \frac{1}{2}c_k(x - x_k)^2 + d_k x\right\}\Bigg|_{x_k}^{x_{k+1}},$$

$$I_k = \frac{1}{4}a_k(x_{k+1} - x_k)^4 + \frac{1}{3}b_k(x_{k+1} - x_k)^3 + \frac{1}{2}c_k(x_{k+1} - x_k)^2 + d_k(x_{k+1} - x_k).$$

If we choose $h_k = x_{k+1} - x_k$, then

$$I_k = \frac{a_k}{4}h_k^4 + \frac{b_k}{3}h_k^3 + \frac{c_k}{2}h_k^2 + d_k h_k. \tag{4-13}$$

Equation (4-13) provides the result of the definite integral for the kth region only. A series of cubic-spline functions therefore must be constructed (with the techniques of Chapter 3) in order to obtain a good approximation for the integral of $f(x)$ over the region [**a**, **b**].

4-4 GAUSS–LEGENDRE INTEGRATION

We have seen that the above integrations are of only a certain accuracy; they depend greatly on the degree of the approximating curve that is chosen. If greater accuracy is required, then either a higher-order degree of polynomial is necessary or the integration interval must be subdivided in a finer manner. The question then presents itself,

> What method of quadrature approximation gives the best answer if a fixed number of function evaluations are to be made?

As an example, suppose that we wish to find the area under the curve $\mathbf{y} = f(x)$ in the interval $-1 \leq x \leq 1$.

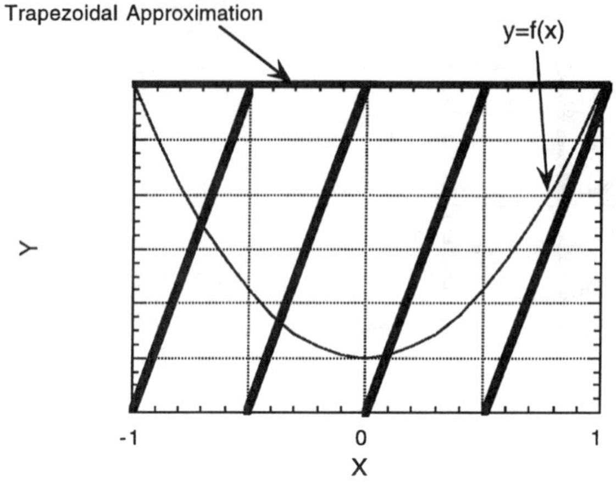

Figure 4-5 Trapezoidal approximation of $y = f(x)$.

We have seen that if we desire a two-point integration formula, the trapezoidal approximation that uses the end points for its evaluation ($x = -1$ and $x = 1$) is chosen. Figure 4-5 shows how the use of the trapezoidal rule over the range $-1 < x < 1$ may calculate the area under the curve (definite integral) inaccurately.

If, instead of demanding that we use the end points of the interval to evaluate our approximating integral, we choose some arbitrary points x_1 and x_2 at which to calculate the integral, then the trapezoidal rule may be applied, as shown in Fig. 4-6. This appears to be a better approximation to the integral than the one that was shown in Fig. 4-5.

Therefore we want to find the best two points, x_1 and x_2, and their corresponding weights, w_1 and w_2, such that

$$\int_{-1}^{1} f(x)\,\mathrm{d}x = w_1 f(x_1) + w_2 f(x_2). \tag{4-14}$$

"Best" means that the choice of x_1 and x_2 is such that the definite integral is the most accurate, when Eq. (4-14) is used, with only two points being used for the evaluation. Equation (4-14) presents the model for a two-point integration formula. It now remains

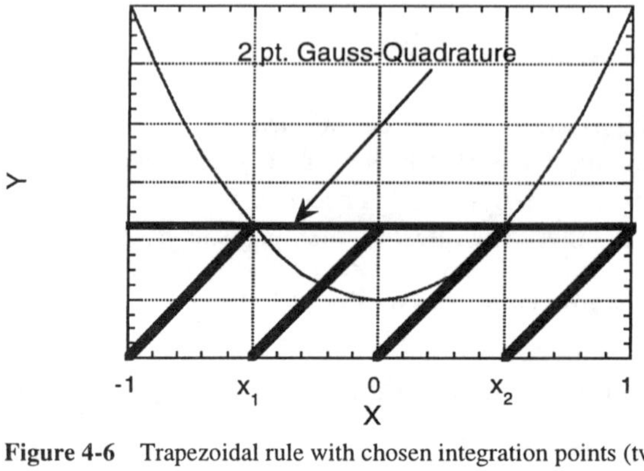

Figure 4-6 Trapezoidal rule with chosen integration points (two-point Gauss quadrature).

to establish a methodology that will allow us, in an orderly manner, to determine the four unknown parameters, w_1, x_1, w_2, and x_2. In Section 4-2, two available data points would yield a quadrature formula of only first-order accuracy (a straight line). Here, however, we are attempting to obtain an integration formula that has third-order accuracy but still uses only two data points. We accomplish this task by using the following strategies.

Since there are four unknowns, the integration would be exact if we choose a cubic polynomial approximation, viz.,

$$f(x) = a_3x^3 + a_2x^2 + a_1x + a_0. \tag{4-14a}$$

We desire to choose w_1, w_2, x_1, x_2 to satisfy the four coefficients a_0, a_1, a_2, a_3. In order to satisfy Eq. (4-14), we integrate the cubic polynomial [Eq. (4-14a)] term by term and equate the results of this integration to an evaluation of Eq. (4-14). This procedure will yield a sufficient number of conditions to determine the location of the integration points and their weight functions. A similar procedure can be used for any integration formula for any number of weight functions and integration points.

To satisfy the constant term in polynomial (4-14a), we choose $f(x) = 1$. Integrating $f(x)$ over the interval and equating the result to Eq. (4-14) with $f(x) = 1$, we achieve

$$\int_{-1}^{1} 1\,\mathrm{dx} = 2 = w_1 + w_2. \tag{4-15}$$

To satisfy the linear term, we choose $f(x) = x$. Then,

$$\int_{-1}^{1} x\,\mathrm{dx} = 0 = w_1x_1 + w_2x_2, \tag{4-16}$$

where we obtain the right-hand side of Eq. (4-16) by using $f(x) = x$ in Eq. (4-14).

The quadratic term in the polynomial (4-14a) may be satisfied by $f(x) = x^2$. Then, in a similar manner,

$$\int_{-1}^{1} x^2\,\mathrm{dx} = \frac{2}{3} = w_1x_1^2 + w_2x_2^2. \tag{4-17}$$

Finally, the cubic term in the polynomial is satisfied by $f(x) = x^3$:

$$\int_{-1}^{1} x^3\,\mathrm{dx} = 0 = w_1x_1^3 + w_2x_2^3. \tag{4-18}$$

Equations (4-15)–(4-18) may be solved simultaneously for

$$w_1 = w_2 = 1, \tag{4-19}$$

$$-x_1 = x_2 = \frac{1}{\sqrt{3}} = 0.57735, \tag{4-20}$$

so that the definite integral of the function $f(x)$ may be approximated by the two-point formula [from Eq. (4-14)], viz.,

$$\int_{-1}^{1} f(x)\,\mathrm{dx} = G[f, 2] = f\left(-\frac{1}{\sqrt{3}}\right) + f\left(\frac{1}{\sqrt{3}}\right). \tag{4-21}$$

Points x_1 and x_2 are called the Gauss quadrature points (or simply the Gauss points). These points are important in the construction of finite elements in Chap. 10.

From the above example, we can conclude that the general N-point Gauss–Legendre rule is exact for polynomial functions of degree $\leq 2N-1$. We can now state the generalized formula for numerical integration using the N-point Gauss quadrature methodology, i.e.,

$$G[f, N] = w_{N,1} f(x_{N,1}) + w_{N,2} f(x_{N,2}) + \cdots + w_{N,N} f(x_{N,N}). \tag{4-22}$$

For a two-point formula, a polynomial of degree 3 was required for evaluating all the weighting functions and all the Gauss point locations. In a similar manner, it can be shown that a three-point formula will require a polynomial of degree 5 and a four-point formula will require a polynomial of degree 7 for evaluating both the weighting coefficients and the location of the Gauss points.

Since the Gauss quadrature rule is valid only when the data lie within the region $[-1, 1]$, the interval of integration $[\mathbf{a}, \mathbf{b}]$ must be modified so that it will encompass the interval $[-1, 1]$. Following the procedure that was used in Chapter 3 for the Chebyshev polynomials, we perform the transformation

$$\mathbf{t} = \frac{a+b}{2} + \frac{b-a}{2}x, \qquad \mathrm{d}\mathbf{t} = \frac{b-a}{2}\,\mathrm{d}x,$$

so that we have $\mathbf{t} = \mathbf{b}$ at $x = 1$ and $\mathbf{t} = \mathbf{a}$ at $x = -1$.

Then, using this transformation,

$$\int_a^b f(t)\,\mathrm{d}t = \int_{-1}^{1} f\left(\frac{a+b}{2} + \frac{b-a}{2}x\right)\frac{b-a}{2}\,\mathrm{d}x, \tag{4-23}$$

we get

$$\int_a^b f(t)\,\mathrm{d}t = \frac{b-a}{2}\sum_{k=1}^{N} w_{N,k} f\left\{\frac{a+b}{2} + \frac{b-a}{2}x_{N,k}\right\}. \tag{4-24}$$

Table 4-1 lists the Gauss points and weight functions for two-, three-, four-, and five-point

Table 4-1 Gauss points and weight approximations

N	Abscissa (x_i)	Weight (w_i)	Error (order)
2	−0.57735	1	$f^{(4)}$
	+0.57735	1	
3	+0.7746	0.556	
	−0.7746	0.556	$f^{(6)}$
	0	0.888	
4	+0.861	0.348	
	−0.861	0.348	$f^{(8)}$
	+0.34	0.652	
	−0.34	0.652	
5	+0.9062	0.237	
	−0.9062	0.237	
	+0.538	0.479	$f^{(10)}$
	−0.538	0.479	
	0	0.569	

approximations. The derivations of these additional formulas are reserved for the problems at the end of this chapter.

4-5 COMPARISON OF METHODOLOGIES

Now that several methods have been presented for the evaluation of definite integrals, we compare several of the above-mentioned techniques against an exact solution. Let us consider the definite integral,

$$\int_{-1}^{1} \frac{dx}{x+2} = \ln(3) \approx 1.09861. \tag{4-25}$$

In this example, we have $f(x) = [1/(x+2)]$. The exact solution in Eq. (4-25) is the benchmark against which the approximations are to be compared.

The exact solution is now compared with the trapezoidal rule, Simpson's rule and the two-point Gauss–Legendre quadrature rules.

First, let us use a simple two-point rule, i.e., the trapezoidal rule with $h = 2$ (since the interval is $-1 \leq x \leq 1$). Then,

$$\begin{aligned} T[f, 2] &= \frac{1}{2} 2[f(-1) + f(1)], \\ T[f, 2] &= 1.0 + 0.333, \\ T[f, 2] &= 1.333. \end{aligned}$$

Comparing this result with the exact solution above, we can calculate the error = 0.23472, or 21.37%.

In order to reduce this error, consider a three-point approximation, i.e., Simpson's rule with $h = 1$:

$$\begin{aligned} S[f, 1] &= \left(\frac{h}{3}\right)(f_0 + 4f_1 + f_2) \\ &= \left(\frac{1}{3}\right)[f(-1) + 4f(0) + f(1)] \\ &= \left(\frac{1}{3}\right)[1.0 + 2.0 + 0.333], \\ S[f, 1] &= 1.11111, \end{aligned}$$

from which we calculate an error = 0.0124, or 1.13%. While this result is far superior to that of the trapezoidal rule, we had to resort to a three-point formula.

We now turn to a two-point Gauss quadrature approach and evaluate the relevant definite integral. We can calculate

$$\begin{aligned} f(-0.57735) &= 0.70291, \\ f(+0.57735) &= 0.387995, \end{aligned}$$

so that from Eq. (4-21), we get

$$G[f, 2] = (1)(0.70291) + (1)(0.387995),$$
$$G[f, 2] = 1.0909,$$

which produces an error of 0.00771, or 0.7%. The strength of the Gauss–Legendre methodology is now readily apparent. We were able to achieve a more accurate result with a two-point Gauss rule than could be achieved with Simpson's rule, which is a three-point approximation.

The author has presented in the Appendixes of this chapter several computer codes that use the techniques that have been presented for numerical integration. These codes were used to obtain the above results. The function that is used in the Appendixes is $f(x) = e^{-x}$. Students may use these codes either to solve some of the problems at the end of this chapter or, with a modification of the function, to evaluate any definite integral. It is hoped, of course, that students will use these algorithms merely as guides to the development of their own personal codes. The codes that are represented in the Appendixes are: Appendix 4.1, **trapezoidal.for**; Appendix 4.2, **simpson.for**; Appendix 4.3, **simpson_3/8.for**; Appendix 4.4, **boole.for**; Appendix 4.5, **spline.for**; Appendix 4.6, **gq.for**; the names of the algorithms describe their functions.

4-6 CONCLUSION

This chapter has dealt with the derivation of quadrature formulas that are useful in many applications in engineering and physics. The formulas apply to the evaluation of only definite integrals. These approximations may be applied to functions that cannot be integrated in closed form, to data that are obtained experimentally, and to the development of finite elements, as is explained in Chapter 10.

We must be careful when applying the results of this chapter. The error from any one of the formulas must be considered as well as the optimum number of grid points that would be required for achieving that level of approximation. In general, although the Gauss quadrature formulas are somewhat complicated in their derivations, they usually produce the most accurate approximation of the integral for a given level of approximation. Therefore the Gauss–Legendre formulas should be the first methodology that is considered in a particular problem.

FURTHER READING

Arden, B. W. and Astill, K. N. (1970). *Numerical Algorithms—Origins and Applications*, Addison-Wesley, Publishing Company, Reading, Massachusetts.

Carnahan, B., Luther, H., and Wilkes, J. (1969). *Applied Numerical Methods*, Wiley, New York.

Desai, C. S. and Abel, J. F. (1972). *Introduction to the Finite Element Method—A Numerical Method for Engineering Analysis*, Van Nostrand Reinhold, New York.

Mathews, J. H. (1987). *Numerical Methods for Computer Science, Engineering, and Mathematics*, Prentice-Hall, Englewood Cliffs, New Jersey.

Pennington, R. H. (1965). *Introductory Computer Methods and Numerical Analysis*, Macmillan, New York.

Press, W. H., Teukolsky, S., Vetterling, W., and Flannery, B. (1986). *Numerical Recipes in FORTRAN*, Cambridge University Press, Cambridge.

Salvadori, M. G. and Baron, M. L. (1952). *Numerical Methods in Engineering*, Prentice-Hall, Englewood Cliffs, New Jersey.

Scheid, F. (1968). *Schaum's Outline of Theory and Problems of Numerical Analysis*, McGraw-Hill, New York.

Stasa, F. L. (1985). *Applied Finite Element Analysis for Engineers*, Holt, Rinehart & Winston, New York.

Zienkiewicz, O. (1971). *The Finite Element Method in Engineering Science*, McGraw-Hill, New York.

PROBLEMS

Where applicable, solve the following problems by writing a computer algorithm. Computer codes are included in Appendixes 4.1–4.6 that can act as templates for the development of your algorithms.

4-1 For the function $y = e^{-x}$ in the interval $[0, \infty]$, analytically evaluate the exact integral. This result will aid in the solution of Problems 4-2–4-8.

4-2 Approximate the area under the curve $y = e^{-x}$ in the interval $[0, \infty]$, by using the trapezoidal rule. What is the optimum number of points that is required for an accuracy of 1%?

4-3 For the function of Problem 4-1, approximate the area under the curve by using Simpson's rule for the same interval of integration. What is the optimum number of points that will achieve a 1% error?

4-4–4-8 Solve Problem 4-2 by using

(*a*) Simpson's 3/8 rule,
(*b*) Boole's quadrature rule,
(*c*) cubic-spline quadrature,
(*d*) two-point Gauss quadrature rule,
(*e*) three-point Gauss quadrature rule.

In each of the above problems, evaluate the optimum number of subintervals that are required for achieving 1% accuracy.

4-9 Use a two-point Gauss–Legendre integration rule to approximate

$$\int_{-1}^{1} \frac{dx}{x+2}.$$

Compare your result with the exact solution.

4-10–4-14 Repeat Problem 4-9 for

(*a*) Simpson's 3/8 rule,
(*b*) Boole's quadrature rule,
(*c*) cubic-spline quadrature,
(*d*) trapezoidal rule,
(*e*) three-point Gauss quadrature rule.

In each of the above problems, evaluate the optimum number of subintervals that are required for achieving 1% accuracy.

4-15 Show that the Gauss–Legendre three-point rule is

$$\int_{-1}^{1} f(x)\,dx = G[f,3] = \frac{5f\left(-\sqrt{\frac{3}{5}}\right) + 8f(0) + f\left(\sqrt{\frac{3}{5}}\right)}{9}.$$

4-16 Using the result from Problem 4-15, approximate

$$\int_{-1}^{1} 5x^4\,dx.$$

Compare your result with the exact solution.

4-17 Using the result from Problem 4-15, approximate

$$\int_1^5 \frac{dt}{t}.$$

Compare your result with the exact solution.

4-18–4-23 For the function $f(x) = [1/(1 + 12x^2)]$, find the area under the curve in the interval $[-5, 5]$ by using

(*a*) Simpson's 3/8 rule,
(*b*) Boole's quadrature rule,
(*c*) cubic-spline quadrature,
(*d*) trapezoidal rule,
(*e*) three-point Gauss quadrature rule,
(*f*) cubic-spline quadrature.

Compare the results with the exact solution and find the optimum number of data points that are required for a 1% error.

4-24–4-29 For the function $f(x) = \sin(\pi/2)x$, find the area under the curve in the interval $[-1, 1]$ by using

(*a*) Simpson's 3/8 rule,
(*b*) Boole's quadrature rule,
(*c*) cubic-spline quadrature,
(*d*) trapezoidal rule,
(*e*) three-point Gauss quadrature rule,
(*f*) cubic-spline quadrature.

Compare the results with the exact solution and find the optimum number of data points that are required for a 1% error.

4-30–4-35 For the function $f(x) = ae^{bx}$, with $\mathbf{a} = 1.0$ and $\mathbf{b} = 5.0$, find the area under the curve in the interval [0, 10] by using

(*a*) Simpson's 3/8 rule,
(*b*) Boole's quadrature rule,
(*c*) cubic-spline quadrature,
(*d*) trapezoidal rule,
(*e*) three-point Gauss quadrature rule,
(*f*) cubic-spline quadrature.

Compare the results with the exact solution and find the optimum number of data points that are required for a 1% error.

4-36–4-41 For the data of Problem 3-9, find the area under the curve by using

(*a*) Simpson's 3/8 rule,
(*b*) Boole's quadrature rule,
(*c*) cubic-spline quadrature,
(*d*) trapezoidal rule,
(*e*) four-point Gauss quadrature rule,
(*f*) cubic-spline quadrature.

Find the optimum number of data points that are required for a 1% error.

4-42–4-47 For the data of Problem 3-11, find the area under the curve by using

(*a*) Simpson's 3/8 rule,
(*b*) Boole's quadrature rule,
(*c*) cubic-spline quadrature,
(*d*) trapezoidal rule,
(*e*) three-point Gauss quadrature rule,
(*f*) cubic-spline quadrature.

Compare the results with the exact solution and find the optimum number of data points that are required for a 1% error.

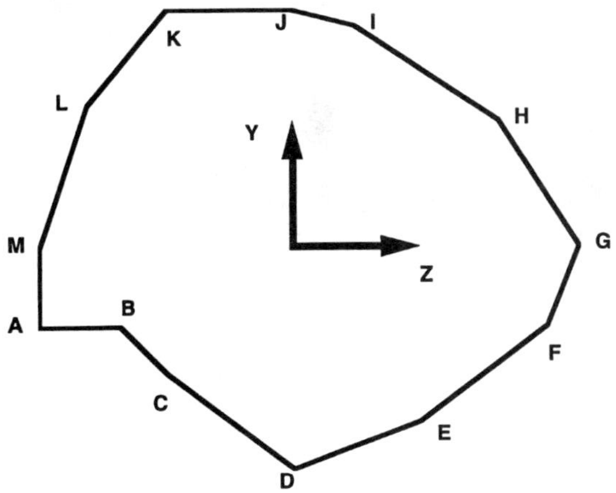

Figure 4-7 Coordinates (z–y): A, (−0.7, −0.2); B, (−0.5, −0.2); C, (−0.2, −0.4); D, (0.0, −0.7); E, (0.4, −0.6); F, (0.6, −0.3); G, (0.7, 0.0); H, (0.6, 0.4); I, (0.2, 0.6); J, (0.1, 0.7); K, (−0.2, 0.7); L, (−0.5, 0.5); M, (−0.7, 0.2).

4-48 Find the area under the curve of Problem 3-15.

4-49 Find the area under the curve of Problem 3-16.

4-50 Find the area under the curve of Problem 3-17.

4-51 Calculate the area of the beam cross section shown in Fig. 4-7.

4-52 Calculate the moment of inertia of the beam cross section shown in Fig. 4–7.

PART TWO

COMPUTATIONAL TOOLS FOR APPLICATIONS

CHAPTER

FIVE

FINITE-DIFFERENCE TECHNIQUES

NOMENCLATURE

$y_{i\pm n}$	nth mesh point to the right (or left) of relevant point
∇	backward differences
δ	central differences
Δ	forward differences
D	differential operator
h	mesh spacing
$O(h)$	order of approximation
p	mesh expander–reducer
y	approximating function

5-1 INTRODUCTION

As previously discussed, the purpose of modeling is to approximate the equations that govern a physical process. This approach leads to the solution of problems involving complex geometries or boundary conditions that are intractable by standard mathematical techniques.

The two most common of the modern modeling techniques are the finite-difference technique, which is presented in this chapter, and the finite-element technique, which is presented in Chap. 10. In all modeling approaches, we assume that the solution to the governing equations can be approximated by a function or by a series of functions. If a sufficiently small region of interest is chosen, then the solution for the dependent variables may be approximated by simple curves. The accuracy of the solution process therefore depends on the complexity of the solution at a point in the region along with the numerical approximation of the governing equation at that point. Decisions as to the modeling size and the complexity of the approximating curve always have to be made. These concerns are addressed in this and subsequent chapters.

In the finite-difference approach to computational mechanics, it is assumed that the governing equations, in the form of differential equations, are known. The role of numerical computation is to obtain accurate values for the dependent variables of these equations for the region of interest. As is seen during these studies, careful numerical modeling is required when the boundary of the region is complex and when the material properties of the solid or the fluid are nonhomogeneous or nonisotropic. In addition, a basic understanding of the physical process being modeled is necessary for several reasons. First, an evaluation of your solution is always required. You must ask the question, "Does this solution make sense?" Second, a knowledge of the physics of the problem will assist modelers in the choice of their numerical approximations. The numerical approximation of the geometric description of the region, called the mesh or discretization, is one of the most important phases of numerical modeling. This point cannot be overemphasized. There is no rule of thumb that can work in all cases to make the proper selection. Therefore experience coupled with a knowledge of the problem is essential to good modeling.

5-2 SIMPLE MODELING FORMULAS

We start with some simple examples and develop some simple finite-difference approximations to the derivatives that appear in differential equation descriptions of physical processes. The simplest method of obtaining expressions for the derivatives consists of substituting an approximating curve for a perceived solution curve and solving for the derivatives. Figure 5-1 shows a one-dimensional solution curve, **y(x)**, which is unknown. The differential equation that governs the process for which **y(x)** is the solution is known.

We wish to approximate the first derivative of the curve **y(x)** at the ith point in the region. We choose the approximating curve,

$$\mathbf{y(x)} = Ax + B, \tag{5-1}$$

as an estimate of the dependent variable at y_i.

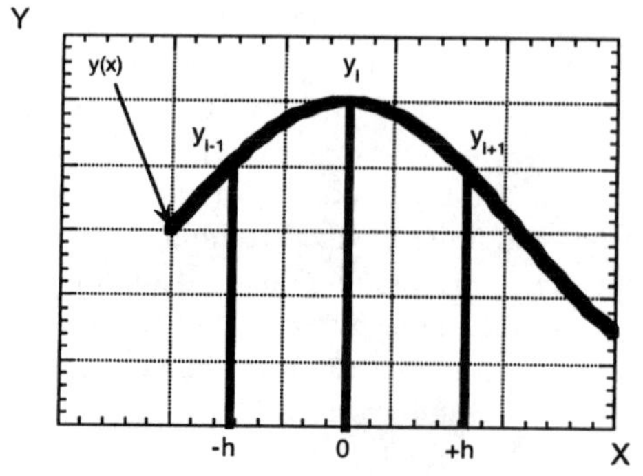

Figure 5-1 Formulation of a first-derivative approximation.

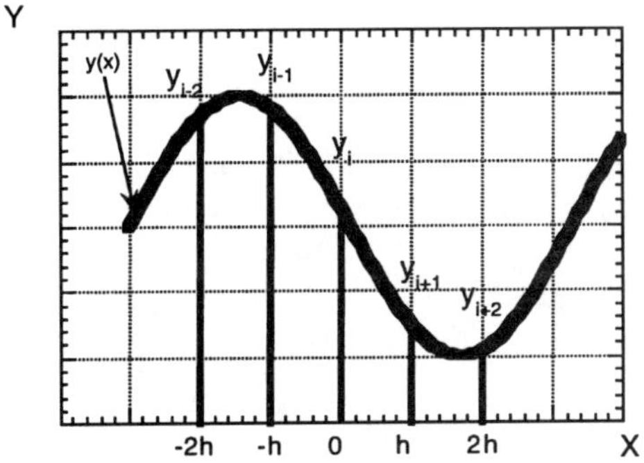

Figure 5-2

Further, we choose a uniform mesh spacing of **h** between the points where we desire the calculation of the derivative. Thus, in order to approximate the first derivative of **y** with respect to **x** at point **i**, we fit the approximating curve, Eq. (5-1), through the surrounding points $\mathbf{x} = -\mathbf{h}$ (called $\mathbf{i} - 1$) and $\mathbf{x} = +\mathbf{h}$ (called $\mathbf{i} + 1$). Thus

$$\mathbf{y}(\mathbf{h}) = Ah + B = y_{i+1}, \tag{5-2}$$

$$\mathbf{y}(-\mathbf{h}) = -Ah + B = y_{i-1}. \tag{5-3}$$

Solving for A from Eqs. (5-2) and (5-3), we get

$$A = \frac{(y_{i+1} - y_{i-1})}{2h} = \mathbf{y}'. \tag{5-4}$$

Thus a simple approximation of the first derivative of **y** with respect to **x** in the differential equation that governs the process at point **i** is given by Eq. (5-4).

The approximation of the second derivative of **y** wth respect to **x** is obtained in a similar manner. Referring to Fig. 5.2, we choose

$$\mathbf{y} = Ax^2 + Bx + C \tag{5-5}$$

as the approximating curve. If we now evaluate Eq. (5-5) at grid points around the ith point, then

$$\mathbf{y}(-\mathbf{h}) = y_{i-1} = Ah^2 - Bh + C, \tag{5-6}$$

$$\mathbf{y}(0) = y_i = C, \tag{5-7}$$

$$\mathbf{y}(\mathbf{h}) = y_{i+1} = Ah^2 + Bh + C. \tag{5-8}$$

If we use Eqs. (5-6)–(5-8) to solve for the second derivative,

$$\mathbf{y}'' = 2A,$$

we get

$$\mathbf{y}'' = \frac{\{y_{i-1} - 2y_i + y_{i+1}\}}{h^2}. \tag{5-9}$$

Equation (5-9) is the approximation of the second derivative. We can continue this derivation process and calculate the third derivative by choosing an approximating function

$$\mathbf{y} = Ax^3 + Bx^2 + Cx + D. \tag{5-10}$$

If we once again use the grid arrangement that is shown in Fig. 5-2 to evaluate the coefficients of Eq. (5-10) and note that the third derivative

$$\mathbf{y}''' = 6A,$$

we get

$$\mathbf{y}(-\mathbf{h}) = y_{i-1} = -Ah^3 + Bh^2 - Ch + D, \tag{5-11}$$

$$\mathbf{y}(0) = y_i = D, \tag{5-12}$$

$$\mathbf{y}(\mathbf{h}) = y_{i+1} = Ah^3 + Bh^2 + Ch + D, \tag{5-13}$$

$$\mathbf{y}(2\mathbf{h}) = y_{i+2} = 8Ah^3 + 4Bh^2 + 2Ch + D, \tag{5-14}$$

so that

$$\mathbf{y}''' = \frac{\{-y_{i-1} + 3y_i - 3y_{i+1} + y_{i+2}\}}{h^3}. \tag{5-15}$$

There are many times when we can best model the physics of the problem by choosing an uneven mesh spacing. In the area of a crack or an irregular boundary, the solution gradient will probably be higher than that in the remainder of the body. This will dictate the need for a finer mesh in that area than would be required in the rest of the region for similar accuracy. In this case, the finite-difference approximation of the second derivative of **y** with respect to **x** may be derived with the aid of Fig. 5-3.

If the mesh spacing to the left of point **i** is **h** and the mesh spacing to the right of point **i** is **ph** (where **p** can be less than or greater than 1), then the second derivative $\mathbf{y}'' = 2A$ can be calculated by the evaluation of Eq. (5-5) at the new mesh points, viz.,

$$\mathbf{y}(-\mathbf{h}) = y_{i-1} = Ah^2 - Bh + C, \tag{5-16}$$

$$\mathbf{y}(0) = y_i = C, \tag{5-17}$$

$$\mathbf{y}(\mathbf{ph}) = y_{i+1} = p^2Ah^2 + pBh + C, \tag{5-18}$$

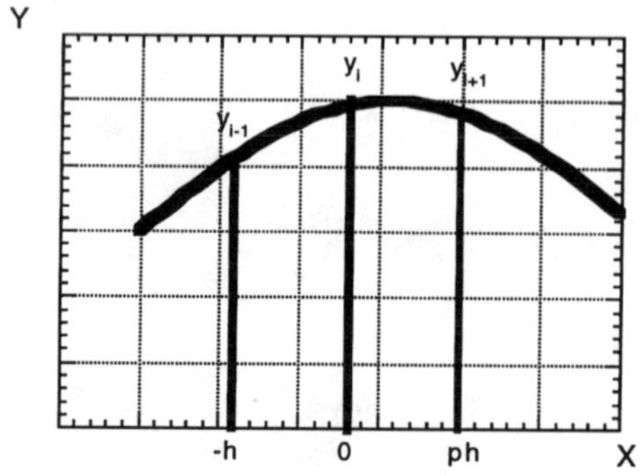

Figure 5-3 Uneven grid spacing.

from which we get

$$\mathbf{y}'' = \frac{2[py_{i-1} - (1+p)y_i + y_{i+1}]}{p(p+1)h^2}, \tag{5-19}$$

which is the approximation of the second derivative, but for an unequal mesh spacing. We can easily show that this equation reduces to Eq. (5-9) when $\mathbf{p} = 1$ (i.e., equal mesh spacing).

While the approach that has just been shown (i.e., choosing a polynomial approximation) will always be successful, the error of the approximation with respect to the mesh spacing **h** is unknown. Therefore the accuracy of the modeling cannot be assessed. In other words, we cannot evaluate how the accuracy of our approximation will converge with the selection of a finer or a coarser mesh. This knowledge is important when we are choosing the order of approximation that will produce the most efficient and robust numerical algorithm.

5-3 ERROR AND MESH CONSIDERATIONS

We now try to derive the differentiation formulas and their error characteristics. We begin by writing a Taylor series for **y** about the origin y_i.

From the Taylor series,

$$\mathbf{y}(\mathbf{x}+\mathbf{h}) = y(x) + hy'(x) + \frac{h^2}{2!}y''(x) + \cdots,$$

or

$$\mathbf{y}(\mathbf{x}+\mathbf{h}) = \sum_{n=0}^{\infty} \frac{h^n}{n!} y^{(n)}(x), \tag{5-20}$$

where

$$y^{(n)} \equiv \frac{\mathrm{d}^n y}{\mathrm{d}x^n},$$

$$y^{(0)}(x) \equiv \mathbf{y}(\mathbf{x}),$$

$$0! = 1,$$

we can now rederive an equation for the first derivative $\mathbf{y}'$ by using the Taylor series. Using the unevenly spaced mesh points $(\mathbf{x} + \mathbf{ph})$ and $(\mathbf{x} - \mathbf{h})$ of Fig. 5-3, we evaluate the Taylor series about point y_i, i.e.,

$$y_{i+1} = y_i + phy_i' + \frac{p^2h^2}{2!}y_i'' + \frac{p^3h^3}{3!}y_i''' + \cdots, \tag{5-21}$$

$$y_{i-1} = y_i - hy_i' + \frac{h^2}{2!}y_i'' - \frac{h^3}{3!}y_i''' + \cdots. \tag{5-22}$$

Subtracting Eq. (5-22) from Eq. (5-21), we get

$$y_{i+1} - y_{i-1} = (p+1)hy_i' + (p^2-1)\frac{h^2}{2!}y_i'' + \cdots. \tag{5-23}$$

We solve for y_i':

$$y_i' = \frac{1}{(p+1)h}(y_{i+1} - y_{i-1}) + (1-p)\frac{h}{2}y_i'' - \cdots. \tag{5-24}$$

The approximate expression for the first derivative then becomes

$$y_i' = \frac{1}{(p+1)h}(y_{i+1} - y_{i-1}) \tag{5-25}$$

with an error

$$(1-p)\frac{h}{2}y_i'' - \frac{1+p^3h^2}{6(1+p)}y_i''' + \cdots, \tag{5-26}$$

which approaches zero as fast as **h** if $\mathbf{p} \neq 1$ and as fast as $\mathbf{h}^2$ if $\mathbf{p} = 1$ (evenly spaced).

Using the Taylor series approach, we can calculate a more accurate formula for the first derivative than Eq. (5-25). If we eliminate y_i'' from both Eqs. (5-21) and (5-22) and subtract, then

$$y_{i+1} = y_i + phy_i' + \frac{p^2h^2}{2}y_i'' + \cdots,$$

$$p^2y_{i-1} = p^2y_i - p^2hy_i' + \frac{p^2h^2}{2}y_i'' + \cdots,$$

from which

$$(y_{i+1} - p^2y_{i-1}) = (1-p^2)y_i + p(1+p)hy_i' + \cdots.$$

We can now express the first derivative as

$$y_i' = \frac{1}{p(p+1)h}[y_{i+1} - (1-p^2)y_i - p^2y_{i-1}] - \frac{1}{6}h^2y_i''', \tag{5-27}$$

which approaches Eq. (5-25) as $\mathbf{p} \to 1$ and whose error approaches zero as $\mathbf{h}^2$ for any **p**. Therefore Eq. (5-27) is a more accurate approximation of $\mathbf{y}'$ than Eq. (5-25).

By using the Taylor series technique, we can obtain several formulas for y_i'' and evaluate their errors.

From Eqs. (5-21) and (5-22), we can eliminate y_i' and obtain the expression for the second derivative; thus

$$y_i'' = \frac{1}{h^2}\frac{2}{p(p+1)}[py_{i-1} - (1+p)y_i + y_{i+1}] + (1-p)\frac{h}{3}y_i''' + \cdots. \tag{5-28}$$

From Eq. (5-28), we see that the error in the second-derivative approximation approaches zero as **h** for $\mathbf{p} \neq 1$ and $\mathbf{h}^2$ for $\mathbf{p} = 1$.

From Fig. 5-4 we can also obtain a more accurate approximation for the second derivative, i.e.,

$$y_i'' = \frac{1}{p(p+1)(p+2)h^2}[p(p^2-1)y_{i-2} - 2(p^3-4p)y_{i-1} + (p^3-7p-6)y_i + 6y_{i+1}], \tag{5-29}$$

which has an error approaching zero as $\mathbf{h}^2$.

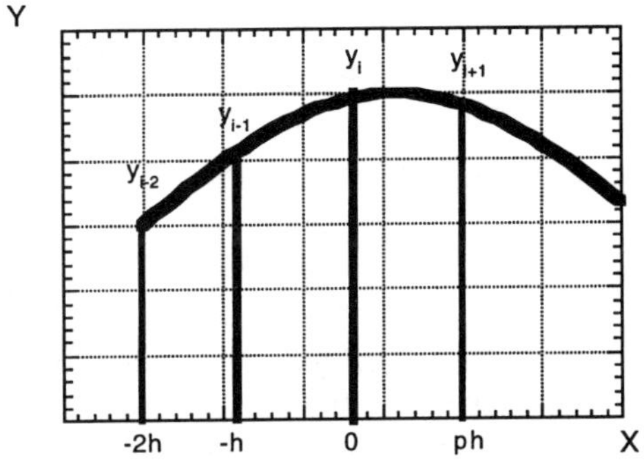

Figure 5-4

When the pivotal points are evenly spaced, the Taylor series technique can be applied symbolically in conjunction with the concept of a difference. This will lead to a set of useful modeling equations for differential equations.

We now adopt the following nomenclature for finite differences: ∇ represents a backward difference, Δ is a forward difference, and δ is a central difference.

5-4 BACKWARD DIFFERENCES

We now derive the finite-difference formulas for backward differences. These serve as approximations to the derivatives that they replace. In the next two sections, we use this approach to write down the equivalent formulas for forward differences and for central differences.

Given the values of the dependent variable

$$y_0, y_1, \ldots, y_{i-2}, y_{i-1}, y_i, y_{i+1}, y_{i+2}, \ldots, y_{n-1}, y_n,$$

evenly spaced by **h**, we define the first backward difference of **y** at point **i** as

$$\nabla y_i \equiv y_i - y_{i-1}. \tag{5-30}$$

The second backward difference of **y** at **i** then becomes

$$\nabla(\nabla y_i) \equiv \nabla^2 y_i = (y_i - y_{i-1}) - (y_{i-1} - y_{i-2}), \tag{5-31}$$

$$\nabla^2 y_i = y_i - 2y_{i-1} + y_{i-2}. \tag{5-32}$$

From this we can infer a generalized formula for successive backward differences, i.e.,

$$\nabla^n y_i \equiv \nabla(\nabla^{n-1} y_i). \tag{5-33}$$

It is easy to verify that the coefficients of the nth backward difference are also the coefficients in the binomial expansion $(a - b)^n$, which are

$$a^n - na^{n-1}b + \left[\frac{n(n-1)}{2}\right] a^{n-2}b^2 + \cdots. \tag{5-34}$$

Using expression (5-34), we can deduce

$$\nabla^3 y_i = y_i - 3y_{i-1} + 3y_{i-2} - y_{i-3}, \tag{5-35}$$

$$\nabla^4 y_i = y_i - 4y_{i-1} + 6y_{i-2} - 4y_{i-3} + y_{i-4}. \tag{5-36}$$

Equations (5-30), (5-32), (5-35), and (5-36) are the first four backward differences of **y** at point **i**. From the definitions of the backward differences, we can state the following properties of backward differences, i.e.,

$$\nabla(y_i + y_j) = \nabla y_i + \nabla y_j = \nabla y_j + \nabla y_i, \tag{5-37}$$

$$\nabla(cy_i) = c\nabla y_i, \tag{5-38}$$

$$\nabla^m(\nabla^n y_i) = \nabla^{m+n} y_i. \tag{5-39}$$

If we define $D \equiv (d/dx)$ as the differential operator, then we have the ability to express the derivatives of a function in terms of backward differences. We now use a methodology that will illustrate this process. We then apply it to some simple derivatives. Finally, we write down expressions for several of the derivatives in terms of forward differences, backward differences, and central differences.

To begin the derivation process, consider the Taylor series expansion,

$$\mathbf{y(x+h)} = y(x) + \frac{h}{1!}y'(x) + \frac{h^2}{2!}y''(x) + \frac{h^3}{3!}y'''(x) + \cdots,$$

$$\mathbf{y(x+h)} = y(x) + \frac{h}{1!}Dy(x) + \frac{h^2}{2!}D^2y(x) + \cdots$$

$$= (1 + hD + \frac{h^2}{2!}D^2 + \frac{h^3}{3!}D^3 + \cdots)y(x).$$

Noting that

$$e^{hD} = 1 + hD + \frac{h^2D^2}{2!} + \cdots,$$

we can write the Taylor series as

$$\mathbf{y(x+h)} = e^{hD}y(x). \tag{5-40}$$

If, as above, we choose $y_{i+1} = y(x_i + h)$ and $y_i = y_i(x_i)$, we derive

$$y_{i+1} = e^{hD}y_i. \tag{5-41}$$

Changing **h** to **−h** and noting that $y(x_i - h) = y_{i-1}$, we obtain

$$\mathbf{y(x-h)} = e^{-hD}y(x)$$

or

$$y_{i-1} = e^{-hD}y_i. \tag{5-42}$$

Then we can derive our first backward-difference formula, i.e., $\nabla y_i = y_i - y_{i-1}$ may be written as

$$\nabla y_i = [1 - e^{-hD}]y_i,$$

or

$$\nabla y_i = \left[hD - \frac{h^2D^2}{2!} + \frac{h^3D^3}{3!} - \cdots\right]y_i,$$

$$\nabla y_i = \left[1 - \frac{hD}{2} + \frac{h^2D^2}{6} - \frac{h^3D^3}{24} + \cdots\right]hDy_i. \tag{5-43}$$

In purely operational form, we see that

$$\nabla = 1 - e^{-hD}. \tag{5-44}$$

We can therefore obtain ∇^2:

$$\nabla^2 = (1 - e^{-hD})^2 = (1 + e^{-2hD} - 2e^{-hD})$$

$$= 1 + \left(1 - 2hD + \frac{4h^2D^2}{2!} - \cdots\right) - 2\left(1 - hD + \frac{h^2D^2}{2!} - \cdots\right)$$

$$\nabla^2 = h^2D^2 - h^3D^3 + \frac{7}{12}h^4D^4 - \cdots. \tag{5-45}$$

Similarly, we can show that

$$\nabla^3 = h^3D^3 - \frac{3}{2}h^4D^4 + \frac{5}{4}h^5D^5. \tag{5-46}$$

For this methodology to be useful in the approximation of differential equations, we must reverse the process in order to obtain expressions for the derivatives of **y** in terms of the differences instead of the differences in terms of the derivatives, as shown. These are the expressions that are used in subsequent chapters to obtain numerical approximations of the governing processes.

Inverting Eq. (5-44), we get

$$e^{-hD} = 1 - \nabla. \tag{5-47}$$

Taking the logarithm of both sides of Eq. (5-47), we obtain a formula that relates the first derivative to the backward difference, viz.,

$$\ln e^{-hD} = -hD = \ln(1 - \nabla) = -\left(\nabla + \frac{\nabla^2}{2} + \frac{\nabla^3}{3} + \frac{\nabla^4}{4} + \cdots\right),$$

so that finally we arrive at

$$hD = h\frac{\mathrm{d}}{\mathrm{d}x} = \nabla + \frac{\nabla^2}{2} + \frac{\nabla^3}{3} + \frac{\nabla^4}{4} + \cdots. \tag{5-48}$$

Equation (5-48) expresses the first derivative in terms of the difference formulas that were derived above. Similar equations can now be derived that express higher-order derivatives in terms of the backward differences.

Taking powers of Eq. (5-48), we get

$$h^2D^2 = \nabla^2 + \nabla^3 + \frac{11}{12}\nabla^4 + \frac{5}{6}\nabla^5 + \cdots, \tag{5-49}$$

$$h^3D^3 = \nabla^3 + \frac{3}{2}\nabla^4 + \frac{7}{4}\nabla^5 + \cdots, \tag{5-50}$$

$$h^4D^4 = \nabla^4 + 2\nabla^5 + \frac{17}{6}\nabla^6 + \cdots. \tag{5-51}$$

Substituting Eqs. (5-45) and (5-46) into Eq. (5-48), we get

$$D = \frac{\nabla}{h} + \frac{1}{h}\frac{\nabla^2}{2} + \frac{1}{h}\frac{\nabla^3}{3} + \cdots, \tag{5-52}$$

$$D = \frac{\nabla}{h} + \frac{1}{h}\left\{h^2 D^2 - h^3 D^3 + \frac{7}{12}h^4 D^4 - \cdots\right\},$$

$$D = \frac{\nabla}{h} + hD^2 - h^2 D^3 + \cdots. \tag{5-53}$$

Truncating Eqs. (5-53), we derive our first significant relationship that will be used in subsequent modeling formulas relating to differential equations, i.e.,

$$D = \frac{\nabla}{h} = \frac{1}{h}\{y_i - y_{i-1}\}. \tag{5-54}$$

Equation (5-54) is a backward-difference approximation for the first derivative in a differential equation and approaches the exact solution (i.e., the error approaches zero) as the mesh spacing **h** [see Eqs. (5-53)]. We refer to this formula as an order **h** approximation to the first derivative.

In order to obtain a more accurate formula for the first derivative, we return to Eq. (5-52), with the use of Eq. (5-46):

$$D = \frac{1}{h}\nabla + \frac{1}{h}\frac{\nabla^2}{2} + \frac{1}{h}\frac{\nabla^3}{3} + \cdots,$$

$$D = \frac{\nabla}{h} + \frac{1}{h}\frac{\nabla^2}{2} + \frac{1}{3h}\left\{h^3 D^3 - \frac{3}{2}h^4 D^4 + \cdots\right\}. \tag{5-55}$$

Truncating Eq. (5-55), we obtain

$$D = \frac{\nabla}{h} + \frac{1}{h}\frac{\nabla^2}{2} \qquad [\text{order } (\mathbf{h}^2)]$$

or

$$D = \frac{1}{2h}\{2\nabla + \nabla^2\}.$$

Substituting the backward-difference formulas for ∇ and ∇^2 [Eqs. (5-30) and (5-32)], we obtain

$$D = \frac{1}{2h}\{2(y_i - y_{i-1}) + (y_i - 2y_{i-1} + y_{i-2})\}$$

or

$$D = \frac{1}{2h}\{3y_i - 4y_{i-1} + y_{i-2}\} \qquad [\text{order } (\mathbf{h}^2)]. \tag{5-56}$$

Thus we have derived two backward-difference approximations for the first derivative of a function with respect to **x**. The first, Eq. (5-54) is accurate to order **h**, while the second, Eq. (5-56) is accurate to order $\mathbf{h}^2$. The second formula is obviously more accurate than the first.

Approximating formulas for the second derivative can be derived in a similar manner. From Eq. (5-49),

$$h^2 D^2 = \nabla^2 + \nabla^3 + \frac{11}{12}\nabla^4 + \cdots,$$

and substituting Eq. (5-46), we obtain

$$D^2 = \frac{\nabla^2}{h^2} + \frac{1}{h^2}\{h^3 D^3 + \cdots\},$$

$$D^2 = \frac{\nabla^2}{h^2} + O(h),$$

or

$$D^2 = \frac{1}{h^2}\{y_i - 2y_{i-1} + y_{i-2}\}. \tag{5-57}$$

Equation (5-57) is a first-order backward-difference approximation to the second derivative.

We can derive a more accurate formulation for the second derivative by carrying one extra term in Eq. (5-49):

$$D^2 = \frac{1}{h^2}\{\nabla^2 + \nabla^3\} + \frac{1}{h^2}\frac{11}{12}\{h^4 D^4 + \cdots\},$$

$$D^2 = \frac{1}{h^2}\{\nabla^2 + \nabla^3\} + O(h^2). \tag{5-58}$$

Substituting the backward-difference formulation [Eqs. (5-32) and (5-35)] into Eqs. (5-58), we obtain

$$h^2 D^2 = (y_i - 2y_{i-1} + y_{i-2}) + (y_i - 3y_{i-1} + 3y_{i-2} - y_{i-3})$$

or

$$h^2 D^2 = 2y_i - 5y_{i-1} + 4y_{i-2} - y_{i-3} + O(h^2). \tag{5-59}$$

As in the case of the first derivative, we have derived two formulas for the second-derivative approximation. In a similar manner, we can derive formulas for all the derivatives. We can also choose the accuracy of the numerical approximation by deriving a formula with the desired accuracy by using the Taylor series approach that has been presented in this section. Table 5-1 shows several backward-difference approximations.

Table 5-1 Backward-difference approximations

Order **h**							
$h\frac{d}{dx}$	=				$-y_{i-1}$	y_i	
$h^2\frac{d^2}{dx^2}$	=			y_{i-2}	$-2y_{i-1}$	y_i	
$h^3\frac{d^3}{dx^3}$	=		$-y_{i-3}$	$3y_{i-2}$	$-3y_{i-1}$	y_i	
$h^4\frac{d^4}{dx^4}$	=	y_{i-4}	$-4y_{i-3}$	$6y_{i-2}$	$-4y_{i-1}$	y_i	
Order $\mathbf{h^2}$							
$2h\frac{d}{dx}$	=				y_{i-2}	$-4y_{i-1}$	$3y_i$
$h^2\frac{d^2}{dx^2}$	=			$-y_{i-3}$	$4y_{i-2}$	$-5y_{i-1}$	$2y_i$
$2h^3\frac{d^3}{dx^3}$	=		$3y_{i-4}$	$-14y_{i-3}$	$24y_{i-2}$	$-18y_{i-1}$	$5y_i$
$h^4\frac{d^4}{dx^4}$	=	$-2y_{i-5}$	$11y_{i-4}$	$-24y_{i-3}$	$26y_{i-2}$	$-14y_{i-1}$	$3y_i$

Table 5-2 Forward-difference approximations

Order h							
$h\frac{d}{dx}$	=	$-y_i$	y_{i+1}				
$h^2\frac{d^2}{dx^2}$	=	y_i	$-2y_{i+1}$	y_{i+2}			
$h^3\frac{d^3}{dx^3}$	=	$-y_i$	$3y_{i+1}$	$-3y_{i+2}$	y_{i+3}		
$h^4\frac{d^4}{dx^4}$	=	y_i	$-4y_{i+1}$	$6y_{i+2}$	$-4y_{i+3}$	y_{i+4}	
Order h²							
$2h\frac{d}{dx}$	=	$-3y_i$	$4y_{i+1}$	$-y_{i+2}$			
$h^2\frac{d^2}{dx^2}$	=	$2y_i$	$-5y_{i+1}$	$4y_{i+2}$	$-y_{i+3}$		
$2h^3\frac{d^3}{dx^3}$	=	$-5y_i$	$18y_{i+1}$	$-24y_{i+2}$	$14y_{i+3}$	$-3y_{i+4}$	
$h^4\frac{d^4}{dx^4}$	=	$3y_i$	$-14y_{i+1}$	$26y_{i+2}$	$-24y_{i+3}$	$11y_{i+4}$	$-2y_{i+5}$

5-5 FORWARD DIFFERENCES

Forward differences are merely the mirror images of backward differences. Therefore we can immediately write the equivalent formulas for the derivative approximations. Forward differences are defined in a manner similar to that of backward differences; thus

$$\Delta y_i \equiv y_{i+1} - y_i, \tag{5-60}$$

$$\Delta^2 y_i = y_{i+2} - 2y_{i+1} + y_i, \tag{5-61}$$

$$\Delta^3 y_i = y_{i+3} - 3y_{i+2} + 3y_{i+1} - y_i. \tag{5-62}$$

Table 5-2 shows these approximations.

5-6 CENTRAL DIFFERENCES

Referring to Fig. 5-5, we define the first central difference as

$$\delta y_i \equiv y\left(x_i + \frac{h}{2}\right) - y\left(x_i - \frac{h}{2}\right), \tag{5-63}$$

$$\delta y_i = y_{i+\frac{1}{2}} - y_{i-\frac{1}{2}}. \tag{5-64}$$

The second central difference is defined as

$$\delta^2 y_i = \delta(\delta y_i) = \left[y_{(i+\frac{1}{2})+\frac{1}{2}} - y_{(i-\frac{1}{2})+\frac{1}{2}}\right] - \left[y_{(i+\frac{1}{2})-\frac{1}{2}} - y_{(i-\frac{1}{2})-\frac{1}{2}}\right],$$

$$\delta^2 y_i = y_{i+1} - 2y_i + y_{i-1}. \tag{5-65}$$

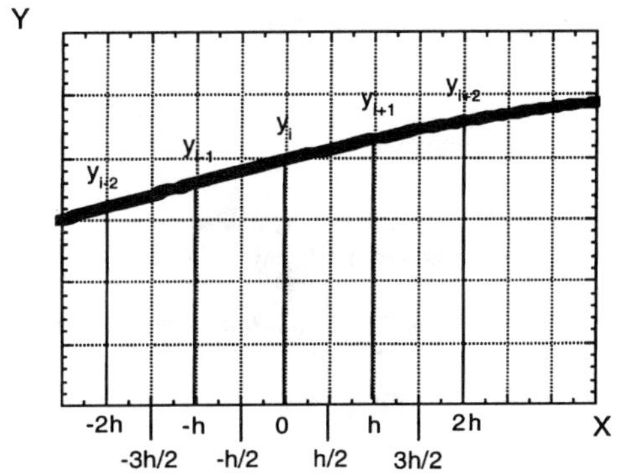

Figure 5-5 Mesh considerations for central-difference approximations.

To eliminate **y** at the intermediate points that appear in all the odd central differences, we define average odd differences. The first averaged difference at **i** becomes

$$\delta y_i = \frac{1}{2}\left(\delta y_{i+\frac{1}{2}} + \delta y_{i-\frac{1}{2}}\right) = \frac{1}{2}[(y_{i+1} - y_i) + (y_i - y_{i-1})] = \frac{1}{2}(y_{i+1} - y_{i-1}). \quad (5\text{-}66)$$

Table 5-3 shows a series of central-difference approximations.

5-7 CONCLUSION

We have developed a technique in this chapter that allows us to calculate finite-difference approximations for any order derivative to any desired accuracy. With these approximations, any differential equation can be approximated to the desired order of accuracy. Several tables have been presented that allow readers to write immediately the difference approximation that they require to simulate their physical process. In the next three

Table 5-3 Central-difference approximations

Accuracy $\mathbf{h}^2$								
$2h\frac{d}{dx}$	$=$		$-y_{i-1}$		y_{i+1}			
$h^2\frac{d^2}{dx^2}$	$=$		y_{i-1}	$-2y_i$	y_{i+1}			
$2h^3\frac{d^3}{dx^3}$	$=$	$-y_{i-2}$	$2y_{i-1}$		$-2y_{i+1}$	y_{i+2}		
$h^4\frac{d^4}{dx^4}$	$=$	y_{i-2}	$-4y_{i-1}$	$6y_i$	$-4y_{i+1}$	y_{i+2}		
Accuracy $\mathbf{h}^4$								
$12h\frac{d}{dx}$	$=$		y_{i-2}	$-8y_{i-1}$		$8y_{i+1}$	$-y_{i+2}$	
$12h^2\frac{d^2}{dx^2}$	$=$		$-y_{i-2}$	$16y_{i-1}$	$-30y_i$	$16y_{i+1}$	$-y_{i+2}$	
$8h^3\frac{d^3}{dx^3}$	$=$	y_{i-3}	$-8y_{i-2}$	$13y_{i-1}$		$-13y_{i+1}$	$8y_{i+2}$	$-y_{i+3}$
$6h^4\frac{d^4}{dx^4}$	$=$	$-y_{i-3}$	$12y_{i-2}$	$-39y_{i-1}$	$56y_i$	$-39y_{i+1}$	$12y_{i+2}$	$-y_{i+3}$

chapters, we use the results of this chapter to formulate computational simulations to problems in engineering.

FURTHER READING

Arden, B. W. and Astill, K. N. (1970). *Numerical Algorithms—Origins and Applications*, Addison-Wesley, Reading, Massachusetts.

Mathews, J. H. (1987). *Numerical Methods for Computer Science, Engineering, and Mathematics*, Prentice-Hall, Englewood Cliffs, New Jersey.

Press, W. H., Teukolsky, S., Vetterling, W., and Flannery, B. (1986). *Numerical Recipes in FORTRAN*, Cambridge University Press, Cambridge.

Salvadori, M. G. and Baron, M. L. (1952). *Numerical Methods in Engineering*, Prentice-Hall, Englewood Cliffs, New Jersey.

Scheid, F. (1968). *Schaum's Outline of Theory and Problems of Numerical Analysis*, McGraw-Hill, New York.

PROBLEMS

5-1 Show that the third derivative of **y** with respect to **x** is given by

$$\frac{d^3 y_i}{dx^3} = \frac{1}{h^3}\{y_i - 3y_{i-1} + 3y_{i-2} - y_{i-3}\}$$

with an accuracy of **h**.

5-2 Derive the backward-difference equation for d^4/dx^4 to an accuracy of **h**.

5-3–5-6 Derive the backward-difference equations for the first four orders of the derivative of **y** with respect to **x** to an accuracy of $\mathbf{h}^2$.

5-7–5-14 Derive the forward-difference equations for the first four orders of the derivatives of **y** with respect to **x** to an accuracy of **h** and to an accuracy of $\mathbf{h}^2$.

5-15–5-22 Repeat Problems 5-7–5-14 for the central-difference approximations.

CHAPTER

SIX

INITIAL-VALUE PROBLEMS

NOMENCLATURE

$f_x = \frac{\partial f}{\partial x}$	derivative of **f** with respect to **x**
$f_0, f_1, f_2, \ldots,$	functional values for Runge–Kutta methods
$y^{(n)}$	nth derivative of **y**
y_n	nth value of the dependent variable
x_n	nth value of the independent variable
Δt	step size
$f(x, y, \ldots, y^{n-1})$	functional form
$f_{xy} = \frac{\partial f}{\partial x \partial y}$	derivative of **f** with respect to **x** and **y**
c	value of the coefficient of viscous damping
h	step size for the independent variable
k	value of the spring constant for the vibrating mass
m	value of the vibrating mass
x, t	independent variables
$y(0)$	dependent variable at $\mathbf{x} = 0$

6-1 INTRODUCTION

Differential equations that describe physical processes come in distinct two types, initial-value problems (which are described in this chapter) and boundary-value problems (which are described in the next two chapters).

In initial-value problems, the solution is generated from a set of initial quantities. The solution process proceeds by incrementing the values of the independent value.

There is no terminal boundary for the independent variables so that no end conditions need to be satisfied.

The independent variable is discretized into specific small units, called increments, at which the dependent variable is to be calculated. The solution strategies that are used toward initial-value problems use the state of the dependent variable at a specific time interval to calculate the state variables at the next time interval. Thus the previous time increment drives the solution for the next increment. It is necessary therefore to calculate an accurate solution for the first few increments (based on the given initial conditions) and to continue the solution process with an algorithm that will accurately predict the remainder of the solution. Also, because of this marching process, error accumulation is probable. The solution algorithm should be written so that this error buildup is minimized.

Many problems in physics and engineering are both initial-value problems and boundary-value problems. These problems, specifically in the area of deformable body mechanics and fluid dynamics, must be treated with a combination of the techniques of this chapter and the next two chapters.

6-2 INITIAL INCREMENTS WITH A TAYLOR SERIES

A robust algorithm that results in an accurate solution to an initial-value problem depends, in part, on the manner by which the solution process is started. There are many techniques that attempt to move the dependent variables away from their initial values into the solution region. Some of these techniques are now examined.

A simple formulation for finding the initial increments in a solution process involves the use of a Taylor series. Consider the nth-order differential equation of the form

$$y^n(x) = f[x, y, y', y'', \ldots, y^{n-1}]$$

with n arbitrary initial conditions given, viz.,

$$y(0) = y_0, \qquad y'(0) = y_0', \ldots, \qquad y^{(n-1)}(0) = y_0^{n-1}.$$

We desire to solve for **y(x)** at points within the region, defined by $x_i = ih$, where **h** is the mesh spacing. To obtain a starting point in the solution process, we use a Taylor series at the given initial values, viz.,

$$y_0^n = f\left[0, y_0, y_0', \ldots, y_0^{n-1}\right]. \tag{6-1}$$

We can calculate higher-order derivatives at the initial point by using the chain rule of differentiation:

$$y_0^{n+1} = \left[\frac{\partial f}{\partial x} + \frac{\partial f}{\partial y}y' + \frac{\partial f}{\partial y'}y'' + \cdots + \frac{\partial f}{\partial y^{n-1}}y^n\right]_{x=0}, \tag{6-2}$$

$$y_0^{n+2} = \left[\frac{\partial^2 f}{\partial x^2} + \frac{\partial^2 f}{\partial y^2}y'^2 + \frac{\partial f}{\partial y}y'' + \cdots\right]_{x=0}. \tag{6-3}$$

We can express the solution **y(x)** near the reference point by using the Taylor series:

$$\mathbf{y(x)} = y_0 + \frac{y_0'}{1!}x + \frac{y_0''}{2!}x^2 + \cdots + . \tag{6-4}$$

This allows the evaluation of **y** at the first few pivotal points, provided that **h** is small.

To illustrate the use of the Taylor series to evaluate the initial solution points, let us consider the differential equation

$$y' = -\frac{0.9}{1+2x}y, \tag{6-5}$$

with the initial condition

$$y(0) = y_0 = 1.$$

We may now calculate the first three derivatives by differentiating Eq. (6-5) and substituting $\mathbf{x} = 0$, i.e.,

$$y'(0) = -0.9[(1+2x)^{-1}y]_{x=0} = -0.9, \tag{6-6}$$

$$y''(0) = -0.9[(1+2x)^{-1}y' - 2(1+2x)^{-2}y]_{x=0}$$

$$= -0.9(y_0' - 2y_0) = +2.61, \tag{6-7}$$

$$y'''(0) = -0.9[(1+2x)^{-1}y'' - 4(1+2x)^{-2}y' + 8(1+2x)^{-3}y]_{x=0}$$

$$= -0.9(y_0'' - 4y_0' + 8y_0) = -12.79. \tag{6-8}$$

Then the Taylor series about $\mathbf{x} = 0$ may be approximated by

$$\mathbf{y(x)} \approx y_0 + \frac{y_0'}{1!}x + \frac{y_0''}{2!}x^2 + \frac{y_0'''}{3!}\cdots + \cdots,$$

or

$$\mathbf{y(x)} = 1 - 0.9x + 1.305x^2 - 2.132x^3 + 3.677x^4. \tag{6-9}$$

Using Eq. (6-9), we can solve for the first few solution points and compare them with the exact solution of Eq. (6-5). Table 6-1 shows this comparison.

We derive the last column in Table 6-1 by solving Eq. (6-5):

$$\frac{dy}{y} = -0.9\frac{dx}{1+2x}$$

for the exact solution

$$\mathbf{y(x)} = (1+2x)^{-0.45}. \tag{6-10}$$

Table 6-1 Taylor series used to calculate initial solution points

x	$y(x)$ (approximate)	$y(x)$ (exact)
0	1.0	1.0
0.02	0.9825	0.9825
0.04	0.9660	0.9660
0.06	0.9503	0.9503
0.08	0.9354	0.9354
0.10	0.9213	0.9212
0.20	0.8610	0.8595
0.30	0.8197	0.8094

Comparing the exact solution at $\mathbf{x} = 0.3$ with the Taylor series at the same point, we see that the approximation is in error by $+1.27\%$

6-3 TAYLOR SERIES METHOD FOR IMPLICIT SYSTEMS

Although the Taylor series method is usually useful only near the starting point, it can nevertheless be used in a limited range. However, when the function $\mathbf{f}$ contains both the dependent variable $\mathbf{y}$ and the independent variable $\mathbf{x}$ explicitly [e.g., $y' = f(y, x)$], then the calculation of the derivatives in the Taylor series becomes somewhat more complex. In this case, the Taylor series may be defined as

$$y(x_k + h) = y(x_k) + hT_N[x_k, y(x_k)], \tag{6-11}$$

where

$$T_N[x_k, y(x_k)] = \sum_{j=1}^{N} \frac{y^{(j)} x_k}{j!} h^{j-1} \tag{6-12}$$

and $y^{(j)}$ is the jth total derivative of the function $\mathbf{f}$ with respect to $\mathbf{x}$.

By using the chain rule of differentiation, we can derive formulas for the derivatives in the Taylor series, i.e.,

$$y'(x) = f, \tag{6-13}$$

which is the differential equation. Differentiating Eq. (6-13) implicitly, we get

$$y''(x) = \frac{\partial f}{\partial x} + \frac{\partial f}{\partial y} y' = f_x + f_y f, \tag{6-14}$$

$$y'''(x) = f_{xx} + 2f_{xy}y' + f_y y'' + f_{yy}[y']^2, \tag{6-15}$$

where the subscripts represent partial derivatives with respect to that variable.

These expressions, once evaluated, may then be used in Eq. (6-12) to form the Taylor series for an implicit expression and therefore provide the basis for the starting point evaluation of a first-order initial-value problem.

6-4 RUNGE–KUTTA METHODS

As seen in the sections above, the difficulty with the Taylor series approach is that the differentiation in most realistic cases may be extremely difficult. This difficulty would lead to the nonuse of this important technique for providing the solution of a differential equation near the initial conditions. Additionally, as the solution process moves away from the initial values, the Taylor series approach becomes more inaccurate.

It would therefore be useful to be able to attain a solution with the accuracy equivalent to a specified number of terms in the Taylor series and yet compute these values by using the weighted sum techniques that were used in numerical integration (which was described in Chapter 4). The techniques that provide this opportunity are called the Runge–Kutta methods. In addition, these methods allow the adjustment of the interval of integration **h** during the solution process.

Since the Runge–Kutta methods approximate the Taylor series to a specified number of terms, there are a variety of algorithm orders. All the various possibilities are not provided in this text. Instead, the derivation of the technique for a specific order is illustrated and the derivation of other levels is left to the readers.

We begin by considering the differential equation,

$$y' = f(x, y)$$

which has the known solution $y = y_n$ at $x = x_n$. We wish to produce an integration formula of the form

$$y_{n+1} = y_n + ak_1 + bk_2 + ck_3 + dk_4 \tag{6-16}$$

at the location $x_{n+1} = x_n + h$, where

$$k_1 = hf(x, y), \tag{6-17}$$

$$k_2 = hf(x + mh, y + mk_1), \tag{6-18}$$

$$k_3 = hf(x + nh, y + nk_2), \tag{6-19}$$

$$k_4 = hf(x + ph, y + pk_3), \tag{6-20}$$

such that Eq. (6-16) becomes identical to the Taylor series through the term in h^4.

Using the techniques that allowed us to derive implicitly Eqs. (6-13)–(6-15), we can express the first four derivatives in the Taylor series, i.e.,

$$y' = f(x, y) = f, \tag{6-21}$$

$$y^{(2)} = f_x + f_y y' = f_x + f_y y = F_1, \tag{6-22}$$

$$y^{(3)} = F_2 + f_y F_1, \tag{6-23}$$

$$y^{(4)} = F_3 + f_y F_2 + 3F_1(f_{xy} + ff_{yy}) + f_y^2 F_1, \tag{6-24}$$

where

$$F_1 = f_x + ff_y, \tag{6-25}$$

$$F_2 = f_{xx} + 2ff_{xy} + f^2 f_{yy}, \tag{6-26}$$

$$F_3 = f_{xxx} + 3ff_{xxy} + 3f^2 f_{xyy} + f^3 f_{yyy}. \tag{6-27}$$

The Taylor series may then be expressed as

$$y(x + h) - y(x) = hf + \frac{1}{2}h^2 F_1 + \frac{1}{6}h^3(F_2 + f_y F_1) + \frac{1}{24}h^4\left[F_3 + f_y F_2 + 3(f_{xy} + ff_{yy})F_1 + f_y^2 F_1\right] + \cdots. \tag{6-28}$$

Equations (6-17)–(6-20) may now be evaluated:

$$k_1 = h_f, \tag{6-29}$$

$$k_2 = h\left[f + mhF_1 + \frac{1}{2}m^2h^2F_2 + \frac{1}{6}m^3h^3F_3 + \cdots\right] \tag{6-30}$$

$$k_3 = h\left\{f + nhF_1 + \frac{1}{2}h^2\left(n^2F_2 + 2npf_yF_1\right) + \frac{1}{6}h^3\left[n^3F_3 + 3m^2nf_yF_2 + 6mn^2(f_{xy} + ff_{yy})F_1\right] + \cdots\right\}, \tag{6-31}$$

$$k_4 = h\left\{f + phF_1 + \frac{1}{2}h^2\left(p^2F_2 + 2npf_yF_1\right) + \frac{1}{6}h^3\left[p^3F_3 + 3n^2pf_yF_2 + 6np^2(f_{xy} + ff_{yy})F_1 + 6mnpf_y^2F_1\right] + \cdots\right\}. \tag{6-32}$$

Substituting k_1–k_4 into Eq. (6-16), we get

$$\begin{aligned} y_{n+1} = {} & y_n + (a+b+c+d)hf + (bm+cn+dp)h^2F_1 \\ & + \frac{1}{2}(bm^2cn^2 + dp^2)h^3F_2 + \frac{1}{6}(bm^3 + cn^3 + dp^3)h^4F_3 \\ & + (cmn + dnp)h^3f_yF_1 + \frac{1}{2}(cm^2n + dn^2p)h^4f_yF_2 \\ & + (cmn^2 + dnp^2)h^4(f_{xy} + ff_{yy})F_1 + dmnph^4f_y^2F_1 + \cdots. \end{aligned} \tag{6-33}$$

Comparing Eq. (6-33) with the Taylor series [Eq. (6-28)], we can express the following conditions:

$$\begin{aligned} a+b+c+d &= 1, \\ bm + cn + dp &= \frac{1}{2}, \\ bm^2 + cn^2 + dp^2 &= \frac{1}{3}, \\ bm^3 + cn^3 + dp^3 &= \frac{1}{4}, \\ cmn + dnp &= \frac{1}{6}, \\ cmn^2 + dnp^2 &= \frac{1}{8}, \\ cm^2n + dn^2p &= \frac{1}{12}, \\ dmnp &= \frac{1}{24}. \end{aligned} \tag{6-34}$$

Since there are eight constraint conditions for the seven unknowns, we choose the usual solution, viz.,

$$\begin{aligned} m &= n = \frac{1}{2}, \\ p &= 1, \\ a &= d = \frac{1}{6}, \\ b &= c = \frac{1}{3}, \end{aligned} \tag{6-35}$$

which leads to

$$\begin{aligned} k_1 &= hf(x, y), \\ k_2 &= hf\left[x + \frac{1}{2}h, y + \frac{1}{2}k_1\right], \\ k_3 &= hf\left[x + \frac{1}{2}h, y + \frac{1}{2}k_2\right], \\ k_4 &= hf\left[x + h, y + k_3\right]. \end{aligned} \tag{6-36}$$

Equations (6-36) along with Eq. (6-16) yield the final expression for use with a fourth-order Runge–Kutta technique, i.e.,

$$y_{n+1} = y_n + \frac{1}{6}(k_1 + 2k_2 + 2k_3 + k_4). \tag{6-37}$$

As an example of the application of the above technique, consider the first-order differential equation,

$$y' = f(x, y) = xy^{1/3}$$

with the initial condition $y(1) = 1$. Choosing $x_0 = 1$ and using a mesh spacing of $\mathbf{h} = 0.1$, we can calculate

$$\begin{aligned} k_1 &= (0.1)f(1, 1) = 0.1, \\ k_2 &= (0.1)f(1.05, 1.05) = 0.10672, \\ k_3 &= (0.1)f(1.05, 1.05336) = 0.10684, \\ k_4 &= (0.1)f(1.1, 1.10684) = 0.11378, \end{aligned}$$

from which we can calculate

$$y_1 = 1 + \frac{1}{6}[0.1 + 0.21344 + 0.21368 + 0.11378] = 1.10682 = y(0.1).$$

We can continue the Runge–Kutta solution process (with a mesh spacing of $\mathbf{h} = 0.1$) to calculate subsequent values of $\mathbf{y}$. Table 6-2 summarizes the results when the Taylor series, the Runge–Kutta solution, and the exact solution of the example differential equation are used.

Table 6-2 Runge–Kutta Methods

x	Taylor	Runge–Kutta	Exact
1.0	1.00	1.00	1.00
2.0	2.82846	2.82843	2.82843
3.0	7.02123	7.02113	7.02113
4.0	14.69710	14.69693	14.69694
5.0	27.00022	26.99998	27.0000

The procedures that are described herein are to be applied to first-order systems only. We can reduce higher-order differential equations, or a system of equations including some high-order members, to a set of first-order equations by making the following transformations. An nth-order equation of the form

$$y^{(n)} = f\left[x, y, y', y'', \dots, y^{(n-1)}\right]$$

can be reduced to a system of simultaneous first-order equations; thus,

$$\begin{aligned} y &= y_0, \\ y' &= y_1, \\ y'' &= y_1' = y_2, \\ &\dots \\ y^{(n)} &= \dots = y_{n-1}' = f(x, y_0, y_1, \dots, y_{n-1}). \end{aligned}$$

This set of equations may then be solved with the techniques of this section.

The Runge–Kutta methods may be used to obtain a solution to a differential equation throughout its entire range of interest. However, these techniques tend to be tedious in their application. More robust methods that are described in the next sections yield accurate solutions within less computational time than either the Taylor or the Runge–Kutta methods. Thus the above techniques are usually used only to obtain a good starting solution near the initial conditions.

6-5 EULER FORWARD-INTEGRATION METHOD

The Taylor series approach of the sections above rapidly loses accuracy as we move away from point $\mathbf{x} = x_0$. Thus this technique is limited to finding the first few solution points near the initial conditions. We now seek a strategy for calculating the rest of the solution at the pivotal mesh points. We refer to these points as incremental points where the solution is required.

An elementary method of step-by-step integration of first-order equations is considered. It is of limited value because of its low accuracy but is useful in illustrating forward-integration processes and does not require the starting of the solution as in the above-described methods. The technique is called the Euler forward-integration method. Consider the first-order differential equation

$$y' = f(x, y)$$

with an initial condition

$$y(0) = y_0.$$

By applying a simple forward-difference formula and evaluating the function at the ith mesh point, we define the Euler forward-integration method by

$$\Delta y_i = y_{i+1} - y_i = f(x_i, y_i)\, h. \tag{6-38}$$

Equation (6-38) states that the change in the value of the dependent variable depends solely on the value of the solution at $[x_i, y_i]$ and the mesh spacing $\mathbf{h}$.

To illustrate this simple technique, let us return to the example of Section 6-2. Then,

$$y' = -\frac{0.9}{1+2x} y$$

with the initial condition $y(0) = y_0 = 1$. We choose an increment for the independent variable, called a step size, of $\mathbf{h} = 0.02$. We establish our integration procedure as

$$y_{i+1} = y_i + h\left\{-\frac{0.9}{1+2x_i}\right\} y_i. \tag{6-39}$$

The solution variable at increment $i+1$ is calculated from the state of the solution at increment i. Using Eq. (6-39), we may begin the solution process by substituting the initial conditions and calculating the dependent variable at $\mathbf{x} = \mathbf{h}$ (also known as x_1), viz.,

$$y_1 = 1 - \frac{0.9(1)(0.02)}{1+2(0)} = 1 - 0.018 = 0.9820.$$

We continue this process in order to calculate the first few solution points. The solutions for these points are shown in Table 6-3.

The solution that is shown in the table was carried out with a step size of $\mathbf{h} = 0.2$. The solution is shown in the table at a selected number of these points.

It can be shown that this solution process has an error of the order of h^2. Comparing the results of the Euler method with the exact solution of the differential equation yields:

at $\mathbf{x} = 0.10$, the error is 0.24%,

at $\mathbf{x} = 0.30$, the error is 2.86%.

Table 6-3 Euler forward-integration technique

x	y_i	f_i	Δy_i
0	1.000	−0.9	−0.0180
0.02	0.9820	−0.8498	−0.0170
0.04	0.9650	−0.8042	−0.0161
0.06	0.9489	−0.7625	−0.0153
0.08	0.9336	−0.7243	−0.0145
0.10	0.9191	−0.6893	−0.0689
0.20	0.8402	−0.5401	−0.0540
0.30	0.7862		

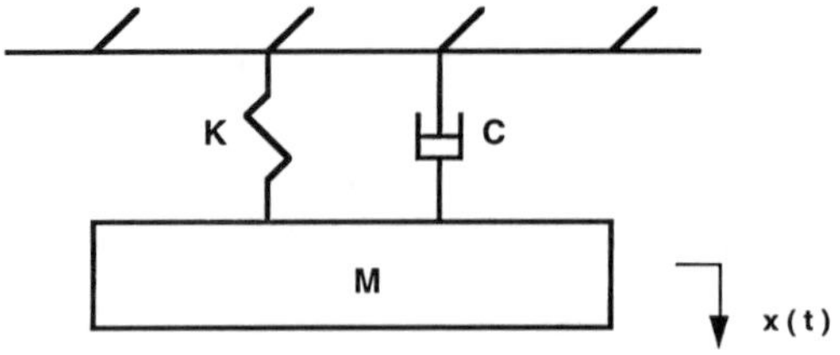

Figure 6-1 Single-degree-of-freedom free vibration of a damped system.

Another important application for which the Euler method has considerable utility is the area of vibratory mechanics. As a second example of the use of this solution method, consider the free vibration of a single-degree-of-freedom, damped mass-spring system. Figure 6-1 shows the physical configuration for this problem. The differential equation of motion for this problem is known to be

$$m\ddot{x} + c\dot{x} + kx = 0, \tag{6-40}$$

subject to the initial conditions $x(0) = x_0$ and $\dot{x}(0) = v(0) = v_0$. In this equation, x is the dependent variable (displacement), t (time) is the independent variable, m is the mass of the vibrating body, c is the coefficient of the viscous damper, and k is the modulus of the spring. Additionally, we have the kinematic relations,

$$v = \frac{dx}{dt} \quad \text{(for the velocity)}, \tag{6-41}$$

$$a = \frac{dv}{dt} \quad \text{(for the acceleration)}. \tag{6-42}$$

We can now approximate Eqs. (6-40)–(6-42) by applying the Euler method, viz.,

$$ma_{i+1} + cv_{i+1} + kx_{i+1} = 0, \tag{6-40a}$$

$$v_i = \frac{x_{i+1} - x_i}{\Delta t}, \tag{6-41a}$$

$$a_i = \frac{v_{i+1} - v_i}{\Delta t}. \tag{6-42a}$$

Equations (6-40a)–(6-42a) are used to solve for the dynamic variables (displacement, velocity, and acceleration) at time increment $i + 1$ when the variables are known at the ith increment. Δt is the size of the step for the independent variable (i.e., the time step). Then, once the Euler method (also called a marching process) has begun, it proceeds with the continuous application of Eqs. (6-40a)–(6-42a). The initial conditions provide x_0 and v_0. The value of a_0 can easily be calculated by the insertion of the initial conditions into Eq. (6-40). Once the initial displacement, velocity, and acceleration are known, the Euler solution method may be applied successfully. Appendix 6-1 contains a computer code, called **explicit.for**, that uses the Euler forward-integration method. A time step of $\Delta t = 0.01$ was used in this solution. Table 6-4 summarizes the results of this analysis. Figure 6-2 plots the actual results.

Table 6-4 Euler forward-integration applied to a single-degree-of-freedom vibrating system

Increment	Time	x	v	a
1	0	1	0	−2.5
50	0.49	0.758	−0.883	−1.01
100	0.99	0.253	−1.02	0.391
150	1.49	−0.176	−0.627	1.07
200	1.99	−0.356	−0.0889	0.978
250	2.49	−0.298	0.283	0.463
300	2.99	−0.123	0.377	−0.0708
350	3.49	0.0429	0.257	−0.364
400	3.99	0.124	0.0626	−0.373
450	4.49	0.115	−0.0862	−0.202
500	4.99	0.0556	−0.137	−0.00233

6-6 MODIFIED EULER METHOD

The Euler method of Section 6-5, although simple to use, is nevertheless highly inaccurate in its application. An inefficiently small step size is required for a fairly accurate solution at the first few steps. Second, since the solution at the $i + 1$st increment depends on the ith increment, there is a natural buildup of the error as the solution process proceeds. It is desirable then to modify the solution that is calculated from the Euler method of Section 6-5. We call this new process the modified Euler method and define this technique as

$$\Delta' y_i = \frac{h}{2}[f(x_i, y_i) + f(x_{i+1}, y_{i+1})]. \tag{6-43}$$

In this method, the error is of the order of h^3. The modified Euler method uses the average result of the known value of the function (at $\mathbf{x} = x_i$) as well as the unknown value of the function (at $\mathbf{x} = x_{i+1}$). Therefore, in order to apply this technique, we must

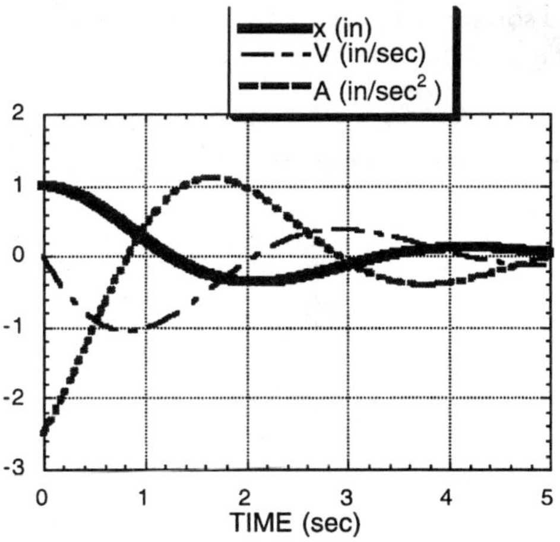

Figure 6-2 Displacement, velocity, and acceleration for a single-degree-of-freedom damped free vibration of the Euler forward-integration method.

first apply the Euler forward-integration method. Using the solution obtained therefrom, we apply the modified Euler method to obtain a more accurate estimate of the solution at increment $i + 1$.

Starting with the Euler method of Section 6-5 for the example problem, we previously obtained

$$x = 0, \qquad y_i = 1.0, \qquad f_i = -0.90, \qquad \Delta y_i = -0.0180,$$

so that

$$y_{i+1} = 1 - 0.0180 = 0.9820.$$

Using y_{i+1}, we can calculate

$$f_{i+1} = -\frac{0.9}{1 + 2(0.02)}(0.982) = -0.8498.$$

Using this result, we now apply the modified Euler method, viz.,

$$\Delta' y_i = 0.01\{-0.9 - 0.8498\} = -0.0175,$$

so that the updated value for y_{i+1} is

$$y'_{i+1} = 1 - 0.0175 = 0.9825.$$

We can now create Table 6-5, which consists of solution steps, by using the Euler forward-integration method, which is then followed by the modified Euler method.

If we once again compare the exact solution with the solution obtained from the modified Euler method, we note that

$$\text{at } \mathbf{x} = 0.10, \text{ the error is } 0.0\%,$$
$$\text{at } \mathbf{x} = 0.30, \text{ the error is } 0.04\%.$$

The modified Euler method may also be applied to the free vibration of the mass, which was described in Section 6-5. In this case, there is no need to perform a two-step process. We discretize Eqs. (6-40)–(6-42) and solve a matrix set. This technique is still a modified Euler method since it uses the average solution at $t = t_i$ and at $t = t_{i+1}$. These

Table 6-5 Modified Euler method

x_i	y_i	f_i	Δy_i	y_{i+1}	f_{i+1}	$\Delta' y_i$
0	1.000	−0.9	−0.018	0.9820	−0.8498	−0.0175
0.02	0.9825	−0.8503	−0.017	0.9655	−0.8046	−0.0165
0.04	0.9660	−0.8050	−0.0161	0.9499	−0.7633	−0.0157
0.06	0.9503	−0.7637	−0.0153	0.9350	−0.7255	−0.0149
0.08	0.9354	−0.7258	−0.0145	0.9209	−0.6907	−0.0142
0.10	0.9212	−0.6909	−0.0691	0.8521	−0.5478	−0.0619
0.20	0.8593	−0.5524	−0.0552	0.8041	−0.4523	−0.0502
0.30	0.8091					

equations may now be discretized with the technique of this section, i.e.,

$$ma_{i+1} + cv_{i+1} + kx_{i+1} = 0, \tag{6-40b}$$

$$\frac{v_{i+1} + v_i}{2} = \frac{x_{i+1} - x_i}{\Delta t}, \tag{6-41b}$$

$$\frac{a_{i+1} + a_i}{2} = \frac{v_{i+1} - v_i}{\Delta t}. \tag{6-42b}$$

Equations (6-40b)–(6-42b) may be put into matrix form for a direct solution algorithm:

$$\begin{bmatrix} m & c & k \\ 0 & \frac{\Delta t}{2} & -1 \\ \frac{\Delta t}{2} & -1 & 0 \end{bmatrix} \begin{bmatrix} a_{i+1} \\ v_{i+1} \\ x_{i+1} \end{bmatrix} = \begin{bmatrix} 0 \\ -x_i - \frac{\Delta t}{2} v_i \\ -v_i - \frac{\Delta t}{2} a_i \end{bmatrix}. \tag{6-44}$$

All quantities on the right-hand side of Eq. (6-44) are known at increment i. Only the $(i + 1)$st incremental quantities are unknown and readily calculated when the coefficient matrix is inverted. It is interesting to note, in this case, that the coefficient matrix is constant, once a choice for Δt is made. Therefore the coefficient matrix need be inverted only once during the entire solution process. The inverted matrix may then be subsequently multiplied by the updated right-hand-side vector to calculate the dynamic variables continually at successive time increments. This strategy will greatly reduce the computational time since only one matrix inversion would then be required. Appendix 6-2 contains a computer code, called **implicit.for**, that solves the free vibration problem by use of the modified Euler method. Table 6-6 summarizes the results of this analysis for a time step $\Delta t = 0.01$. Figure 6-3 plots the results.

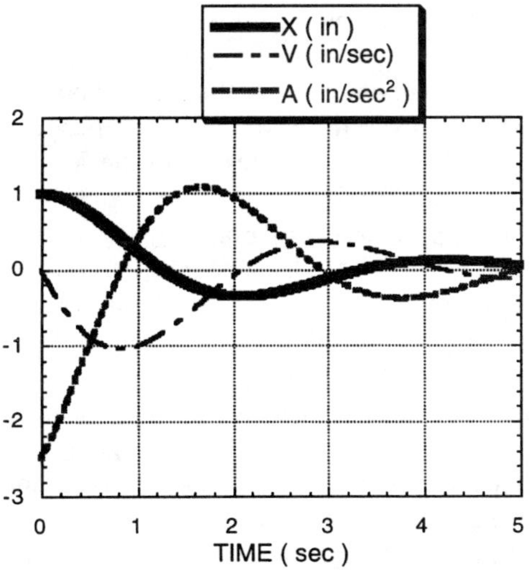

Figure 6-3 Displacement, velocity, and acceleration for a single-degree-of-freedom damped free vibration of the modified Euler method.

Table 6-6 Modified Euler method solution of vibration problem

Increment	Time	x	v	A
1	0	1	0	−2.5
50	0.49	0.756	−0.875	−1.01
100	0.99	0.255	−1.01	0.375
150	1.49	−0.168	−0.623	1.04
200	1.99	−0.346	−0.0961	0.961
250	2.49	−0.292	0.268	0.463
300	2.99	−0.123	0.364	−0.0557
350	3.49	0.0367	0.252	−0.344
400	3.99	0.117	0.0666	−0.358
450	4.49	0.111	−0.0771	−0.2
500	4.99	0.0554	−0.128	−0.0103

6-7 OPTIMUM STEP-SIZE CONSIDERATIONS

It is obvious from the examples of Sections 6-5 and 6-6 that the modified Euler method is more accurate than the Euler forward-integration method. Thus, for the same choice of time step Δt, we obtain a better solution by using the former rather than the latter approach. However, once the choice of a strategy is made, the question of the value of the step size that will produce the most accurate solution while using the least amount of time steps becomes important. This value is called the optimum step size since it provides the most accurate solution with the least cost.

Before considering a methodology for determining this optimum step, note that the best step is a function of the solution strategy that has already been selected. In other words, there is an optimum step size for the Euler forward, and a different optimum step size for the modified Euler, solution techniques. We must decide which solution algorithm is to be used before seeking the best increment for the independent variable.

Referring to Figs. 6-2 and 6-3, we note that maximum value of the velocity occurs near $t = 1$ s (100 time steps for a Δt of 0.01 s). We choose this value of the independent variable to illustrate an empirical (i.e., trial-and-error) method for determining the desired value of Δt.

To produce an example of this trial-and-error approach, let us arbitrarily decide that we are choosing the more accurate modified Euler method. The code, **implicit.for**, in Appendix 6-2 is used for this study. Since we desire the value of the velocity at 1 s, we run the code for several choices of Δt and record the value of the velocity at this time. Table 6-7 shows the results of these runs and Fig. 6-4 plots the results.

It is obvious from Table 6-7 that the optimum step size for this problem is $\Delta t = 0.01$. Any larger choice produces a value for the velocity at $t = 1$ that will change as the increment is reduced. Any smaller choice of the increment will not produce a significant change in the solution and would therefore use computer time without benefit.

In every initial-value problem, an inspection of the solution for a given step size will yield a favorable solution point where the comparisons of the preceding paragraph may be made. It is the opinion of the author that this trial-and-error method for determining

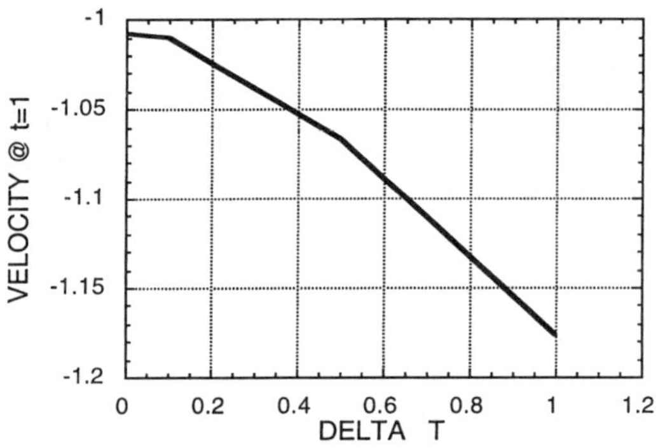

Figure 6-4 Study to determine the optimum step size.

optimum discretation size is better than any analytical approach that can be developed. Also, this approach will work for any initial-value problem and does not depend on any *a priori* knowledge of the solution.

6-8 MILNE METHOD

We have thus far explored relatively simple methods for the integration of initial-value problems. We now explore the Milne method, which is a higher-order integration method that yields accurate results for a reasonable choice of **h**. However, the function **f** must be known at several previous pivotal points in order to apply this technique. Therefore it can be used only after the solution has progressed. To use the Milne method, we start the solution process with one of the above-illustrated methods. After several initial increments are calculated, we turn our attention to the new approach.

We start, once again, with the basic differential equation $y' = f(x, y)$, where the values of the dependent variable (and therefore the function) are known at $[x_0, x_1, x_2, x_3]$. Specifically,

$$[y_0, y_1, y_2, y_3]$$

or

$$[f_0, f_1, f_2, f_3]$$

is known at these points.

Table 6-7 Optimum step-size consideration

Δt	Increment	Time (s)	Velocity
1	1	1	−1.176
0.5	2	1	−1.067
0.1	10	1	−1.0109
0.05	20	1	−1.009
0.01	100	1	−1.008
0.005	200	1	−1.008

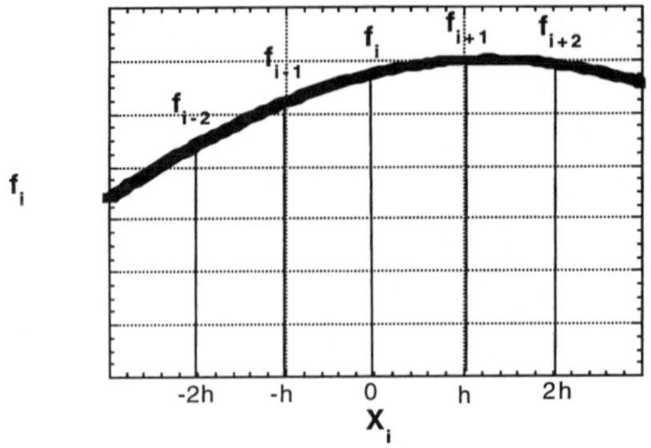

Figure 6-5 Data point setup for the Milne method.

We desire to calculate y_{i+1} by using the formula

$$y_{i+1} - y_{i-3} = \int_{x_{i-3}}^{x_{i+1}} f(x, y)\,\mathrm{d}x = \frac{4h}{3}(2f_{i-2} - f_{i-1} + 2f_i) + \frac{14}{15}h^5 f^{IV}(\xi). \quad (6\text{-}45)$$

The above formula is the Milne method. This formula states that the dependent variable at the $(i + 1)$st point is calculated from the values of the dependent variable at the four previous points.

In order to understand this technique fully, we derive Eq. (6-45). Figure 6-5 shows the graphical relationship of the points that surround point i. We assume a parabola of the form

$$y = ax^2 + bx + c. \quad (6\text{-}46)$$

Then, substituting the known points around point i, we get

$$\begin{aligned} f_{i+1} &= ah^2 + bh + c, \\ f_i &= c, \\ f_{i-1} &= ah^2 - bh + c, \end{aligned}$$

from which we can derive

$$a = \frac{f_{i+1} - 2f_i + f_{i-1}}{2h^2},$$

$$b = \frac{f_{i+1} - f_i}{2h},$$

$$c = f_i.$$

Then the integral,

$$\int_{-2h}^{2h} f(x)\,\mathrm{d}x = \frac{16h^3}{3}a + 4hc,$$

becomes, on substitution,

$$= \frac{8h}{3}\{f_{i+1} - 2f_i + f_{i-1}\} + 4h\{f_i\},$$

Table 6-8 Use of Milne integration method

x	Milne	Exact
0		0
0.3		0.05
0.6		0.222
0.9		0.56
1.2	1.119	1.119
1.5	1.982	1.981
1.8	3.249	3.249
2.1	5.066	5.066
2.4	7.622	7.623

or finally,

$$\int_{-2h}^{2h} f(x)\,\mathrm{d}x = \frac{4h}{3}\{2f_{i+1} - f_i + 2f_{i-1}\}. \tag{6-47}$$

Returning to the differential equation, $y' = f(x, y)$, and shifting one step to the left, we find that Eq. (6-47) becomes

$$y_{i+1} - y_{i-3} = \int_{x_{i-3}}^{x_{i+1}} f(x, y)\,\mathrm{d}x = \frac{4h}{3}(2f_{i-2} - f_{i-1} + 2f_i). \tag{6-48}$$

The y_{i+1} that is calculated from the above formula is used in $f(x, y)$ to find f_{i+1}, which can then be used in Simpson's 1/3 rule, i.e.,

$$y_{i+1} = y_{i-1} + \frac{h}{3}\{f_{i-1} + 4f_i + f_{i+1}\} \tag{6-49}$$

to obtain a more accurate formula for y_{i+1}. Equation (6-49) has an accuracy of the order of h^5, as shown by Eq. (6-45).

The Milne method is also known as a predictor–corrector technique since it allows the prediction of the solution based on the solution at several previous points. We can also use it iteratively by successively applying Eq. (6-48) until two successive values for y_{i+1} are within a small convergence criterion.

As an example of the use of the Milne method, let us consider the differential equation $y' = x + y$ with an initial condition $y_0 = 0$. The exact solution to this equation is $y = e^x - (1 + x)$. Table 6-8 compares the solution of this differential equation with the Taylor series for the initial solution points, the Milne method for successive points, and the exact solution as a comparative tool.

6-9 CONCLUSION

In this chapter several techniques have been described that can be brought to bear in order to solve initial-value problems as they occur in physics and engineering, such as the vibration of a mass-spring dashpot system. The methods, which can be used to solve the class of problems that are known as initial-value problems, are considerable. This chapter has illustrated the Taylor series, the Runge–Kutta method, the Euler forward-integration method, the modified Euler method, and the Milne method. The latter is a predictor–corrector technique that can be used to accelerate the solution process with no loss in accuracy. These techniques are but a small portion of the algorithms that are available for the integration of differential equations that are subject to initial conditions. This chapter does not pretend to delineate all the techniques that are available for this solution process. Reference to the section entitled Further Reading, at the end of this chapter, will enhance students' use of these techniques. The purpose of this chapter was to expose the readers to the more relevant solution techniques that are used to solve initial-value differential equations. It is hoped that this chapter will motivate readers to select the solution process that is most efficient for their particular application.

FURTHER READING

Arden, B. W. and Astill, K. N. (1970). *Numerical Algorithms—Origins and Applications*, Addison-Wesley, Reading, Massachusetts.

Mathews, J. H. (1987). *Numerical Methods for Computer Science, Engineering and, Mathematics*, Prentice-Hall, Englewood Cliffs, New Jersey.

Rao, S. S. (1990). *Mechanical Vibrations*, Addison-Wesley, Reading, Massachusetts. Thomsen, W. T. (1972). *Theory of Vibrations, with Applications*, Prentice-Hall, Englewood Cliffs, New Jersey.

Scheid, F. (1968). *Schaum's Outline of Theory and Problems of Numerical Analysis*, McGraw-Hill, New York.

PROBLEMS

6-1 Solve the equation $y' = t^2 - y$ over [0, 0.2] with $y(0) = 1$. Compare the resulting solution with

$$y(t) = -e^{-t} + t^2 - 2t + 2.$$

Find the optimum size of the numerical integration interval Δt. Justify your choice. Use the Taylor series approach.

6-2 Resolve Problem 6-1 by using the Euler forward-integration method. Explain your choice of the optimum step size.

6-3 Resolve Problem 6-1 by using the modified Euler method. Explain your choice of the optimum step size.

6-4 An undamped spring-mass system, shown in Fig. 6-6, is driven by a force $P = P_0 \sin \omega t$. The differential equation of motion for this case is

$$m\ddot{x} + kx = P_0 \sin \omega t,$$

where k is the spring stiffness and has a value of 100, m is the mass and has a value of 1, ω is the forcing frequency and has a value of 8, and P_0 is the magnitude of the driving force and has a value of

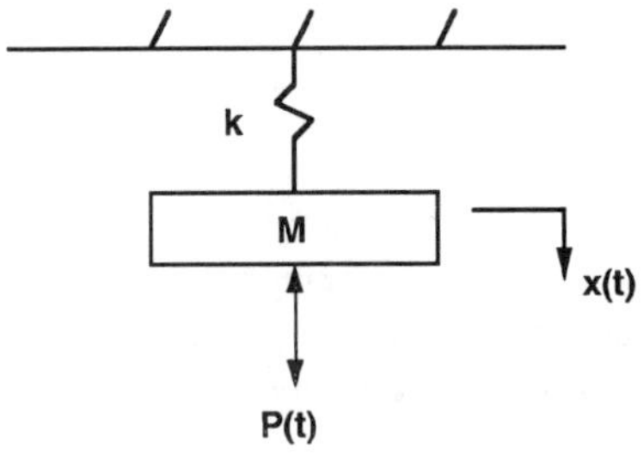

Figure 6-6 Single-degree-of-freedom forced vibration of a damped system.

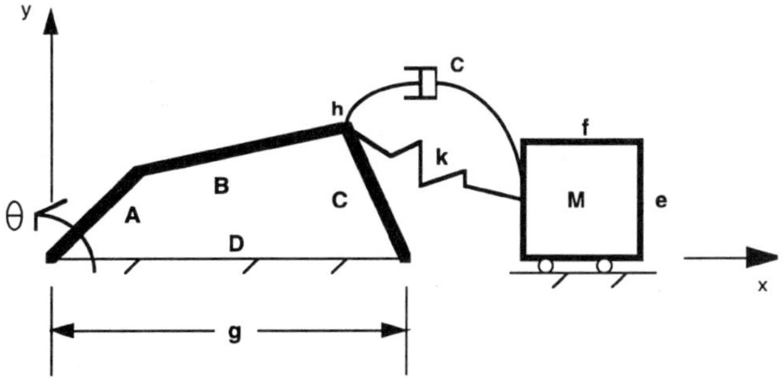

Figure 6-7 Vibration of a four-bar linkage.

5. Numerically integrate this differential equation and solve for the displacement, the velocity, and the acceleration. Compare with the exact result, if you can. Use the Euler forward-integration method.

6-5 Solve Problem 6-4 using the modified Euler method.

6-6 Consider a projectile that is fired straight up and falls straight down. If air resistance is proportional to velocity, the initial-value problem for the velocity is

$$v' = -32 - \left(\frac{K}{M}\right) v, \qquad v(0) = v_0.$$

Let $v_0 = 160$ ft./s and $K/M = 0.1$. Use the Euler forward-integration method to solve for the velocity during the time interval [0, 30]. Calculate the displacement $y(t)$. Discuss your choice for the time step in the numerical integration of the differential equation. Derive the exact solution, if you can, and compare the numerical results with the exact solution.

6-7 Solve Problem 6-6 using the modified Euler method.

6-8 A four-bar linkage, shown in Fig. 6-7, is used to vibrate the mass m by means of a spring and a dashpot that is attached to both the linkage and to the center of the vertical dimension on the mass. The mass is constrained to move in the x direction only. The following parameters apply:

$$\begin{aligned}
M &= 2.0\,\text{lb. s}^2/\text{in.},\\
k &= 10.0\,\text{lb./in.},\\
c &= 7.0\,\text{lb. s/in.},\\
A &= 1.0\,\text{in.},\\
B &= 6.0\,\text{in.},\\
C &= 3.0\,\text{in.},\\
D &= 4.0\,\text{in.},\\
e &= 0.5\,\text{in.},\\
f &= 0.5\,\text{in.},\\
g &= 20.0\,\text{in. (initial position of the mass)},\\
\theta &= \pi/6 + (\pi/12)\sin(\pi/2)t.
\end{aligned}$$

The mass is initially at rest and the spring is unstretched when the angular velocity, $\omega = \dot{\theta}$, is applied to the linkage at bar $\boldsymbol{A}$.

(a) Perform a kinematic analysis on the linkage to determine the displacement to both the spring and the dashpot.

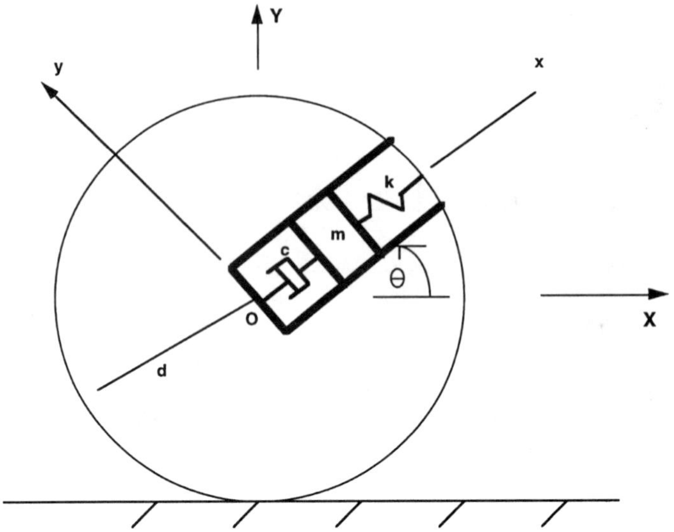

Figure 6-8 Vibrating mass in a moving coordinate system.

(b) Calculate the resulting motion of the mass. A complete description of the motion includes the complete time history of the displacement x, the velocity v, the acceleration a, the spring force F_s, and the dashpot force F_d. For this section, use a time increment of $\Delta t = 0.005$ s.
(c) Find the optimum Δt for the above problem. Justify your choice.

Use the Euler forward-integration method.

The computer code that can solve this problem is shown in Appendix 6-3, in a code called **Four-Bar.for**.

6-9 Solve Problem 6-8 using the modified Euler method.

6-10 The wheel shown in Fig. 6-8 rolls without slipping. Mounted on this wheel is a mass m that slides in a frictionless slot and is attached to a nonlinear damper and to a nonlinear spring. The spring and the dashpot are governed by

$$F_s = k_1([x - x_0]|x - x_0| + k_2|x - x_0|),$$
$$F_d = c_1 x|x| + c_2 x,$$

respectively, where x_0 is the initial position of the mass at rest.

The angle θ is shown. Initially, θ is zero and the wheel and the mass are at rest. The system is then given an excitation as shown in Fig. 6-9.

For this system, a dynamic analysis yields the following differential equation of motion:

$$m\ddot{x} - m\omega^2 x + md\ddot{\theta} \cos\theta = -F_s - F_d.$$

For values $k_1 = 10$ lb./in.^{2}, $d = 5$ in., $c_1 = 1.265$ lb. sec^2/in.^{2}, $m = 1$ lb. sec^2/in., $x_0 = 2.5$ in., $v_0 = 0.0$, $k_2 = 15$ lb./in., and $c_2 = 1.500$ lb. sec/in.

(a) Derive the above equation of motion for m.
(b) Calculate and plot the displacement, velocity, and acceleration of the mass as a function of time. Calculate and plot the spring force and the dashpot force as functions of time.
(c) Explain concisely how your program operates.
(d) Explain how you arrived at the optimum time step for your program. Use equations or graphs to prove that your time selection is the best. Quantify your time step.

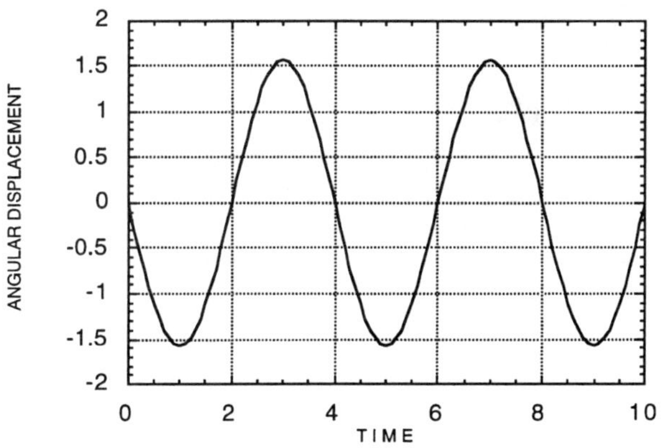

Figure 6-9 Angular displacement of the ball with respect to time. ($\theta = -\frac{\pi}{2}\sin\{\frac{\pi}{2}t\}$).

(e) If it were your design goal to limit the maximum displacement of the mass relative to the wheel, would you increase or decrease the overall damping?

Make a general statement about how you would decide what to do if your system were linear (i.e., $k_1 = c_1 = 0$). Prove your conclusion.
Use the Euler forward-integration method.
The code, **ball.for**, in Appendix 6-4 was written to solve Problem 6-10.

6-11 Solve Problem 6-10 using the Modified Euler method.

6-12 Derive the second-order Runge–Kutta formula. Start with

$$y_{n+1} \approx y_n + h(af_0 + bf_1),$$

where

$$f_0 = f(x_n, y_n),$$
$$f_1 = f(x_n + ph, y_n + qhf_0).$$

Obtain relations and values for a, b, p, and q. If $b = 1$, show that your result is the modified Euler formula, i.e.,

$$y_{n+1} = y_n + hf\left[x_n + \frac{1}{2}h,\ y_n + \frac{1}{2}hf(x_n, y_n)\right].$$

6-13 The equation of motion for a rocket in vertical flight is given by

$$\frac{(M_0 - m_f t)}{g_0}\frac{\mathrm{d}u}{\mathrm{d}t} = -cu^2 - \frac{(M_0 - m_f t)}{g_0}g + \frac{m_f V_j}{g_0},$$

where M_0 is the initial mass of the rocket; fuel is 20,000 lbm; t is the time in seconds; m_f is the rate of fuel burned, which is 200 lbm/s, V_j is the velocity of the combustion products that leave the rocket nozzle, which is $= 12{,}000$ ft/s, g_0 is the mass conversion factor, which is 32.2 [(lbm-ft)/(lbf-s^2)]; u is the rocket velocity in feet per second; c is the drag coefficient for air resistance, which is 0.002 [(lbf-s^2)/(ft^2)]; and g is the acceleration of gravity, which is 32.2 ft/s.

Write a third-order Runge–Kutta program to determine the trajectory and the velocity as functions of time. Assume that there is enough fuel for a 10-s burn (2000 lbm). Continue this problem with $m_f = 0$ for the remainder of the ascent when $u = 0$. Determine the value of the time for maximum height. Determine the maximum height that is attained by the rocket.

6-14 Use the Euler forward-integration method to solve Problem 6-13.

6-15 Use the modified Euler method to solve Problem 6-13.

CHAPTER

SEVEN

ORDINARY DIFFERENTIAL EQUATIONS

NOMENCLATURE

y'	first derivative of y with respect to x
y''	second derivative of y with respect to x
f_i	value of the function at the ith mesh point
h	mesh spacing
x	independent variable
y	dependent variable
$y(0)$	value of y at $x = 0$
$y(N)$	value of y at $x = N$

7-1 INTRODUCTION

We have used the techniques that were described in Chap. 5 to evaluate the solution of initial-value problems. We now turn our attention to the class of problems that are known as ordinary boundary-value problems. In Chap. 8, the solution of partial differential equations will be explored. Ordinary differential equations are solved by the creative use of the finite-difference method, which reduces the differential equations to the evaluation of the roots of a system of simultaneous algebraic equations. It is shown that central differences should be used where convenient, while forward and backward differences should be used only near the boundaries of the region.

Since the basic methodology has been developed in the preceding chapters, this section is concerned with the solution of ordinary differential equations and the solution strategies that are used.

We must remember, however, that in ordinary boundary-value problems, the independent variable is bounded within a region, and therefore must satisfy a specific boundary value.

7-2 ITERATIVE SOLUTIONS OF BOUNDARY-VALUE PROBLEMS

To begin the study of this class of problems, let us consider a simple ordinary differential equation. In Chap. 5, we showed that we can approximate the second derivative by

$$y'' = \frac{1}{h^2}\{f_{i+1} - 2f_i + f_{i-1}\}, \tag{7-1}$$

so that the differential equation,

$$y'' + y^2 = 0, \tag{7-2}$$

subject to the boundary conditions

$$y(0) = 2,$$

$$y(1) = 0,$$

becomes, when approximated by Eq. (7-1),

$$y_{i+1} - 2y_i + y_{i-1} + h^2 y_i^2 = 0 \tag{7-3}$$

or

$$y_{i+1} = 2y_i - h^2 y_i^2 - y_{i-1}. \tag{7-4}$$

Since the region of interest is [0, 1] (see the boundary conditions), let us choose to divide the region (i.e., discretize the region) by choosing a coarse mesh, i.e., $\mathbf{h} = 1/4$. Equation (7-4) then becomes

$$y_{i+1} = 2y_i - \left(\frac{y_i}{4}\right)^2 - y_{i-1}. \tag{7-5}$$

In general, we would set up a matrix set of simultaneous algebraic equations in order to solve for the values of y at the mesh points that are created by the discretization parameter $\mathbf{h}$. These points are located at $x_i = (i - 1)\mathbf{h}$. Before the matrix process is illustrated, let us attempt to solve Eq. (7-5) by a simple but inefficient method of successive updates.

In order to start the solution process, let us make an initial guess for the value of the dependent variable y at the second mesh point. Let us choose $y_1 = 1.5$. This will be considered the initial iterate for the solution process that will be carried out by trial and error (or iteration).

We are given the value of the dependent variable at $x_0 = 0\,[y(0) = 2]$, and we have estimated the value of the solution at the next iteration point $x_1[y(1) = 1.5]$. With the successive application of Eq. (7-5), we can calculate the values of y at all the grid points until we reach the other boundary at $x = 1$. We then compare the calculated value of y at this boundary with the given boundary condition. Any deviation in the two values at

Table 7-1 Results of iterative process

		Iterate			
		$y_1 = 1.5$	$y_1 = 2.0$	$y_1 = 1.70$	$y_1 = 1.6825$
i	x_i	y_i	y_i	y_i	y_i
0	0	2.0	2.0	2.0	2.0
1	0.25	1.50	2.0	1.70	1.6825
2	0.50	0.86	1.75	1.22	1.1881
3	0.75	0.17	1.31	0.65	0.6055
4	1.00	−0.52	0.76	0.05	0.000

this point results in an error and causes us to modify the estimate of y_1 (at point x_1). This iteration process is then continued until convergence is achieved at the boundary that is represented by $x = x_n$. Convergence occurs when the calculated solution at $x = x_n$ matches the boundary condition $y = 0$. Table 7-1 shows the results of this process.

Four iterations were necessary in order to obtain a solution that converged to the given boundary condition.

7-3 MATRIX SOLUTIONS OF LINEAR BOUNDARY-VALUE PROBLEMS

7-3.1 Example in Heat Transfer

The method described in Section 7-2 becomes cumbersome when the order of the differential equation is higher than 3 or 4. Also, since the difference equation (7-5) is linear, a matrix formulation of the solution is readily applicable. Further, it is shown that when the matrix approach is used, no iteration is required for linear problems. Two examples of using the matrix approach are now illustrated in order to expose the readers to the techniques that may be used in the solution of linear ordinary differential equations.

As our first example, let us consider a one-dimensional heat conduction problem. Figure 7-1 shows a thin rod whose ends are held at fixed temperatures and is submersed in a medium whose temperature distribution is known and varies in the x-direction. We require that the steady-state temperature within the rod be calculated. Conduction heat transfer takes place within the rod while a convective heat transfer process occurs between the rod and the surrounding medium. For this case, the differential equation that governs the temperature distribution within the rod is

$$\frac{d^2T}{dx^2} - \frac{2k_1}{kR}T = -\frac{2k_1}{kR}T_\infty, \tag{7-6}$$

where k is the thermal conductivity of the wire, k_1 is the convection coefficient of the wire, x is the distance along the wire axis, $T_\infty(x)$ is the temperature of the surrounding medium, $T(x)$ is the temperature distribution within the wire, and R is the radius of the wire.

Equation (7-6) is an ordinary differential equation since x is the only independent variable.

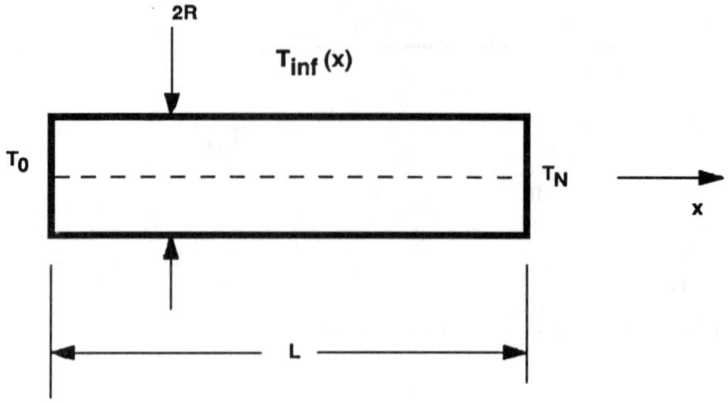

Figure 7-1 Conduction within a thin rod.

The temperature boundary conditions at the ends of the rod are

$$T(0) = T_0, \qquad T(L) = T_N. \tag{7-7}$$

It is normal practice in heat transfer problems to normalize the variables so that the solution to the problem can be made applicable to a wide range of varying geometric conditions. When the current problem is nondimensionalized with

$$z = \frac{x}{L}, \qquad v(z) = \frac{T(x)}{T_N}, \qquad F(z) = \frac{T_\infty(x)}{T_\infty(0)}, \tag{7-8}$$

Eq. (7-6) becomes

$$v'' - \frac{2k_1L^2}{kR}v = -\frac{2k_1L^2}{kR}\frac{T_\infty(0)}{T_N}F(z) \tag{7-9}$$

and the boundary conditions become

$$v(0) = \frac{T_0}{T_N}, \qquad v(1) = 1. \tag{7-10}$$

If we assign values to the geometric and the thermal parameters before attempting to approximate Eq. (7-9) by the finite-difference technique, i.e.,

$$L = 100\,\text{cm},$$

$$R = 1\,\text{cm},$$

$$k = 1\,\text{cal/sec}\,^\circ\text{C}\,\text{cm}^2/\text{cm},$$

$$k_1 = 6 \times 10^{-4}\left(\frac{3}{2}z + \frac{1}{3}\right)\,\text{cal/sec}\,^\circ\text{C}\,\text{cm}^2,$$

$$v_0 = 0.0,$$

$$\frac{T_\infty(0)}{T_N} = 1,$$

$$F(z) = e^z,$$

the differential equation becomes

$$v'' - 2(9z + 2)v = -2(9z + 2)e^z \tag{7-11}$$

with boundary conditions,

$$v(0) = 0, \qquad v(1) = 1. \tag{7-12}$$

It is to be noted that although the coefficients of the terms in Eq. (7-11) are nonlinear, the differential equation itself is still linear.

To begin the approximation process, the region (rod) is divided into N equal parts of width $h = 1/N$. Using central differences to approximate the second derivative in Eq. (7-11), we get

$$v_{i+1} - 2v_i + v_{i-1} - 2h^2(9z_i + 2)v_i = -2h^2(9z_i + 2)e^{z_i}. \tag{7-13}$$

If we apply this difference equation [Eq. (7-13)] at $i = 1, 2, \ldots, n-1$, we will derive $n-1$ linear algebraic equations that can be solved by the matrix inversion techniques of Chap. 1. As an example, let us choose a mesh size of $\mathbf{h} = 1/4$. The difference equations, as represented by Eq. (7-13), then become, at $i = 1$, $(z = 0.25)$,

$$v_2 - 2.53v_1 + v_0 = -0.682, \tag{7-14}$$

at $i = 2$, $(z = 0.50)$,

$$v_3 - 2.81v_2 + v_1 = -1.34, \tag{7-15}$$

and at $i = 3$, $(z = 0.75)$,

$$v_4 - 3.09v_3 + v_2 = -2.32 \tag{7-16}$$

or

$$\begin{bmatrix} 1.0 & 0.0 & 0.0 & 0.0 & 0.0 \\ 1.0 & -2.53 & 1.0 & 0.0 & 0.0 \\ 0.0 & 1.0 & -2.81 & 1.0 & 0.0 \\ 0.0 & 0.0 & 1.0 & -3.09 & 1.0 \\ 0.0 & 0.0 & 0.0 & 0.0 & 1.0 \end{bmatrix} \begin{bmatrix} v_0 \\ v_1 \\ v_2 \\ v_3 \\ v_4 \end{bmatrix} = \begin{bmatrix} 0.0 \\ -0.682 \\ -1.34 \\ -2.32 \\ 1.0 \end{bmatrix}.$$

Equations (7-14)–(7-16) are a set of simultaneous equations that, when $v_0 = 0.0$ and $v_4 = 1.0$ are applied, yields the solution for the interior temperatures within the wire, i.e., v_1, v_2, and v_3. Table 7-2 shows the values of the temperatures while Fig. 7-2 displays the results graphically.

The preceding approximation was carried out, for demonstration purposes, with a very coarse grid. The value of an optimum n (or $\mathbf{h}$) is a parameter that must be determined

Table 7-2 Temperature distribution in a thin rod

Point	x	v
1	0	0
2	0.25	0.7753
3	0.5	1.2802
4	0.75	1.4855
5	1	1

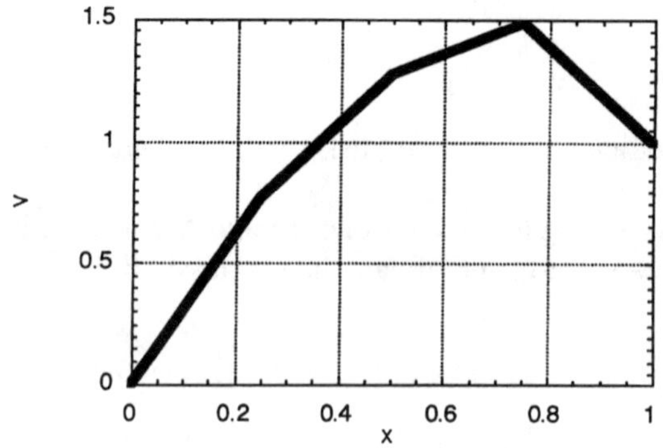

Figure 7-2 Temperature distribution in a thin rod.

in this problem. In other words, students must determine the value of **h** that will yield an approximate solution to the differential equation that approaches the exact solution. This is usually done by trial and error as illustrated in Chap. 6.

Appendix 7-1 contains a FORTRAN code, **rod.for**, that was used to solve this problem.

7-3.2 Example in Structural Mechanics

We now turn our attention to an ordinary boundary-value problem in the area of structural mechanics. The differential equation, which governs the deflection of an elastic beam on an elastic foundation of modulus k, is given by

$$y^{IV} + \left(\frac{k}{EI}\right)y = \left(\frac{q}{EI}\right). \tag{7-17}$$

The geometry of this problem is shown in Fig. 7-3. In this problem the parameters have the following values: modulus of the foundation $k = 2{,}000$ psi, modulus of elasticity of the beam $E = 30 \times 10^6$ psi, moment of inertia of the beam $I = 3 \times 10^3$ in.^{4}, uniformly distributed load $q = 40{,}000$ lb/in., length of the beam $L = 120$ in., x is the distance along the beam, and y is the deflection of the beam.

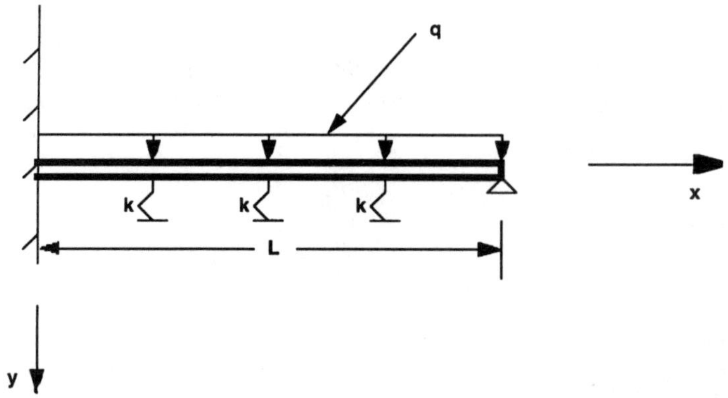

Figure 7-3 Elastic beam on an elastic foundation.

This differential equation is linearly similar to the one that governed the heat transfer in Subsection 7-3.1. Here, however, the boundary conditions are mixed. The values of the dependent variable and one of its derivatives are given at the boundaries of the beam. For this geometric configuration, the end conditions are

$$y(0) = y'(0) = 0 \qquad \text{(the deflection and the slope vanish at } x = 0\text{)},$$
$$y(L) = y''(L) = 0 \qquad \text{(the deflection and the moment vanish at } x = L\text{)}.$$

It is required that we calculate the deflection $y(x)$.

We begin the analysis by creating a central difference approximation of differential equation (7-17). Using the techniques of Chap. 5, we find that the differencing process yields

$$y_{i-2} - 4y_{i-1} + \left\{6 + \frac{h^4k}{EI}\right\} y_i - 4y_{i+1} + y_{i+2} = \frac{h^4q}{EI}, \tag{7-18}$$

where h, the size of the discretization parameter, is defined by $h = [L/(N-1)]$ and N is the number of mesh points in the region. Before proceeding to approximate the boundary conditions by finite-difference techniques, we find that a study of Eq. (7-18) reveals some interesting concerns.

Because of the $(i-2)$ term in Eq. (7-18), the index for i cannot be less than 3. Therefore, the first and second equations must come from the two boundary conditions on the left-hand side of the beam. Similarly, because of the $(i+2)$ term in Eq. (7-18), the index i cannot exceed $(N-2)$. Therefore equations $(N-1)$ and N must come from the two boundary conditions at the right end of the beam.

We now turn our attention to the two boundary conditions at the left end of the beam, i.e., $y(0) = y'(0) = 0$. The boundary condition for the deflection yields

$$y_1 = 0. \tag{7-19}$$

The second equation that comes from the derivative boundary condition requires a bit of care in its selection. One would be tempted to choose a forward finite-difference approximation of the lowest order, as described in Chap. 5. This, however, would lead to a physically impossible result, i.e., $y_2 = y_1$, which must be rejected. Selecting the next higher order of approximation for the forward difference of the first derivative {i.e., $[O(h^2)]$} and noting that $y_1 = 0$, for the second equation we obtain

$$4y_2 - y_3 = 0. \tag{7-20}$$

We now turn our attention to the two boundary conditions at the right end of the beam. The deflection boundary condition at the right end of the beam yields

$$y_N = 0, \tag{7-21}$$

which forms the Nth equation, while an approximation of the second derivative by the backward-difference technique yields

$$-y_{N-3} + 4y_{N-2} - 5y_{N-1} = 0 \tag{7-22}$$

when one also substitutes Eq. (7-21). Equation (7-22) is the $(N-1)$st equation in the solution set.

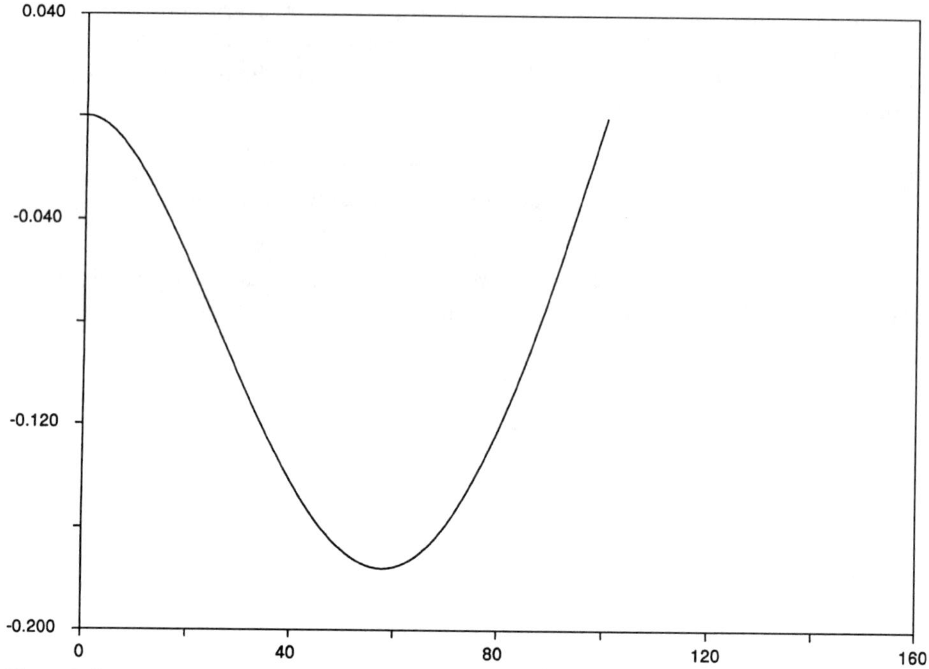

Figure 7-4 Deflection of an elastic beam on an elastic foundation.

The algorithm that will solve the elastic beam on an elastic foundation can now be written. Equations (7-19)–(7-22) yield four equations while the cycling of Eq. (7-18) through the index range of $i = 3, \ldots, N-2$ yields the remaining $(N-4)$ equations for the N unknown dependent variables at the mesh points.

If one arbitrarily chooses $N = 11$ $(h = 12)$, then the above approximation produces the solution shown in Table 7-3. Figure 7-4 plots the results of this analysis.

The above solution was carried out with a FORTRAN code. This code is in Appendix 7-2 under the title of **beam.for**.

Once again, one must determine whether the solution is physically reasonable. Readers must also determine the optimum size of the discretization parameter h. This latter study is left to the readers in Problem 7-3.

7-4 MATRIX SOLUTIONS OF NONLINEAR BOUNDARY-VALUE PROBLEMS

This section briefly describes the methods that may be brought to bear on nonlinear ordinary boundary-value problems. We demonstrated in Chap. 2 that the solution of nonlinear algebraic equations requires that a fair amount of creativity be brought to the solution process and the ultimate success of the process relies heavily on the choice of the initial guess of the solution. These difficulties are compounded when one approaches nonlinear differential equations. It is beyond the scope of a basic textbook to develop procedures

Table 7-3 Deflection of an elastic beam on an elastic foundation

I	*x*	*y*	*I*	*x*	*y*
1	0.000	0.000000000	47	51.111	−0.170872435
2	1.111	−0.000237606	48	52.222	−0.172373012
3	2.222	−0.000950426	49	53.333	−0.173613220
4	3.333	−0.002111159	50	54.444	−0.174588516
5	4.444	−0.003693014	51	55.556	−0.175294891
6	5.556	−0.005669716	52	56.667	−0.175729290
7	6.667	−0.008015480	53	57.778	−0.175888553
8	7.778	−0.010705049	54	58.889	−0.175770357
9	8.889	−0.013713658	55	60.000	−0.175372884
10	10.000	−0.017017059	56	61.111	−0.174694702
11	11.111	−0.020591503	57	62.222	−0.173734888
12	12.222	−0.024413738	58	63.333	−0.172493219
13	13.333	−0.028461041	59	64.444	−0.170969784
14	14.444	−0.032711193	60	65.556	−0.169165179
15	15.556	−0.037142470	61	66.667	−0.167080507
16	16.667	−0.041733619	62	67.778	−0.164717168
17	17.778	−0.046463989	63	68.889	−0.162077382
18	18.889	−0.051313307	64	70.000	−0.159163684
19	20.000	−0.056261919	65	71.111	−0.155979067
20	21.111	−0.061290599	66	72.222	−0.152527317
21	22.222	−0.066380650	67	73.333	−0.148812339
22	23.333	−0.071513884	68	74.444	−0.144838780
23	24.444	−0.076672561	69	75.556	−0.140611738
24	25.556	−0.081839480	70	76.667	−0.136136651
25	26.667	−0.086997963	71	77.778	−0.131419793
26	27.778	−0.092131853	72	78.889	−0.126467541
27	28.889	−0.097225450	73	80.000	−0.121287040
28	30.000	−0.102263555	74	81.111	−0.115885668
29	31.111	−0.107231550	75	82.222	−0.110271536
30	32.222	−0.112115227	76	83.333	−0.104453154
31	33.333	−0.116900854	77	84.444	−0.098439537
32	34.444	−0.121575207	78	85.556	−0.092240267
33	35.556	−0.126125589	79	86.667	−0.085865311
34	36.667	−0.130539864	80	87.778	−0.079325199
35	37.778	−0.134806290	81	88.889	−0.072630979
36	38.889	−0.138913617	82	90.000	−0.065794148
37	40.000	−0.142851114	83	91.111	−0.058826789
38	41.111	−0.146608457	84	92.222	−0.051741481
39	42.222	−0.150175929	85	93.333	−0.044551216
40	43.333	−0.153544381	86	94.444	−0.037269570
41	44.444	−0.156704903	87	95.556	−0.029910581
42	45.556	−0.159649253	88	96.667	−0.022488829
43	46.667	−0.162369639	89	97.778	−0.015019344
44	47.778	−0.164858624	90	98.889	−0.007517706
45	48.889	−0.167109594	91	100.000	0.000000000
46	50.000	−0.169116139			

fully for nonlinear differential equations. However, for the sake of completeness, a brief presentation is given here.

Consider the nonlinear differential equation

$$2xy'' + (y')^2 - 4y = 0 \tag{7-23}$$

subject to the boundary conditions

$$y(0) = 1, \qquad y(10) = 121. \tag{7-24}$$

We apply the standard finite-difference technique to the solution of this problem. However, since the equation is nonlinear, the matrix that will be created will depend, in part, on the unknown solution values. Therefore, as explained in Chap. 2, we have to use an iterative approach in order to achieve the solution. The iterative approach that is used in this section is the one that was illustrated in Chap. 2, i.e., the matrix Newton–Raphson method.

For the sake of comparison only, we can show that the exact solution to Eq. (7-23) is

$$y = x^2 + 2x + 1 = (x + 1)^2. \tag{7-25}$$

We discretize the region of interest [0, 10] into N nodal points such that $h = [1/(N - 1)]$. Using a standard central-difference approach, we can approximate Eq. (7-23), viz.,

$$2x_i\left\{\frac{1}{h^2}[y_{i-1} - 2y_i + y_{i+1}]\right\} + \left\{\frac{1}{4h^2}\right\}(y_{i+1} - y_{i-1})^2 - 4y_i = 0. \tag{7-26}$$

In order to linearize Eq. (7-26), we now define the variables y_i^*. The starred variables are estimates of the solution at the ith point and must be updated as the solution process proceeds. Equation (7-26) then becomes

$$y_{i-1}\{8x_i - y_{i+1}^* + y_{i-1}^*\} + y_i\{-16x_i - 16h^2\} + y_{i+1}\{8x_i + y_{i+1}^* - y_{i-1}^*\} = 0. \tag{7-27}$$

If we evaluate Eq. (7-27) for ($i = 2, \ldots, N-1$) and append the boundary conditions $y(0) = 1$ and $y(10) = 121$, we form a system of N simultaneous equations with which to calculate the values of y. However, since we used estimated values of the y's to form the coefficient matrix, the calculated values may not be accurate. Using the residual vector approach of Chap. 2, an iteration can be constructed to achieve the final result.

Table 7-4(a) shows the results of this approach when only four mesh points are used. Table 7-4(b) shows the results of the analysis when 11 mesh points are used. Figure 7-5(a)

Table 7-4(a) Four-point solution of a nonlinear differential equation

Point	x	y (exact)	y (approximate)
1	0	1.0000	1.0000
2	3.33	18.7489	22.3015
3	6.67	58.8289	60.9509
4	10	121.0000	121.0000

Table 7-4(b) 11-point solution of a nonlinear differential equation

Point	x	y(exact)	y(approximate)
1	0	1.0000	1.0000
2	1	4.0000	6.7624
3	2	9.0000	11.8229
4	3	16.0000	19.1383
5	4	25.0000	28.1606
6	5	36.0000	38.9874
7	6	49.0000	51.6466
8	7	64.0000	66.1598
9	8	81.0000	82.5441
10	9	100.0000	100.8134
11	10	121.0000	121.0000

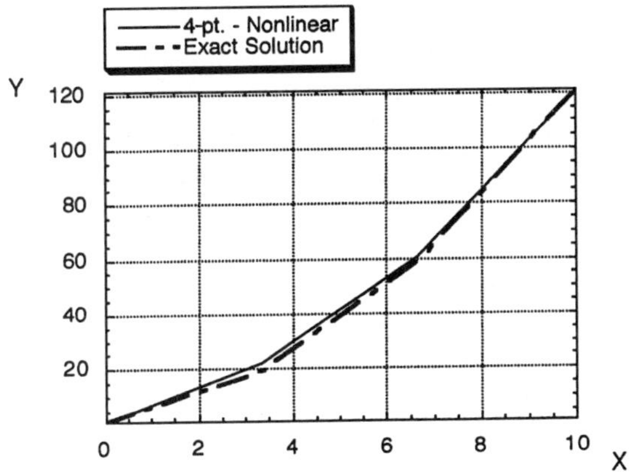

(*a*)

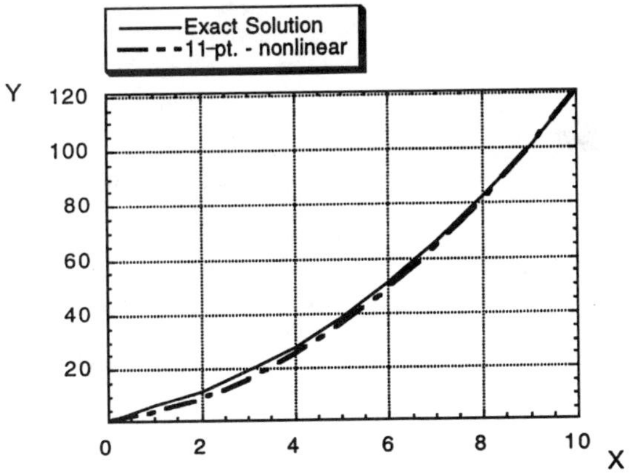

(*b*)

Figure 7-5 (a) Four-point solution, (b) 11-point solution of a nonlinear differential equation.

displays the results for the four-mesh-point approximation, and Fig. 7-5(b) shows the 11-mesh-point case. The above solution was achieved with the aid of the FORTRAN program **nonlinear.for**, which is given in Appendix 7-3.

Both of the above solutions were carried out until the dependent variable converged within a tolerance of 0.1%. Although a small convergence tolerance was used, the numerical simulation deviates considerably from the exact solution. In other words, as explained above, the solution has satisfied the difference equations but has not necessarily solved the differential equation. The accuracy of the modeling depends on both the convergence tolerance and the numerical approximation that is chosen. The results from this example should alert readers to the difficulties that accompany the solution of nonlinear differential equations.

7-5 CONCLUSION

This chapter has brought together the techniques that were described in previous chapters in order to approximate the solution of ordinary boundary-value problems. The differential equations were discretized with finite differences (central differences, when possible). This approach yields a set of simultaneous algebraic equations for the values of the dependent variables at mesh points that are chosen within the region of interest. The approximations of the boundary conditions yield a set of equations that is added to the above set to form a complete set of matrix equations that can then be solved by the techniques of Chap. 1.

As was illustrated in this chapter, there is a fundamental thought process that must be used when doing any numerical modeling. Foremost, one must always question whether the resultant solution describes the physical process that is being modeled. Second, one must question the accuracy of the solution by determining whether the optimum size of the discretization parameter was used.

While these techniques in this chapter demonstrated how differential equations may be approximated, the applications in engineering and physics are three dimensional. The physical laws are then represented by partial differential equations. These are discussed in Chap. 8.

FURTHER READING

Arden, B. W. and Astill, K. N. (1970). *Numerical Algorithms—Origins and Applications*, Addison-Wesley, Reading, Massachusetts.

Kardestuncer, H. (1974). *Elementary Matrix Analysis of Structures*, McGraw-Hill, New York.

Mathews, J. H. (1987). *Numerical Methods*, Prentice-Hall, Englewood Cliffs, New Jersey.

Press, W. H., *et al.* (1986). *Numerical Recipes in FORTRAN*, Cambridge University Press, New York.

Rao, S. S. (1990). *Mechanical Vibrations*, Addison-Wesley, Reading, Massachusetts.

Salvadori, M. G. and Baron, M. L. (1952). *Numerical Methods in Engineering*, Prentice-Hall, Englewood Cliffs, New Jersey.

PROBLEMS

7-1 The exact solution to the differential equation

$$y'' + 4y = 4x^2 + 2$$

within the interval [0, 1], subject to the boundary conditions

$$y(0) = 0,$$
$$y(1) = 1,$$

is $y = x^2$. Approximate this equation using finite differences with a mesh size of $\mathbf{h} = 1$ and then with a mesh size of $\mathbf{h} = 0.333$. Compare your solution with the exact solution. Attempt to determine the optimum mesh spacing that will converge with the approximation that you have chosen for this problem.

7-2 Determine the values of y at the pivotal points of the interval [0, 1], if y satisfies the boundary-value problem

$$y''' + 2y = 12x^2 + 2,$$

subject to the boundary conditions

$$y(0) = 0,$$
$$y(1) = y'(1) = 0.$$

Use finite differences and determine the optimum mesh spacing.

7-3 Determine the optimum step size for the beam on an elastic foundation that is described in subsection 7-3.2.

7-4 Determine the optimum step size for heat conduction in the rod problem that is described in subsection 7-3.1.

7-5 A prismatic bar whose length is 15 in. and whose cross section is a unit square is shown in Fig. 7-6. This bar will be loaded as a tensile bar so that all stresses vanish except for the normal stress in the z direction.

The bar is made of steel whose material properties are modulus of elasticity $E = 30 \times 10^6$ psi, density $\rho = 0.278$ lb/in., Poisson ratio $\nu = 0.293$, and thermal conductivity $\alpha = 6.7 \times 10^{-6}$ in./(in./deg F).

The bar is loaded by application of a uniform displacement of 0.015 in. to the right end. The left end is restrained so that only normal tensile stresses are produced in the z direction. All other stress components vanish.

This problem is to be solved by using the finite-difference technique.

A complete solution for this problem involves the description of the stresses, the strains, and the displacements in the entire body.

Review of the governing equations for a tensile bar:

$$\varepsilon_{12} = \varepsilon_{13} = \varepsilon_{23} = 0, \qquad \sigma_{12} = \sigma_{13} = \sigma_{23} = 0,$$

$$\varepsilon_{11} = -\nu\varepsilon_{33} = -\nu\frac{du_3}{dx_3} = \text{normal strain in the } x_1 \text{ direction},$$

$$\varepsilon_{22} = -\nu\varepsilon_{33} = -\nu\frac{du_3}{dx_3} = \text{normal strain in the } x_2 \text{ direction},$$

$$\varepsilon_{33} = \frac{du_3}{dx_3} = \text{normal strain in the } x_3 \text{ direction},$$

$$\Delta = \varepsilon_{11} + \varepsilon_{22} + \varepsilon_{33} = \text{dilatation},$$

$$\sigma_{11} = \lambda\Delta + 2G\varepsilon_{11},$$

$$\sigma_{22} = \lambda\Delta + 2G\varepsilon_{22},$$

$$\sigma_{33} = \lambda\Delta + 2G\varepsilon_{33},$$

where modulus of rigidity $G = [E/2(1+\nu)]$ psi, Lame constant $\lambda = \{E\nu/[(1+\nu)(1-2\nu)]\}$ psi, u_3 is the displacement of a point in the z direction (in inches), ε is the strain, and σ is the stress (in pounds per square inch).

Governing equation of equilibrium:

$$\frac{d\sigma_{33}}{dx_3} = 0.$$

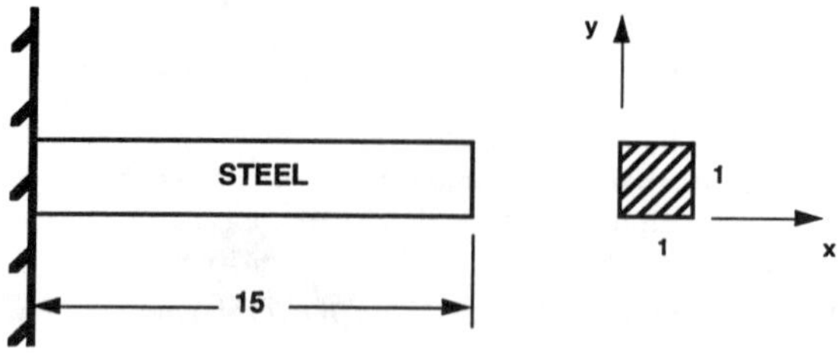

Figure 7-6 Stress analysis of a tensile bar.

The solution to this problem is contained in a FORTRAN program, called **tensile.for**, and is given in Appendix 7-4. Students are encouraged to produce their own codes and should use Appendix 7-4 only as a guide.

7-6 A prismatic bar whose length is 15 in. and whose cross section is a unit square is shown in Fig. 7-7. This bar will be subjected to the following thermal environment. The left end of the bar is to be held at 70°F. The right end of the bar is to be held at 1000°F. All the other surfaces are adiabatic (i.e., no heat transfer occurs across these boundaries). Only conduction heat transfer within the bar need be considered.

The bar is a bimetallic strip composed of steel and aluminum, as shown. For steel, the thermal properties are $\lambda = 9.4$ (Btu/h ft. °F) $= 2.176 \times 10^{-4}$ Btu/s in. °F), $c = 0.11$ Btu/lb$_m$ °F, $\rho = 488$ lb$_m$/ft.$^3 = 0.282$ lb$_m$/in.$^3 = 7.3057 \times 10^{-4}$ lb$_f$/in.^{3}, $a = k/c = 0.15$ ft.^{2}/h.

For aluminum, the thermal properties are $\lambda = 119$ Btu/h ft. °F $= 2.754 \times 10^{-3}$ Btu/s in. °F, $c = 0.208$ Btu/lb$_m$ °F, $\rho = 169$ lb$_m$/ft.$^3 = 9.78 \times 10^{-2}$ lb$_m$/in.$^3 = 2.534 \times 10^{-4}$ lb$_f$/in.^{3}, $a = k/c = 3.33$ ft.^{2}/h.

This problem is to be solved by using the finite-differences technique. Only the steady-state temperature field is required within the bar.

A complete solution for this problem involves the description of the temperatures in the entire body.

For steady-state heat conduction,

$$\frac{\partial}{\partial x_1}\left(\lambda \frac{\partial T}{\partial x_1}\right) + \frac{\partial}{\partial x_2}\left(\lambda \frac{\partial T}{\partial x_2}\right) + \frac{\partial}{\partial x_3}\left(\lambda \frac{\partial T}{\partial x_3}\right) = 0,$$

where $\lambda(x, y, z)$ is conductivity, $T(x, y, z)$ is temperature, and x_1, x_2, and x_3 are the independent coordinate directions.

The solution to this problem is given in Appendix 7-5 in a FORTRAN code, called **heat.for**. Students are encouraged to produce their own codes and should use Appendix 7-5 only as a guide.

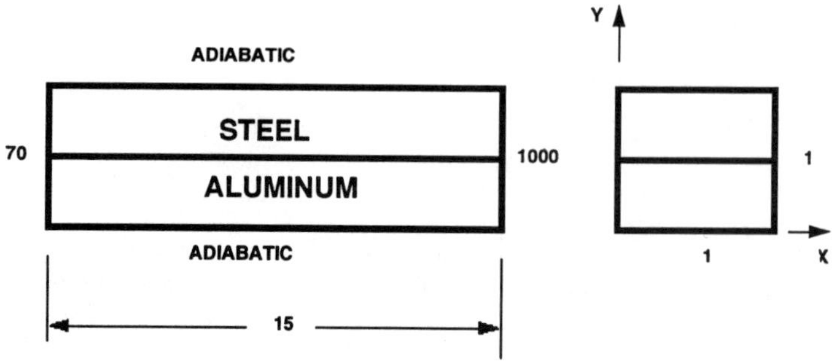

Figure 7-7 Heat conduction in bimetallic strip.

CHAPTER

EIGHT

PARTIAL DIFFERENTIAL EQUATIONS

NOMENCLATURE

ρ	density
γ	specific heat ratio
ρ_0	stagnation density
ψ	stream function
φ	velocity potential
c	speed of wave propagation
$\mathbf{h}$	mesh size
K	conductivity
$\mathbf{k}$	mesh size
p	pressure
S	tension
T	temperature
t	time
u	displacement in a solid or velocity in fluid
V	flow velocity
v	displacement in a solid or velocity in fluid
w	displacement in a solid or velocity in fluid
x	spatial independent variable
y	spatial independent variable
z	displacement

8-1 INTRODUCTION

Partial differential equations occur most frequently when one is modeling physical phenomena in the real three-dimensional world of engineering and physics. In these cases, the dependent variables may be functions of space and time. Thus, in a problem that has both spatial and temporal variations of the unknown variables, the partial differential equation that describes the phenomenon may possess as many as four independent variables.

This chapter illustrates some of the techniques that are used to solve these equations. We see that there are no additional simulation strategies required for partial differential equations than were already delineated for initial-value and for ordinary boundary-value problems. Forward-, backward-, and central-difference techniques may be applied to this class of problems as long as the basic nature of a partial derivative is not violated. This statement is explained in Section 8-3.

8-2 CLASSIFICATION OF PARTIAL DIFFERENTIAL EQUATIONS

A typical second-order partial differential equation, with a dependent variable $u(x, y)$, may be written as

$$A(x, y)\frac{\partial^2 u}{\partial x^2} + B(x, y)\frac{\partial^2 u}{\partial x \partial y} + C(x, y)\frac{\partial^2 u}{\partial y^2} + f\left(x, y, u, \frac{\partial u}{\partial x}, \frac{\partial u}{\partial y}\right) = 0. \qquad (8\text{-}1)$$

This equation is linear in the second-order terms, but the term

$$f\left(x, y, u, \frac{\partial u}{\partial x}, \frac{\partial u}{\partial y}\right)$$

may be either linear or nonlinear. Additionally, the coefficients A, B, and C may be variable. This equation is considered quasi-linear.

We may classify partial differential equations by the value of the term $B^2 - 4AC$. As is explained fully in Chapter 12,

when $B^2 - 4AC < 0$, Eq. (8-1) is called an elliptic differential equation;
when $B^2 - 4AC = 0$, Eq. (8-1) is called a parabolic differential equation;
when $B^2 - 4AC > 0$, Eq. (8-1) is called a hyperbolic differential equation.

An example of an elliptic equation is the Laplace equation, which appears in the field of linear elasticity, viz.,

$$\nabla^2 u = \frac{\partial^2 u}{\partial x^2} + \frac{\partial^2 u}{\partial y^2} = 0,$$

where $u(x, y)$ is the displacement field.

An example of a parabolic equation occurs in the transient conduction of heat, i.e.,

$$K\frac{\partial^2 T}{\partial x^2} = \frac{\partial T}{\partial t},$$

where $T(x, t)$ is the temperature, K is the conductivity, x is the spatial variable, and t is the temporal variable.

An example of a hyperbolic equation occurs in the area of deformable body dynamics (wave propagation), viz.,

$$c^2 \frac{\partial^2 u}{\partial x^2} = \frac{\partial^2 u}{\partial t^2},$$

where $u(x, y)$ is the displacement field and c is the speed of wave propagation.

Since the existence of hyperbolic differential equations in solid mechanics and in fluid mechanics can result in discontinuities of the dependent variable, a discussion of hyperbolic equations is reserved for Chapter 12, where they are thoroughly discussed and used in several examples. This chapter is reserved for the delineation of techniques that are associated with elliptic and parabolic differential equations.

8-3 MODELING OF PARTIAL DIFFERENTIAL EQUATIONS

We model partial derivatives in the same way that we modeled ordinary derivatives, i.e., backward, forward, and central differences. Figure 8-1 shows the geometric configuration for this modeling. Since finite-difference approximations were derived with a Taylor series (see Chapter 5), the existence of all the derivatives of the dependent variable is assumed. Thus, in hyperbolic equations in which some of the derivatives can become indeterminate, the above-mentioned modeling techniques would fail. This chapter is concerned only with equations that satisfy the criteria on which the finite-difference modeling is based.

We begin our illustration of the simulation by considering the scalar Laplacian of the variable $z(x, y)$, i.e.,

$$\nabla^2 z = \frac{\partial^2 z}{\partial x^2} + \frac{\partial^2 z}{\partial y^2} = D_x^2 + D_y^2. \tag{8-2}$$

The terms in Eq. (8-2) can be modeled with a central-difference strategy:

$$\frac{\partial^2 z_i}{\partial x^2} = \frac{1}{h^2}\{z_R - 2z_i + z_L\}, \tag{8-3}$$

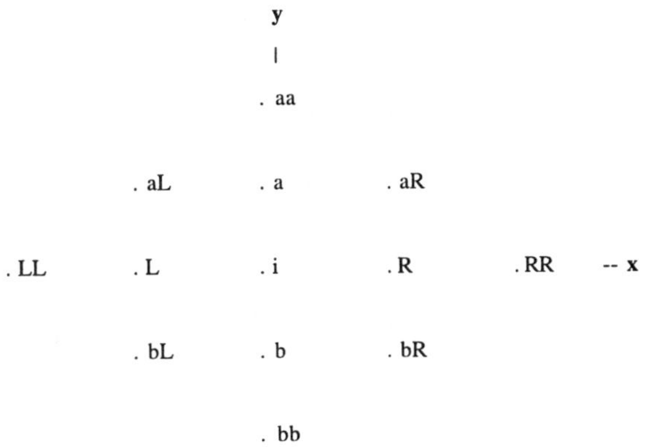

Figure 8-1 Configuration for the modeling of partial differential equations.

$$\frac{\partial^2 z_i}{\partial y^2} = \frac{1}{k^2}\{z_a - 2z_i + z_b\}. \tag{8-4}$$

Then,

$$h^2k^2\nabla^2 z_i = k^2\{z_R - 2z_i + z_L\} + h^2\{z_a - 2z_i + z_b\}. \tag{8-5}$$

If **h** = **k** (i.e., equal mesh spacing in both the x and the y directions),

$$h^2\nabla^2 z_i = z_a + z_b + z_R + z_L - 4z_i. \tag{8-6}$$

The mixed second partial derivative of z can be similarly modeled by a central-difference scheme. Note that the second derivative is the first derivative of the first derivative, i.e.,

$$\frac{\partial^2 z_i}{\partial x \partial y} = \frac{\partial}{\partial y}\frac{\partial z_i}{\partial x} = \frac{\partial}{\partial y}\{z_R - z_L\}\frac{1}{2h}, \tag{8-7}$$

or

$$\frac{\partial^2 z_i}{\partial x \partial y} = [\{z_{aR} - z_{bR}\} - \{z_{aL} - z_{bL}\}]\frac{1}{4h^2}, \tag{8-8}$$

$$4h^2\frac{\partial^2 z_i}{\partial x \partial y} = z_{aR} - z_{bR} - z_{aL} + z_{bL}. \tag{8-9}$$

8-4 EXAMPLE IN HEAT CONDUCTION

All the theoretical development has been done to model partial differential equations by finite differences. It remains only to demonstrate how these strategies may be used in physical problems. The problem in this section and the problem in Section 8-5 close our discussion of finite-difference modeling techniques.

Consider the thin plate of Fig. 8-2, which is subjected to temperature boundary conditions, as shown. Only steady-state heat conduction in the plate is to be considered. The differential equation that governs the steady-state temperature $T(x, y)$ in the plate is Laplace's relationship:

$$\nabla^2 T = \frac{\partial^2 T}{\partial x^2} + \frac{\partial^2 T}{\partial y^2} = 0. \tag{8-10}$$

We assume that the boundaries of the plate, of length L, are held at the temperatures that are shown.

To begin the discretization process, we divide the plate into an equally spaced grid by choosing **h** $= [L/(N-1)]$, where N is the number of mesh points in the x and the y directions. It is not necessary to construct the same number of mesh points in each direction. We have chosen this condition for convenience. There are therefore N^2 points in the rectangular grid. Since the temperatures are to be evaluated at all the grid points where they are not specified, our modeling will yield a system of N^2 simultaneous equations to be solved by the techniques of Chapter 1. Additionally, we require a method of numbering the grid points so that we can match the temperature with the physical point.

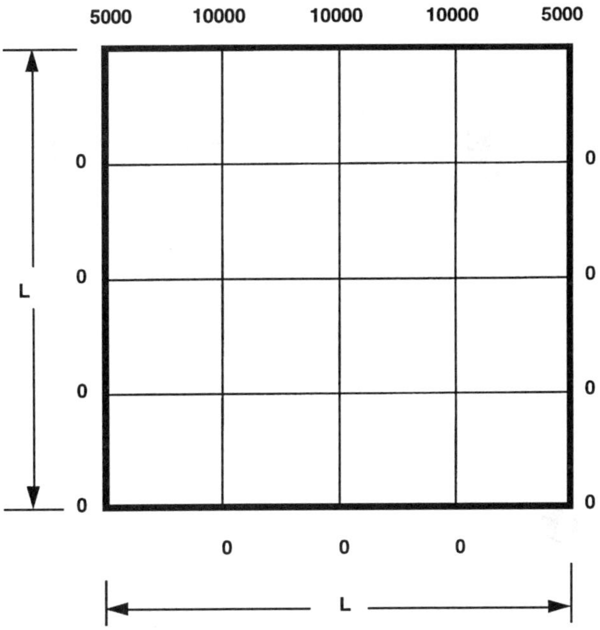

Figure 8-2 Heat conduction in a plate.

We choose the index i to locate the **x** coordinate of the point (i.e., the rows) and we choose j to locate the **y** coordinate of the point (i.e., the columns). Then we can adopt the following numbering scheme for the grid points. The point called $i$pos is found by the formula

$$i\text{pos} = (j - 1) * N + i.$$

Using a modified form of Eq. (8-6), we can approximate Eq. (8-10) by

$$h^2\nabla^2 z_i = z_{(i,\,j+1)} + z_{(i,\,j-1)} + z_{(i+1,\,j)} + z_{(i,\,j-1)} - 4z_i = 0. \tag{8-11}$$

Equation (8-11) forms the basis for the construction of the matrix that is used in the solution of the temperatures. Since the partial differential equation is linear, no iteration will be required. However, a consideration of the optimum step size must still be made. Appendix 8.1 contains the code, **plate.f**, that solves this problem and may be used by readers to solve other thermal and geometric configurations. For the example in this section, we arbitrarily choose a square plate whose side is 11 in. long.

In any numerical simulation, the question of mesh size becomes important. Therefore, in this problem, we have solved the heat conduction problem for mesh sizes of $N = 3$ and $N = 11$. Tables 8-1 and 8-2 show the steady-state temperature distributions in the plate for a 3×3 and an 11×11 grid, respectively. Figure 8-3 shows a contour plot of the temperatures for the coarse (3×3) mesh, while Fig. 8-4 shows the temperature distribution in the same plate. Similar results from the fine-grid approximation are shown in Figs. 8-5 and 8-6. If we were to look merely at Figs. 8-3 and 8-4 and compare these with Figs 8-5 and 8-6, we might conclude that the 3×3 grid results are less

Table 8-1 Steady-state temperature distributions in a plate for a 3 × 3 grid

I	x	y	T
1	0.00	0.00	0.00
2	5.50	0.00	0.00
3	11.00	0.00	0.00
4	0.00	5.50	0.00
5	5.50	5.50	2500.00
6	11.00	5.50	0.00
7	0.00	11.00	5000.00
8	5.50	11.00	10000.00
9	11.00	11.00	5000.00

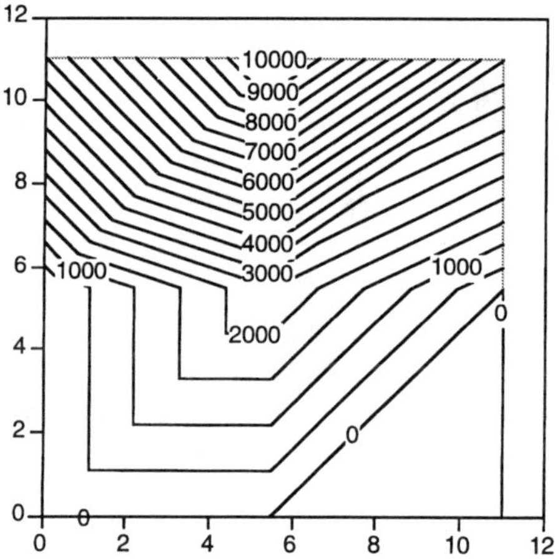

Figure 8-3 Temperature contours (3 × 3 grid).

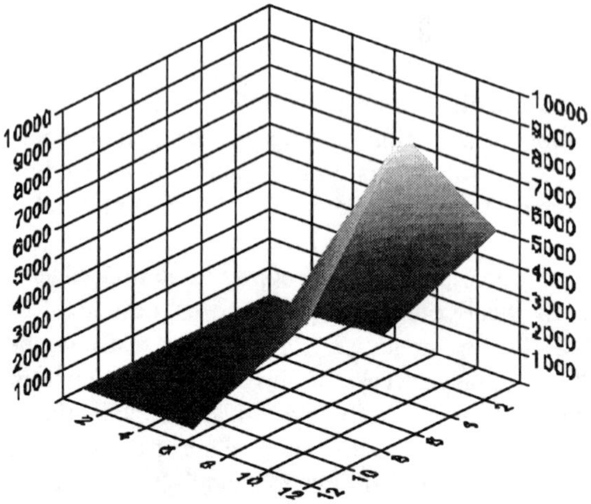

Figure 8-4 Temperature distribution in a plate (3 × 3 grid).

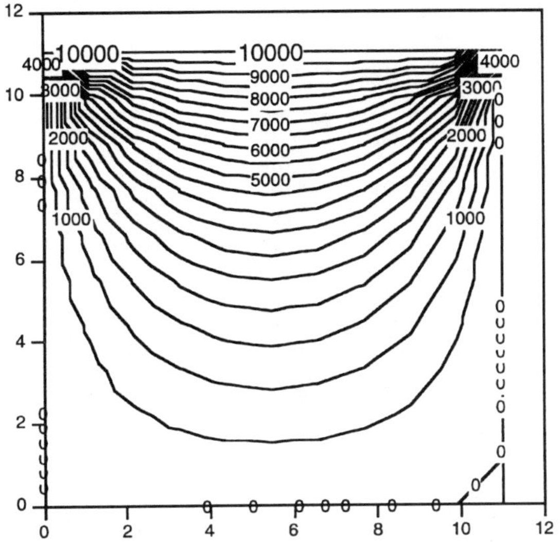

Figure 8-5 Temperature contours (11 × 11 grid).

accurate, but are acceptable. However, if we look at Tables 8-1 and 8-2, which represent the actual output values from the computer solutions, we immediately realize that Figs. 8-3 and 8-4 do not truly represent the temperatures within the plate. Table 8-1 has only one solution point (point 5) that was actually produced by the code **plate.f**. The distribution in Figs. 8-3 and 8-4 are the plotting package's guesses for the contours within the plate. This condition produces a false sense that a coarse grid can actually yield an acceptable solution. Table 8-2, which is produced from the 11 × 11 mesh simulation,

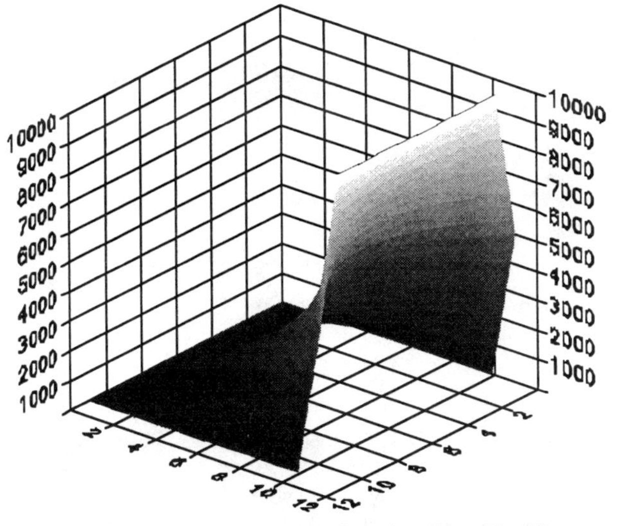

Figure 8-6 Temperature distribution in a plate (11 × 11 grid).

Table 8-2 Steady-state temperature distributions in a plate for an 11 × 11 grid

I	*x*	*y*	*T*	*I*	*x*	*y*	*T*
1	0.00	0.00	0.0	51	6.60	4.40	1690.12
2	1.10	0.00	0.00	52	7.70	4.40	1453.18
3	2.20	0.00	0.00	53	8.80	4.40	1070.81
4	3.30	0.00	0.00	54	9.90	4.40	569.99
5	4.40	0.00	0.00	55	11.00	4.40	0.00
6	5.50	0.00	0.00	56	0.00	5.50	0.00
7	6.60	0.00	0.00	57	1.10	5.50	828.87
8	7.70	0.00	0.00	58	2.20	5.50	1542.04
9	8.80	0.00	0.00	59	3.30	5.50	2071.96
10	9.90	0.00	0.00	60	4.40	5.50	2392.99
11	11.00	0.00	0.00	61	5.50	5.50	2500.00
12	0.00	1.10	0.00	62	6.60	5.50	2392.99
13	1.10	1.10	110.75	63	7.70	5.50	2071.96
14	2.20	1.10	209.93	64	8.80	5.50	1542.04
15	3.30	1.10	287.75	65	9.90	5.50	828.87
16	4.40	1.10	337.16	66	11.00	5.50	0.00
17	5.50	1.10	354.07	67	0.00	6.60	0.00
18	6.60	1.10	337.16	68	1.10	6.60	1203.44
19	7.70	1.10	287.75	69	2.20	6.60	2196.51
20	8.80	1.10	209.93	70	3.30	6.60	2899.63
21	9.90	1.10	110.75	71	4.40	6.60	3309.87
22	11.00	1.10	0.00	72	5.50	6.60	3443.93
23	0.00	2.20	0.00	73	6.60	6.60	3309.88
24	1.10	2.20	233.07	74	7.70	6.60	2899.63
25	2.20	2.20	441.24	75	8.80	6.60	2196.51
26	3.30	2.20	603.91	76	9.90	6.60	1203.44
27	4.40	2.20	706.81	77	11.00	6.60	0.00
28	5.50	2.20	741.95	78	0.00	7.70	0.00
29	6.60	2.20	706.81	79	1.10	7.70	1788.37
30	7.70	2.20	603.91	80	2.20	7.70	3140.95
31	8.80	2.20	441.24	81	3.30	7.70	4020.16
32	9.90	2.20	233.07	82	4.40	7.70	4502.96
33	11.00	2.20	0.00	83	5.50	7.70	4655.95
34	0.00	3.30	0.00	84	6.60	7.70	4502.96
35	1.10	3.30	380.27	85	7.70	7.70	4020.16
36	2.20	3.30	718.04	86	8.80	7.70	3140.95
37	3.30	3.30	979.84	87	9.90	7.70	1788.37
38	4.40	3.30	1144.22	88	11.00	7.70	0.00
39	5.50	3.30	1200.12	89	0.00	8.80	0.00
40	6.60	3.30	1144.22	90	1.10	8.80	2809.10
41	7.70	3.30	979.84	91	2.20	8.80	4558.76
42	8.80	3.30	718.04	92	3.30	8.80	5537.10
43	9.90	3.30	380.27	93	4.40	8.80	6025.86
44	11.00	3.30	0.00	94	5.50	8.80	6173.97
45	0.00	4.40	0.00	95	6.60	8.80	6025.86
46	1.10	4.40	569.99	96	7.70	8.80	5537.10
47	2.20	4.40	1070.81	97	8.80	8.80	4558.76
48	3.30	4.40	1453.18	98	9.90	8.80	2809.10
49	4.40	4.40	1690.12	99	11.00	8.80	0.00
50	5.50	4.40	1770.09	100	0.00	9.90	0.00

contd.

Table 8-2 (*Continued*).

I	x	y	T	I	x	y	T
101	1.10	9.90	4889.25	112	1.10	11.00	10000.00
102	2.20	9.90	6747.91	113	2.20	11.00	10000.00
103	3.30	9.90	7543.60	114	3.30	11.00	10000.00
104	4.40	9.90	7889.42	115	4.40	11.00	10000.00
105	5.50	9.90	7988.20	116	5.50	11.00	10000.00
106	6.60	9.90	7889.42	117	6.60	11.00	10000.00
107	7.70	9.90	7543.60	118	7.70	11.00	10000.00
108	8.80	9.90	6747.90	119	8.80	11.00	10000.00
109	9.90	9.90	4889.25	120	9.90	11.00	10000.00
110	11.00	9.90	0.00	121	11.00	11.00	5000.00
111	0.00	11.00	5000.00				

has many points that are calculated as a result of using the code. Therefore Figs. 8-5 and 8-6 are better representations of the temperature solutions and less a representation of the robustness of the plotting package.

We have, by no means, found the optimum grid spacing for this problem. In order to determine this optimum discretization, we must continue to subdivide the region (i.e., the plate) and recalculate the solution. When the mesh size produces a solution that is almost identical (to an arbitrarily chosen convergence tolerance) to the one from the last (and coarser) grid, then we conclude that the optimum spacing has been achieved for the chosen finite-difference approximation to the differential equation.

A note of caution must be introduced here. Even if the optimum spacing is determined, this does not guarantee that the solution models the physical behavior of the plate. The optimum spacing refers to only the finite-difference model that was chosen to simulate the differential equation. If the physical description of the problem is modeled by a poor or crude finite-difference technique, then the solution to the actual physical problem will never be achieved, even with a fine mesh.

8-5 EXAMPLE IN STRUCTURAL MECHANICS

Consider the thin plate of Fig. 8-7, which is subjected to a uniform tension S and a uniformly distributed pressure load p, as shown. The differential equation that governs the deflection z in the plate is given by

$$\frac{\partial^2 z}{\partial x^2} + \frac{\partial^2 z}{\partial y^2} + \frac{p}{S} = 0, \tag{8-12}$$

which is called Poisson's equation.

We assume that the boundaries of the plate of length L are fixed (i.e., $z = 0$), as shown.

As in Section 8-4, we discretize the plate surface and number the points with the *i*pos method.

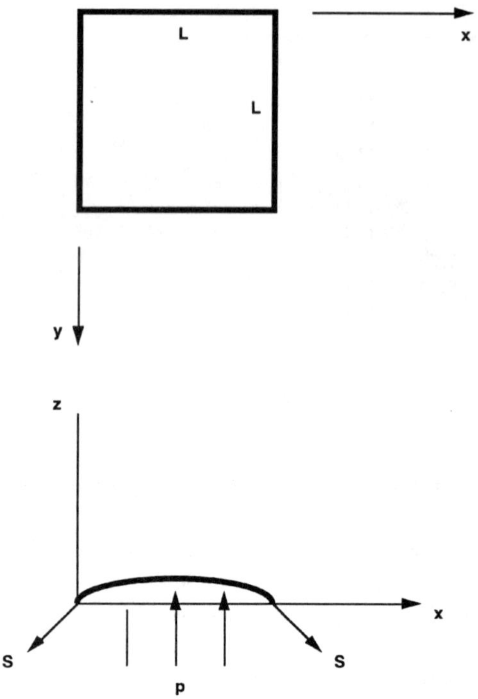

Figure 8-7 Membrane under pressure and tension.

Using a modified form of Eq. (8-12), we get

$$\left\{\frac{1}{h^2}\right\}\{z_{i,j+1} + z_{i,j-1} + z_{i+1,j} + z_{i-1,j} - 4z_i\} + \frac{p_i}{S_i} = 0. \tag{8-13}$$

If we arbitrarily choose the expressions

$$S_i = k,$$
$$p_i = m$$

(i.e., uniform tension and pressure), Eq. (8-13) becomes

$$\{z_{i,j+1} + z_{i,j-1} + z_{i+1,j} + z_{i-1,j} - 4z_i\} = -h^2\frac{m}{k}. \tag{8-14}$$

Equation (8-14), which is similar to Eq. (8-11) but with a right-hand side, forms the basis for the construction of the matrix that is used in the solution. Since this partial differential equation is linear, no iteration is required. However, the consideration of the optimum step size is still required. Appendix 8.2 contains the code, **membrane.f**, that solves this problem and may be used by readers to solve other geometric configurations. A square membrane 11 in. long was chosen as a trial example. Two separate grids were chosen for the discretization of the region (i.e., a coarse 7 × 7 grid and a finer 15 × 15 grid). Table 8-3, shows the solution of the simulation for the coarse grid while Table 8-4 is the solution with the fine grid. Figure 8-8 shows the displacement

Table 8-3 Solution of the simulation for the coarse grid

I	*x*	*y*	*u*	*I*	*x*	*y*	*u*
1	0	0	0	26	7.3333	5.5	4.6923
2	1.8333	0	0	27	9.1667	5.5	3.0769
3	3.6667	0	0	28	11	5.5	0
4	5.5	0	0	29	0	7.3333	0
5	7.3333	0	0	30	1.8333	7.3333	2.8077
6	9.1667	0	0	31	3.6667	7.3333	4.25
7	11	0	0	32	5.5	7.3333	4.6923
8	0	1.8333	0	33	7.3333	7.3333	4.25
9	1.8333	1.8333	1.9038	34	9.1667	7.3333	2.8077
10	3.6667	1.8333	2.8077	35	11	7.3333	0
11	5.5	1.8333	3.0769	36	0	9.1667	0
12	7.3333	1.8333	2.8077	37	1.8333	9.1667	1.9038
13	9.1667	1.8333	1.9038	38	3.6667	9.1667	2.8077
14	11	1.8333	0	39	5.5	9.1667	3.0769
15	0	3.6667	0	40	7.3333	9.1667	2.8077
16	1.8333	3.6667	2.8077	41	9.1667	9.1667	1.9038
17	3.6667	3.6667	4.25	42	11	9.1667	0
18	5.5	3.6667	4.6923	43	0	11	0
19	7.3333	3.6667	4.25	44	1.8333	11	0
20	9.1667	3.6667	2.8077	45	3.6667	11	0
21	11	3.6667	0	46	5.5	11	0
22	0	5.5	0	47	7.3333	11	0
23	1.8333	5.5	3.0769	48	9.1667	11	0
24	3.6667	5.5	4.6923	49	11	11	0
25	5.5	5.5	5.1923				

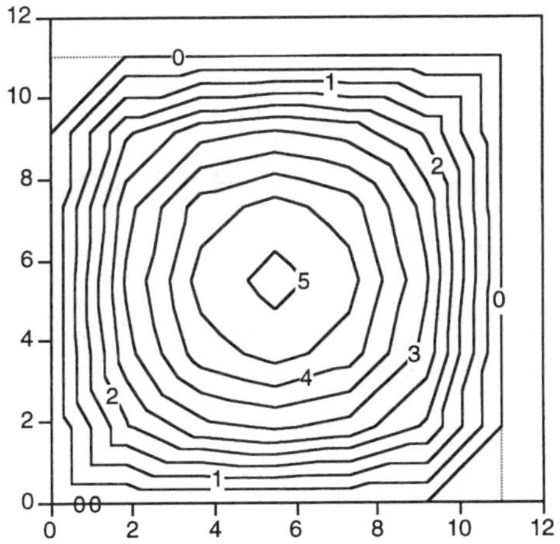

Figure 8-8 Displacement contours (7 × 7 grid).

Table 8-4 Solution of the simulation for the fine grid

I	*x*	*y*	*u*	*I*	*x*	*y*	*u*
1	0.0000	0.0000	0.0000	50	3.1429	2.3571	17.2839
2	0.7857	0.0000	0.0000	51	3.9286	2.3571	19.0050
3	1.5714	0.0000	0.0000	52	4.7143	2.3571	19.9982
4	2.3571	0.0000	0.0000	53	5.5000	2.3571	20.3228
5	3.1429	0.0000	0.0000	54	6.2857	2.3571	19.9982
6	3.9286	0.0000	0.0000	55	7.0714	2.3571	19.0050
7	4.7143	0.0000	0.0000	56	7.8571	2.3571	17.2839
8	5.5000	0.0000	0.0000	57	8.6429	2.3571	14.7282
9	6.2857	0.0000	0.0000	58	9.4286	2.3571	11.1724
10	7.0714	0.0000	0.0000	59	10.2143	2.3571	6.3749
11	7.8571	0.0000	0.0000	60	11.0000	2.3571	0.0000
12	8.6429	0.0000	0.0000	61	0.0000	3.1429	0.0000
13	9.4286	0.0000	0.0000	62	0.7857	3.1429	7.3402
14	10.2143	0.0000	0.0000	63	1.5714	3.1429	13.0069
15	11.0000	0.0000	0.0000	64	2.3571	3.1429	17.2839
16	0.0000	0.7857	0.0000	65	3.1429	3.1429	20.3954
17	0.7857	0.7857	2.9935	66	3.9286	3.1429	22.5068
18	1.5714	0.7857	4.9870	67	4.7143	3.1429	23.7306
19	2.3571	0.7857	6.3749	68	5.5000	3.1429	24.1314
20	3.1429	0.7857	7.3402	69	6.2857	3.1429	23.7306
21	3.9286	0.7857	7.9789	70	7.0714	3.1429	22.5068
22	4.7143	0.7857	8.3440	71	7.8571	3.1429	20.3954
23	5.5000	0.7857	8.4629	72	8.6429	3.1429	17.2839
24	6.2857	0.7857	8.3440	73	9.4286	3.1429	13.0069
25	7.0714	0.7857	7.9789	74	10.2143	3.1429	7.3402
26	7.8571	0.7857	7.3402	75	11.0000	3.1429	0.0000
27	8.6429	0.7857	6.3749	76	0.0000	3.9286	0.0000
28	9.4286	0.7857	4.9870	77	0.7857	3.9286	7.9789
29	10.2143	0.7857	2.9935	78	1.5714	3.9286	14.2313
30	11.0000	0.7857	0.0000	79	2.3571	3.9286	19.0050
31	0.0000	1.5714	0.0000	80	3.1429	3.9286	22.5068
32	0.7857	1.5714	4.9870	81	3.9286	3.9286	24.8964
33	1.5714	1.5714	8.5797	82	4.7143	3.9286	26.2859
34	2.3571	1.5714	11.1724	83	5.5000	3.9286	26.7418
35	3.1429	1.5714	13.0069	84	6.2857	3.9286	26.2859
36	3.9286	1.5714	14.2313	85	7.0714	3.9286	24.8964
37	4.7143	1.5714	14.9343	86	7.8571	3.9286	22.5068
38	5.5000	1.5714	15.1636	87	8.6429	3.9286	19.0050
39	6.2857	1.5714	14.9343	88	9.4286	3.9286	14.2313
40	7.0714	1.5714	14.2313	89	10.2143	3.9286	7.9789
41	7.8571	1.5714	13.0069	90	11.0000	3.9286	0.0000
42	8.6429	1.5714	11.1724	91	0.0000	4.7143	0.0000
43	9.4286	1.5714	8.5797	92	0.7857	4.7143	8.3440
44	10.2143	1.5714	4.9870	93	1.5714	4.7143	14.9343
45	11.0000	1.5714	0.0000	94	2.3571	4.7143	19.9982
46	0.0000	2.3571	0.0000	95	3.1429	4.7143	23.7306
47	0.7857	2.3571	6.3749	96	3.9286	4.7143	26.2859
48	1.5714	2.3571	11.1724	97	4.7143	4.7143	27.7749
49	2.3571	2.3571	14.7282	98	5.5000	4.7143	28.2638

contd.

Table 8-4 ***(Continued).***

I	x	y	u	I	x	y	u
99	6.2857	4.7143	27.7749	147	8.6429	7.0714	19.0050
100	7.0714	4.7143	26.2859	148	9.4286	7.0714	14.2313
101	7.8571	4.7143	23.7306	149	10.2143	7.0714	7.9789
102	8.6429	4.7143	19.9982	150	11.0000	7.0714	0.0000
103	9.4286	4.7143	14.9343	151	0.0000	7.8571	0.0000
104	10.2143	4.7143	8.3440	152	0.7857	7.8571	7.3402
105	11.0000	4.7143	0.0000	153	1.5714	7.8571	13.0069
106	0.0000	5.5000	0.0000	154	2.3571	7.8571	17.2839
107	0.7857	5.5000	8.4629	155	3.1429	7.8571	20.3954
108	1.5714	5.5000	15.1636	156	3.9286	7.8571	22.5068
109	2.3571	5.5000	20.3228	157	4.7143	7.8571	23.7306
110	3.1429	5.5000	24.1314	158	5.5000	7.8571	24.1315
111	3.9286	5.5000	26.7418	159	6.2857	7.8571	23.7306
112	4.7143	5.5000	28.2638	160	7.0714	7.8571	22.5068
113	5.5000	5.5000	28.7638	161	7.8571	7.8571	20.3954
114	6.2857	5.5000	28.2638	162	8.6429	7.8571	17.2839
115	7.0714	5.5000	26.7418	163	9.4286	7.8571	13.0069
116	7.8571	5.5000	24.1315	164	10.2143	7.8571	7.3402
117	8.6429	5.5000	20.3228	165	11.0000	7.8571	0.0000
118	9.4286	5.5000	15.1636	166	0.0000	8.6429	0.0000
119	10.2143	5.5000	8.4629	167	0.7857	8.6429	6.3749
120	11.0000	5.5000	0.0000	168	1.5714	8.6429	11.1724
121	0.0000	6.2857	0.0000	169	2.3571	8.6429	14.7282
122	0.7857	6.2857	8.3440	170	3.1429	8.6429	17.2839
123	1.5714	6.2857	14.9343	171	3.9286	8.6429	19.0050
124	2.3571	6.2857	19.9982	172	4.7143	8.6429	19.9982
125	3.1429	6.2857	23.7306	173	5.5000	8.6429	20.3228
126	3.9286	6.2857	26.2859	174	6.2857	8.6429	19.9982
127	4.7143	6.2857	27.7749	175	7.0714	8.6429	19.0050
128	5.5000	6.2857	28.2638	176	7.8571	8.6429	17.2839
129	6.2857	6.2857	27.7749	177	8.6429	8.6429	14.7282
130	7.0714	6.2857	26.2859	178	9.4286	8.6429	11.1724
131	7.8571	6.2857	23.7306	179	10.2143	8.6429	6.3749
132	8.6429	6.2857	19.9982	180	11.0000	8.6429	0.0000
133	9.4286	6.2857	14.9343	181	0.0000	9.4286	0.0000
134	10.2143	6.2857	8.3440	182	0.7857	9.4286	4.9870
135	11.0000	6.2857	0.0000	183	1.5714	9.4286	8.5797
136	0.0000	7.0714	0.0000	184	2.3571	9.4286	11.1724
137	0.7857	7.0714	7.9789	185	3.1429	9.4286	13.0069
138	1.5714	7.0714	14.2313	186	3.9286	9.4286	14.2313
139	2.3571	7.0714	19.0050	187	4.7143	9.4286	14.9343
140	3.1429	7.0714	22.5068	188	5.5000	9.4286	15.1636
141	3.9286	7.0714	24.8964	189	6.2857	9.4286	14.9343
142	4.7143	7.0714	26.2859	190	7.0714	9.4286	14.2313
143	5.5000	7.0714	26.7418	191	7.8571	9.4286	13.0069
144	6.2857	7.0714	26.2859	192	8.6429	9.4286	11.1724
145	7.0714	7.0714	24.8964	193	9.4286	9.4286	8.5797
146	7.8571	7.0714	22.5068	194	10.2143	9.4286	4.9870

Table 8-4 (*Continued*).

I	*x*	*y*	*u*	*I*	*x*	*y*	*u*
195	11.0000	9.4286	0.0000	211	0.0000	11.0000	0.0000
196	0.0000	10.2143	0.0000	212	0.7857	11.0000	0.0000
197	0.7857	10.2143	2.9935	213	1.5714	11.0000	0.0000
198	1.5714	10.2143	4.9870	214	2.3571	11.0000	0.0000
199	2.3571	10.2143	6.3749	215	3.1429	11.0000	0.0000
200	3.1429	10.2143	7.3402	216	3.9286	11.0000	0.0000
201	3.9286	10.2143	7.9789	217	4.7143	11.0000	0.0000
202	4.7143	10.2143	8.3440	218	5.5000	11.0000	0.0000
203	5.5000	10.2143	8.4629	219	6.2857	11.0000	0.0000
204	6.2857	10.2143	8.3440	220	7.0714	11.0000	0.0000
205	7.0714	10.2143	7.9789	221	7.8571	11.0000	0.0000
206	7.8571	10.2143	7.3402	222	8.6429	11.0000	0.0000
207	8.6429	10.2143	6.3749	223	9.4286	11.0000	0.0000
208	9.4286	10.2143	4.9870	224	10.2143	11.0000	0.0000
209	10.2143	10.2143	2.9935	225	11.0000	11.0000	0.0000
210	11.0000	10.2143	0.0000				

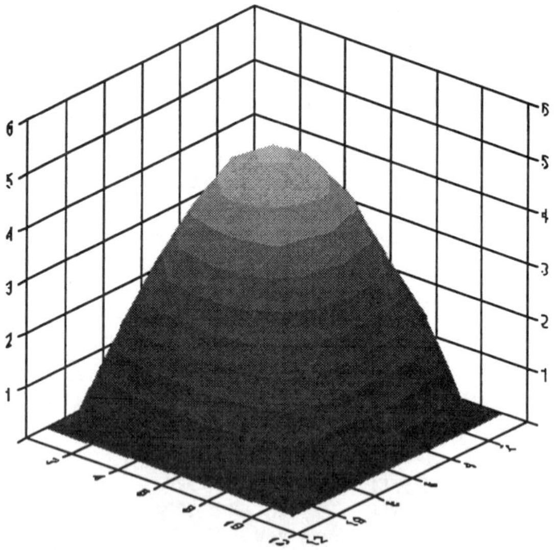

Figure 8-9 Displacement distribution in a membrane (7 × 7 grid).

contours in the plate for the coarse grid, and Fig. 8-9 shows a plot of the displacement distribution. Figures 8-10 and 8-11 are the equivalent representations for the fine grid.

The discussion of mesh spacing, optimum sizing, etc., follows directly from the one that was given for heat conduction in a plate. One would think that since the coarse grid (i.e., 7 × 7) and the fine grid (i.e., 11 × 11) are better approximations to the difference equations than those that were used in the heat conduction problem of

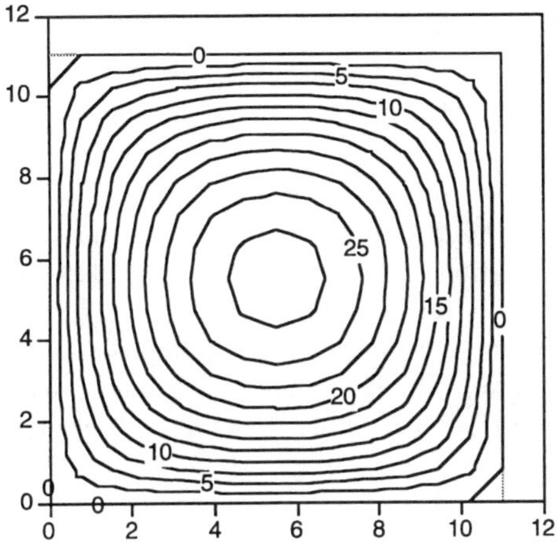

Figure 8-10 Displacement contours (15 × 15 grid).

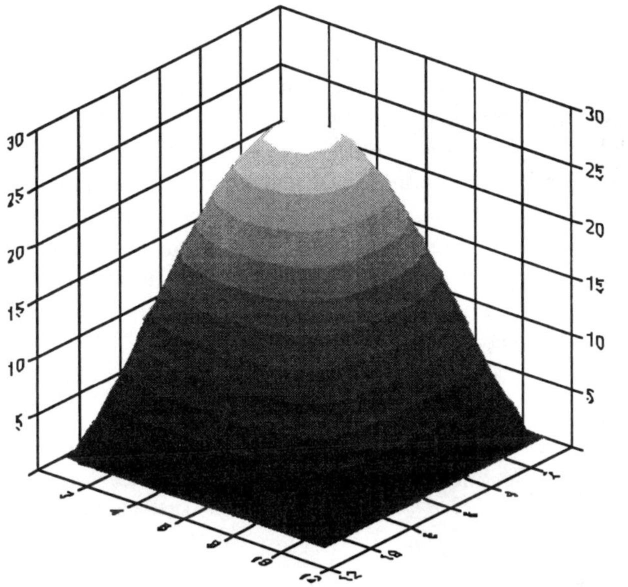

Figure 8-11 Displacement distribution in a membrane (15 × 15 grid).

Section 8-4, the two results in this case would closely match. This, however, would be an extremely erroneous conclusion. Figure 8-8, which was obtained from Table 8-3, shows a maximum deflection in the membrane of 5 in. (actually 4.7 in. in the table). Figure 8-10, however, shows a maximum membrane deflection of 25 in. (actually 28.76 in. in Table 8-4). The conclusion that must be made is that, for this problem, the optimum mesh spacing has not nearly ben achieved.

Problems 8-1 and 8-2 request that students resolve the two example problems in this chapter, by using the supplied codes, in order to determine the optimum grids in each case. The caution regarding the difference equations versus the differential equations that was given at the end of Section 8-4 is still valid.

8-6 CONCLUSION

This chapter has demonstrated the use of the finite-difference techniques as they are applied to the simulation of partial differential equations. The methods that were derived in Chapter 5 and presented in Chapters 6 and 7 are directly applicable here.

The two example problems that are solved in this chapter are linear and therefore require no additional methodology for their solution. The techniques of Chapter 1 will invert the matrix that is generated by the discretization process. Nonlinear examples were not discussed since their application would require complex solution algorithms because of the nonlinearity of the generated matrix. These ideas must be considered to be beyond the scope of an introductory textbook in computational mechanics and should be included only in an advanced presentation of modeling techniques.

This chapter concludes the entire discussion of the use of vectorial approaches toward the simulation of problems in physics and engineering. As noted in the introduction of the textbook, an alternative approach to the derivation and the solution of physical problems lies in the use of energy methods. These techniques are scalar in nature and therefore only one equation is required for describing a physical process, as opposed to the vectorial approach, which requires an equation for each coordinate direction. Energy approaches are theoretically based on the field of mathematics that is known as the calculus of variations. Chapter 9 introduces the use of these mathematical ideas to the development of the physical equations of engineering. A new solution strategy emerges from this development and leads directly to the modeling techniques known as the finite-element methods. These methods are then explored in Chapter 10.

FURTHER READING

Mathews, J. H. (1987). *Numerical Methods*, Prentice-Hall, Englewood Cliffs, New Jersey.

Salvadori, M. G. and Baron, M. L. (1952). *Numerical Methods in Engineering*, Prentice-Hall, Englewood Cliffs, New Jersey.

Schreier, S. (1982). *Compressible Flow*, Wiley, New York.

Sokolnikoff, I. S. (1956). *Mathematical Theory of Elasticity*, McGraw-Hill, New York.

PROBLEMS

8-1 Using the computer code **plate.f**, which is presented in Appendix 8.1, resolve the heat conduction problem that is presented in Section 8-4 until the optimum mesh spacing is achieved. Compare this solution with those that are presented in Tables 8-1 and 8-2.

8-2 Using the computer code **membrane.f**, which is presented in Appendix 8.2, resolve the membrane problem that is presented in Section 8-5 until the optimum mesh spacing is achieved. Compare this solution with those that are presented in Tables 8-3 and 8-4.

8-3 Consider the thin plate of Fig. 8-7, which is subjected to a variable tension S and a distributed pressure load p. The differential equation that governs the deflection z in the plate is given by

$$\frac{\partial^2 z}{\partial x^2} + \frac{\partial^2 z}{\partial y^2} + \frac{p}{S} = 0.$$

We assume that the boundaries of the plate of length L are fixed at $z = 0$ and that

$$S_i = kz_i^2,$$
$$p_i = e^{-z_i}.$$

Solve this nonlinear problem for the displacement distribution in the plate. Use the value of **k** that was given above in the chapter.

Hint: Because of the nonlinearity of the differential equation, an iteration process will be required for a solution.

8-4 Solve Problem 8-3 until the optimum grid spacing is achieved. Compare all solutions from the various meshes to prove that your discretization is optimal.

8-5 In the theory of elasticity, the differential equation that governs the torsion of a noncircular prismatic cylinder is

$$\frac{\partial^2 \psi}{\partial x^2} + \frac{\partial^2 \psi}{\partial y^2} = -2,$$

where ψ is the stress function that vanishes along the boundary of the region. The torsional stresses in the cylinder are given by

$$\tau_{zx} = \mu\alpha \frac{\partial \psi}{\partial y},$$
$$\tau_{zy} = -\mu\alpha \frac{\partial \psi}{\partial x},$$

where μ is the modulus of rigidity and α is the angle of twist per unit length. Using the values

$$\mu = 11.54 \text{ ksi},$$
$$\alpha = 0.1,$$

calculate the stresses in an elliptical cylinder whose major axis is 6 in. and whose minor axis is 3 in.

8-6 Solve the torsional problem for a cylinder of square cross section whose side is equal to seven. Use all equations and parameters from Problem 8-5.

8-7 Solve the torsional problem for a cylinder that has the cross section of an equilateral triangle whose sides are each equal to four. Use all equations and parameters from Problem 8-5.

8-8 A beam that is subjected to bending by terminal couples M is shown in Fig. 8-12. The numerical values for this problem are $M = 100$ in. lb (bending moment), $I = 1.0$ in.^{4} (moment of inertia), $E = 30$ ksi (modulus of elasticity), and $\sigma = 0.3$ (Poisson's ratio).

The bending stress in the beam is given by

$$\tau_{zz} = -\frac{M}{I}x,$$

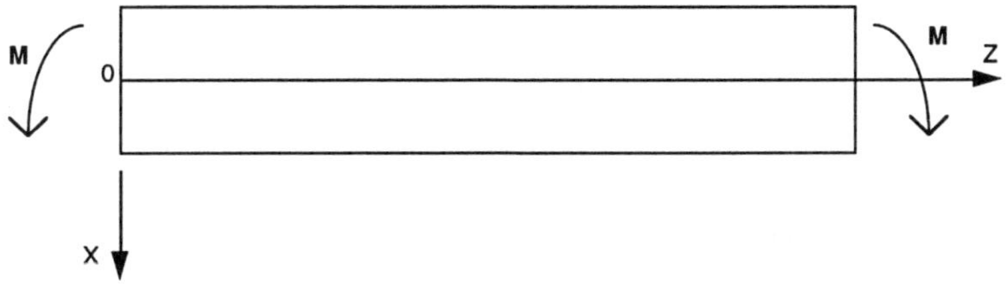

Figure 8-12 Beam loaded by terminal couples.

while the equations that govern the displacements u, v, and w within the beam are

$$\frac{\partial u}{\partial x} = \frac{\sigma M}{EI}x,$$

$$\frac{\partial v}{\partial y} = \frac{\sigma M}{EI}x,$$

$$\frac{\partial w}{\partial z} = -\frac{M}{EI}x,$$

$$\frac{\partial v}{\partial x} + \frac{\partial u}{\partial y} = 0,$$

$$\frac{\partial w}{\partial y} + \frac{\partial v}{\partial x} = 0,$$

$$\frac{\partial u}{\partial z} + \frac{\partial w}{\partial x} = 0.$$

Find the displacements and the stresses in the beam.

8-9 The beam of Fig. 8-13 hangs under the influence of its own weight. The top of the beam is fixed in the z direction. The density of the beam material $\rho = 0.278$ lb/in.^{3} while the acceleration of gravity $g = 386.4$ in./s^2. Poisson's ratio $\sigma = 0.3$, and the modulus of elasticity $E = 30$ ksi. The equations that govern the displacements u, v, and w in the x, y, and z directions, respectively, are

$$\frac{\partial u}{\partial x} = \frac{\partial v}{\partial y} = -\frac{\sigma \rho g z}{E},$$

$$\frac{\partial w}{\partial z} = \frac{\rho g z}{E},$$

$$\frac{\partial v}{\partial x} + \frac{\partial u}{\partial y} = 0,$$

$$\frac{\partial w}{\partial y} + \frac{\partial v}{\partial z} = 0,$$

$$\frac{\partial u}{\partial z} + \frac{\partial w}{\partial x} = 0,$$

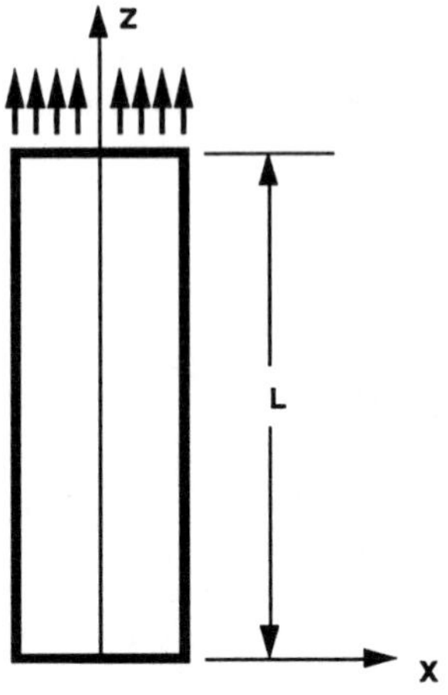

Figure 8-13 Beam hanging under the influence if its own weight.

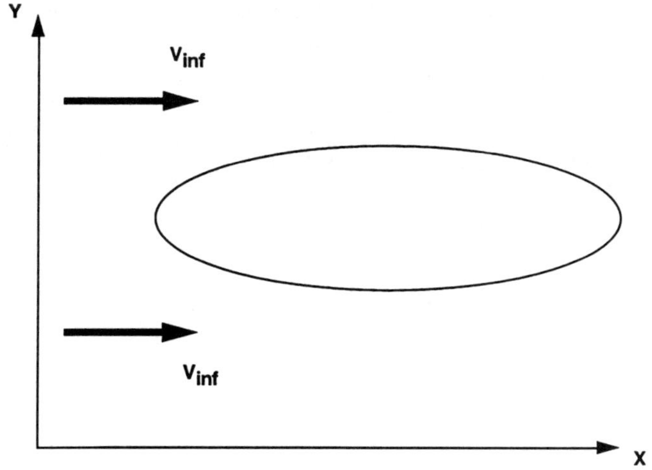

Figure 8-14 Flow over an airfoil.

while the stress in the bar is

$$\tau_{zz} = \rho g z$$

Find the stress and displacement distribution in the bar.

8-10 A square plate of side a is built in on all sides and is subjected to a uniform distribution q. The governing differential equation for the transverse displacement w is given by

$$\nabla^4 w = \left(\frac{\partial^4}{\partial x^4} + 2\frac{\partial^4}{\partial x^2 \partial y^2} + \frac{\partial^4}{\partial y^4} \right) w = \frac{q}{D},$$

with the boundary conditions

$$w = \frac{\partial w}{\partial n} = 0,$$

where

$$D = \frac{E h^3}{12(1 - \nu^2)}$$

is the flexural rigidity, n is the outward normal to the edge of the plate, $q = 1{,}000$ lb/in.^{2} is the distributed load, $E = 30$ ksi is the modulus of elasticity, $h = 0.1$ in. is the plate thickness, and $\nu = 0.3$ is Poisson's ratio. Calculate the displacement distribution $w(x, y)$.

8-11 Figure 8-14 shows an airfoil in a stream whose far-field velocity is V_∞. The stream function ψ is governed by

$$\frac{\partial^2 \psi}{\partial x^2} + \frac{\partial^2 \psi}{\partial y^2} - \frac{\partial \psi}{\partial x}\left(\ln \frac{\rho}{\rho_0} \right) - \frac{\partial \psi}{\partial y}\left(\ln \frac{\rho}{\rho_0} \right) = 0,$$

where ρ is the fluid density and ρ_0 is the stagnation density.

The density ratio is given by

$$\frac{\rho}{\rho_0} = \left[1 - \frac{\gamma - 1}{2} \frac{\left(\frac{\rho}{\rho_0} \psi_y \right)^2 + \left(\frac{\rho}{\rho_0} \psi_x \right)^2}{c_0^2} \right]^{\frac{1}{\gamma - 1}},$$

where γ is the gas constant and c_0 is the speed of sound in the fluid. The velocity in the fluid is given by

$$V^2 = \left(\frac{\rho_0}{\rho}\psi_y\right)^2 + \left(\frac{\rho_0}{\rho}\psi_x\right)^2.$$

The velocity must vanish on the surface of the airfoil while reaching the far-field value $V = V_\infty$. Using the numerical values of $\rho_0 = 0.075$ lb/ft.^{3}, $\gamma = 1.4$, $c_0 = 1128$ ft./s, and $V_\infty = 1000$ ft./s, find the density and velocity distribution in the fluid.

8-12 If Problem 8-11 were expressed in terms of the velocity potential φ in the fluid, then the governing equation becomes

$$\varphi_{xx} + \varphi_{yy} = \frac{1}{c^2}\varphi_x^2\varphi_{xx} + \frac{1}{c^2}\varphi_y^2\varphi_{yy} - \frac{1}{c^2}\varphi_x\varphi_y 2\varphi_{xy},$$

where

$$c^2 = c_0^2 - \frac{\gamma - 1}{2}\left(\varphi_x^2 + \varphi_y^2\right).$$

The fluid velocities u and v are defined by

$$\frac{\rho u}{\rho_0} = \psi_y,$$

$$\frac{\rho v}{\rho_0} = -\psi_x.$$

Using the numerical values of Problem 8-11, calculate the velocity potential distribution in the fluid.

CHAPTER

NINE

ENERGY METHODS AND MINIMIZATION

NOMENCLATURE

$\bar{y}$	approximation curve in energy formulations
η	arbitrary curve
y^n	nth derivative of y with respect to x
$u_x, u_y, \ldots$	partial derivatives of u with respect to coordinate directions
ε	small perturbation parameter or emissivity
σ	Stefan–Boltzmann constant
ε_{ij}	strain component in a structure
τ_{ij}	stress component in a structure
A	cross-sectional area
E	modulus of elasticity of a material
e	elongation of a string
F	functional expression
$f(x)$	arbitrarily distributed load
h	coefficient of convective heat transfer
I	moment of inertia for beam bending
$I(y)$	extremum expression
k_i	thermal conductivity in the ith coordinate direction
L	length of a beam or string
P	surface perimeter of a body
Q	heat source
q^s	surface heat transfer
s	arc length of a curve
T	tension in a string or temperature in a body
T_a	ambient temperature

T_r	receiver temperature for radiation heat transfer
U	strain energy in a structure
u	dependent variable
V	potential energy within a structure
W	work done by the external forces on a body
x	coordinate direction
y	coordinate direction and deflection of a structure

9-1 INTRODUCTION

In the preceding chapters we examined the area of physics that is known as vector mechanics. A free-body diagram, representing a portion of the physical process, was analyzed. Force summations (in structural mechanics) or balance procedures (in heat transfer and fluid flow) were applied to the body in the three coordinate directions. These summations then resulted in differential equations that described the relevant processes.

In order to obtain numerical solutions to these differential equations, finite-difference procedures were applied to the governing laws. These techniques have been discussed in preceding chapters.

The above numerical procedures, while efficient and robust, introduced several difficulties in their application, i.e.,

1. There is difficulty in applying the difference techniques at nonrectangular boundaries.
2. The expression of the boundary conditions becomes difficult when
   (a) the boundary is not rectangular, or
   (b) a derivative boundary condition needs to be applied.
   This may occur, for example when a heat flux is applied to the boundary of a body undergoing heat transfer.
3. The choice of a nonrectangular approximating grid creates difficulty in the expression of derivatives that are evaluated with respect to the coordinate directions.

It is therefore fair to ask whether there is an alternative method of expressing these physical laws in a manner that will either eliminate, or at least greatly reduce, the above-mentioned problems in the numerical simulation of these processes. Is there, for example, a numerical method that can be derived that is not dependent on the shape of the physical boundary? Can the boundary conditions be expressed in a manner that is more suited to the physics of the problem? For example, in structural mechanics, can the forces and the displacements within the body and on the boundary of the body be simply expressed?

The answers to these questions bring us to the area of physics known as energy techniques. These procedures are scalar and therefore do not involve directional quantities in their derivation. This chapter, then, is devoted to the introduction of energy techniques. It is demonstrated how these techniques are derived, how they are used in mechanics, and how they lead to the numerical solution of the physical quantities that are associated with

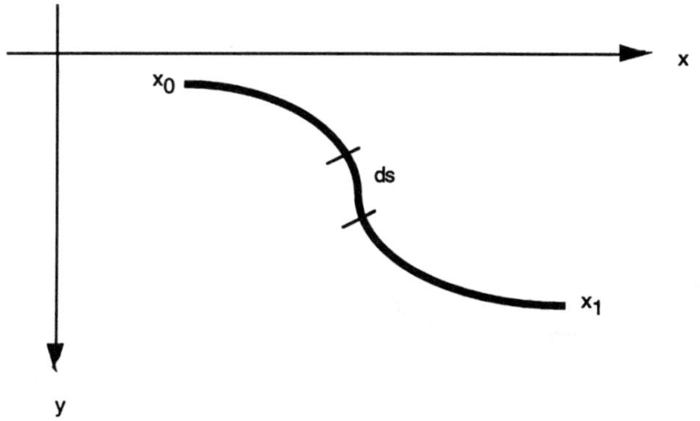

Figure 9-1 Arbitrary curve that passes through the fixed points x_0 and x_1.

a particular problem. The derivation of energy functions, finally, leads to the introduction of the finite-element techniques in Chapter 10.

9-2 CALCULUS OF VARIATIONS

The basis for all energy techniques that are presented in this textbook is the calculus of variations. This mathematical technique is based on the ability to find an expression, called a functional, that, when minimized, results in the governing physical laws for the appropriate problem.

To introduce the calculus of variations, let us turn our attention to a simple problem, i.e., that of determining the shortest distance between two points. Figure 9-1 shows an arbitrary curve that is constrained to pass through the two fixed points x_0 and x_1. We would like to determine the curve that represents the shortest distance between these two points.

The length of the entire curve is

$$s = \int_{x_0}^{x_1} \mathrm{d}s = \int_{x_0}^{x_1} \sqrt{1 + y'^2}\, \mathrm{d}x \tag{9-1}$$

or

$$I(y) = \int_{x_0}^{x_1} \sqrt{1 + y'^2}\, \mathrm{d}x. \tag{9-2}$$

The reason that this integral is expressed as $I(y)$, even though it is a definite integral, is because the value of the definite integral in Eq. (9-2) depends on the curve $y(x)$ that is chosen. We must now develop a methodology to minimize an expression such as $I(y)$, which is called a functional.

Consider the functional $I(y) = F(x, y, y')$. We seek the curve $y(x)$ that will minimize $I(y)$. For each choice of $y(x)$ [say $\bar{y}(x)$] in Eq. (9-2), the definite integral will produce a number that will be the distance between points x_0 and x_1. We seek the curve $y(x)$ that will yield the lowest number. To begin this process, Fig. 9-2 shows a family of curves such that

$$\bar{y}(x) = y(x) + \varepsilon\eta(x), \tag{9-3}$$

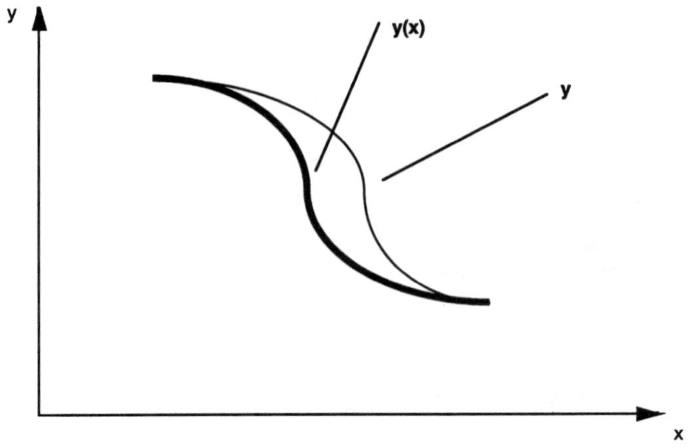

Figure 9-2 Family of curves in which $\bar{y}(x) = y(x) + \varepsilon\eta(x)$.

where ε is a small parameter. $\eta(x)$ must be chosen so that all curves pass through x_0 and x_1 (i.e., $\eta(x)$ must be zero at these points). We define the variation of y, δy, as

$$\bar{y} - y = \varepsilon\eta(x) = \text{variation of } y = \delta y. \tag{9-4}$$

We want to choose $y(x)$ such that it minimizes $I(y)$ and satisfies the relation

$$I(\bar{y}) \geq I(y). \tag{9-5}$$

Therefore,

$$\left.\frac{\mathrm{d}I(\bar{y})}{\mathrm{d}\varepsilon}\right]_{\varepsilon\to 0} = \left.\frac{\mathrm{d}I(y+\varepsilon\eta)}{\mathrm{d}\varepsilon}\right]_{\varepsilon\to 0} = \delta I = 0. \tag{9-6}$$

This is a necessary condition since at $\varepsilon = 0$, $\bar{y} = y$ and $I(y)$ is a minimum. Figure 9-3 shows this condition.

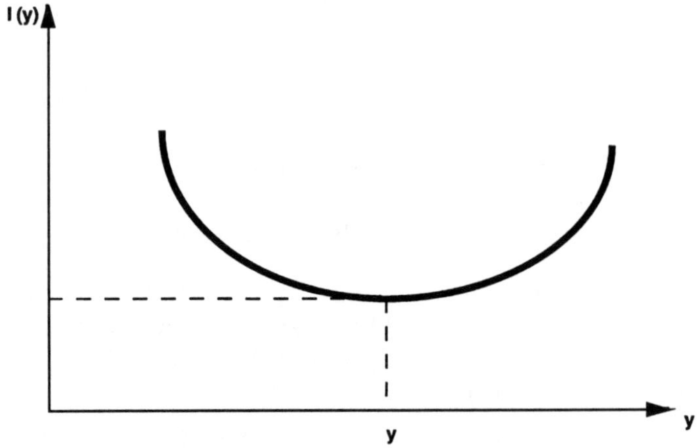

Figure 9-3 Condition of Eq. (9-6).

To calculate δI

$$\delta I = \int_{x_0}^{x_1} \frac{\mathrm{d}F}{\mathrm{d}\varepsilon}\bigg]_{\varepsilon\to 0} \mathrm{d}x = \int_{x_0}^{x_1} \left(\frac{\partial F}{\partial \bar{y}}\frac{\mathrm{d}\bar{y}}{\mathrm{d}\varepsilon} + \frac{\partial F}{\partial \bar{y}'}\frac{d\bar{y}'}{\mathrm{d}\varepsilon}\right)\bigg]_{\varepsilon\to 0} \mathrm{d}x \tag{9-7}$$

$$= \int_{x_0}^{x_1} (F_y\eta + F_{y'}\eta')\,\mathrm{d}x \qquad (\bar{y}=y \quad \text{at} \quad \varepsilon=0). \tag{9-8}$$

Integrating the second term in Eq. (9-8), by parts, yields

$$\int_{x_0}^{x_1} \eta' F_{y'}\,\mathrm{d}x = \eta F_{y'}\bigg]_{x_0}^{x_1} - \int_{x_0}^{x_1} \eta \frac{\mathrm{d}F_{y'}}{\mathrm{d}x}\,\mathrm{d}x, \tag{9-9}$$

but

$$\eta F_{y'}\big]_{x_0}^{x_1} = 0,$$

since $\eta(x) = 0$ at the end points.

$$\therefore \quad \delta I = \int_{x_0}^{x_1} \left(F_y - \frac{\mathrm{d}F_{y'}}{\mathrm{d}x}\right) \eta\,\mathrm{d}x = 0. \tag{9-10}$$

Since $\eta(x)$ is arbitrary,

$$F_y - \frac{\mathrm{d}F_{y'}}{\mathrm{d}x} = 0. \tag{9-11}$$

Equation (9-11) is called the Euler equation of the original variational integral. The satisfaction of Eq. (9-11) for $y(x)$ yields the curve that minimizes the original functional and, in this example, calculates the least distance between points x_0 and x_1.

In Eq. (9-11) we have used the abbreviations

$$F_y = \frac{\partial F}{\partial y}(x, y', \text{const.}),$$

$$F_{y'} = \frac{\partial F}{\partial y'}(x, y, \text{const.}).$$

Returning to the previous example, we can show that

$$F = \sqrt{1 + y'^2},$$

$$\frac{\partial F}{\partial y} = 0,$$

$$\frac{\partial F}{\partial y'} = \{1 + y'^2\}^{-\frac{1}{2}} y'.$$

This yields a Euler equation:

$$\frac{\mathrm{d}}{\mathrm{d}x}\left(\frac{\partial F}{\partial y'}\right) = 0 \qquad \text{or} \qquad \frac{\partial F}{\partial y'} = C_1. \tag{9-12}$$

Thus,

$$\frac{y'}{\sqrt{1 + y'^2}} = C_1,$$

$$y'^2 = C_2,$$

$$y' = C_3 = \text{const.}$$

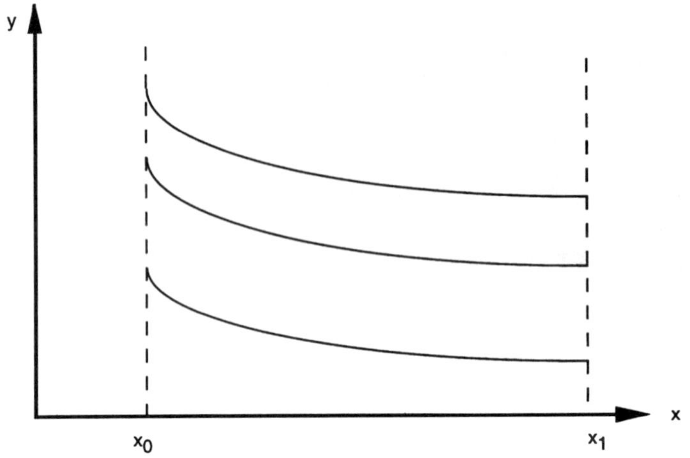

Figure 9-4

This is, of course, the equation of a straight line. We have thus demonstrated that the minimum distance between two points occurs when the curve $y(x)$ is chosen as a straight line. This is known to be the correct answer.

Before we proceed to more complex functionals, it would be instructive to determine what additional conditions, if any, must be placed on the minimization process if the values of the ordinates at points x_0 and x_1 (i.e., y_0 and y_1) were not fixed in space. Figure 9-4 shows the same geometry as that of the previous problem. Now, however, the end points can move in the y direction.

Returning to Eq. (9-9), we still derive the same Euler equation, i.e., Eq. (9-12). Now, however, we must also satisfy the additional condition

$$\frac{\partial F}{\partial y'} = 0 \quad \text{at} \quad x = x_0 \quad \text{and} \quad x = x_1. \tag{9-13}$$

Equation (9-13) is called the natural boundary condition.

To summarize the important results of this section, we note that

1. If the dependent variable (e.g., the displacement in structural problems) is known on the boundary, then the minimization of the functional leads to the solution of a differential equation (Euler's equation).
2. If, on the other hand, the dependent variable is not explicitly defined at a particular point of the boundary (e.g., the tractions are known) then an additional condition must be included in the solution process, i.e., the natural boundary condition.

9-3 HIGHER-ORDER FUNCTIONALS

The functional of Section 9-2 is perhaps the simplest expression that one would encounter in mathematics or physics. It is therefore instructive to delineate more complex functionals along with their Euler's equations and their natural boundary conditions. The derivations of these results are left to the readers. If this were a textbook in mathematics rather than computational mechanics, we might develop the calculus of variation in greater depth. Since our goal is to apply these techniques to physical problems, only a few relevant cases are considered at this time. When these procedures are reintroduced in the next chapters in order to develop the finite-difference procedures, the relevant energy functional is listed for that case.

Case I: Higher derivatives

$$I(y) = \int_{x_0}^{x_1} F\left(x, y, y', y'', \ldots, y^{(n)}\right) \mathrm{d}x.$$

Euler's equation becomes

$$F_y - \frac{\mathrm{d}}{\mathrm{d}x}(F_{y'}) + \frac{\mathrm{d}^2}{\mathrm{d}x^2}(F_{y''}) - \cdots + (-1)^n \frac{\mathrm{d}^n}{\mathrm{d}x^n} F_{y^{(n)}} = 0, \tag{9-14}$$

where F_y is an abbreviation for $\partial F/\partial y$.

Case II: One dependent and two independent variables

$$I(u) = \iint_R F(x, y, u, u_x, u_y, u_{xx}, u_{xy}, u_{yy})\, \mathrm{d}x\, \mathrm{d}y.$$

The Euler equation becomes

$$F_u - \frac{\partial F_{u_x}}{\partial x} - \frac{\partial F_{u_y}}{\partial y} + \frac{\partial^2 F_{u_{xx}}}{\partial x^2} + \frac{\partial^2 F_{u_{yy}}}{\partial y^2} + \frac{\partial^2 F_{u_{xy}}}{\partial x \partial y} = 0. \tag{9-15}$$

Case III: Two dependent and two independent variables

$$I(u, v) = \iint_A F(x, y, u, v, u_x, u_y, v_x, v_y)\, \mathrm{d}x\, \mathrm{d}y + \oint_S \phi(s, u, v, u_S, v_S)\, \mathrm{d}S$$

Since there are two dependent variables, the Euler equations become

$$F_u - \frac{\partial F_{u_x}}{\partial x} - \frac{\partial F_{u_y}}{\partial y} = 0,$$

$$F_v - \frac{\partial F_{v_x}}{\partial x} - \frac{\partial F_{v_y}}{\partial y} = 0, \tag{9-16}$$

and the natural boundary conditions are

$$\begin{aligned} F_{u_x} \frac{\mathrm{d}y}{\mathrm{d}s} - F_{u_y} \frac{\mathrm{d}x}{\mathrm{d}s} + \phi_u - \frac{\mathrm{d}}{\mathrm{d}s}\left[\phi_{u_s}\right] &= 0, \\ F_{v_x} \frac{\mathrm{d}y}{\mathrm{d}s} - F_{v_y} \frac{\mathrm{d}x}{\mathrm{d}s} + \phi_v - \frac{\mathrm{d}}{\mathrm{d}s}\left[\phi_{v_s}\right] &= 0. \end{aligned} \tag{9-17}$$

In Chapter 10, Eqs. (9-16) and (9-17) form the basis for the development of a two-dimensional structural finite-element methodology.

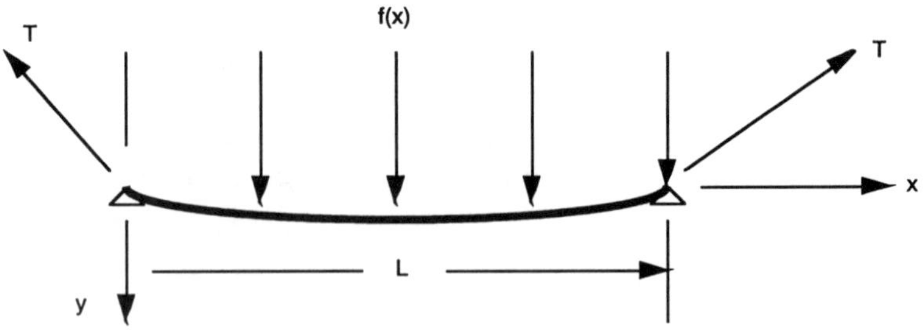

Figure 9-5 String under tension T and loaded by a lateral distribution of forces $f(x)$.

9-4 EXAMPLES IN STRUCTURAL ANALYSIS

We now turn our attention to examples in structural analysis. We study how the calculus of variations can be applied to these problems and form the basis of an energy approach for solution methodology. We must identify a functional that, when minimized by the techniques of the preceding sections, yields a condition (differential equation) that will allow us to obtain the significant state variables for the structural problem. This functional for structural problems is the potential energy. If this were a textbook in mathematics, a complete proof of the previous statement would be necessary. However, in this text, we demonstrate the validity of the potential energy theorem by applying the above-mentioned concepts to several simple problems.

Problem I: String under tension and normal loading Figure 9-5 shows a string that is under a tension T and is loaded by a lateral distribution of forces $f(x)$. The string can elongate as well as deflect.

To calculate the elongation in the string, we calculate the length of the string in its loaded position and compare it with the undeformed length, viz.,

$$e = \int_0^L (\mathrm{d}s - \mathrm{d}x) = \int_0^L \left[\sqrt{1 + y'^2} - 1\right] \mathrm{d}x;$$

but

$$(1 + y'^2)^{\frac{1}{2}} - 1 = 1 + \frac{1}{2}y'^2 + \cdots - 1 \cong \frac{1}{2}y'^2,$$

so that the elongation in the string becomes

$$e = \frac{1}{2}\int_0^L y'^2\, \mathrm{d}x.$$

The potential energy in a structural member is

$$V = U - W,$$

where U is the strain energy in the structure and W is the work done by the external forces on the body.

For the string the potential energy becomes

$$V = U - W = \frac{T}{2}\int_0^L y'^2\, \mathrm{d}x - \int_0^L f(x) y\, \mathrm{d}x = \int_0^L \left[\frac{1}{2}Ty'^2 - f(x)y\right] \mathrm{d}x. \quad (9\text{-}18)$$

This expression can now be considered a functional of the form

$$I(y) = \int_0^L F(x, y, y')\,dx,$$

which is of the same form as that of the functional in Section 9-2. Thus the Euler equation that was derived in that section may be applied to the string problem:

$$\frac{\partial F}{\partial y} = -f(x),$$

$$\frac{\partial F}{\partial y'} = Ty'$$

$$\frac{d}{dx}(Ty') = Ty''.$$

Equation (9-11) becomes for this problem

$$Ty'' + f(x) = 0. \tag{9-19}$$

Equation (9-19) is the governing differential equation for this problem. The function $y(x)$ that satisfies this equation is the solution to the problem of a transversely loaded string that is under tension.

Problem II: Simply supported beam with a transverse loading The next example is a simply supported beam with a distributed transverse loading. This configuration is shown in Fig. 9-6. Once again, we will apply the Potential Energy Theorem. Using Bernoulli's beam theory, we can define the following quantities, viz.,

$$W = \int_A T_i u_i\,dA = \int_0^L f(x) y\,dx,$$

$$M = EIy'' \quad \text{(Euler–Bernoulli beam condition)},$$

$$\tau_{xx} = \frac{My}{I} \quad \text{(definition of bending stress)},$$

$$\varepsilon_{xx} = \frac{\tau_{xx}}{E} = \frac{My}{EI} \quad \text{(stress–strain relation)},$$

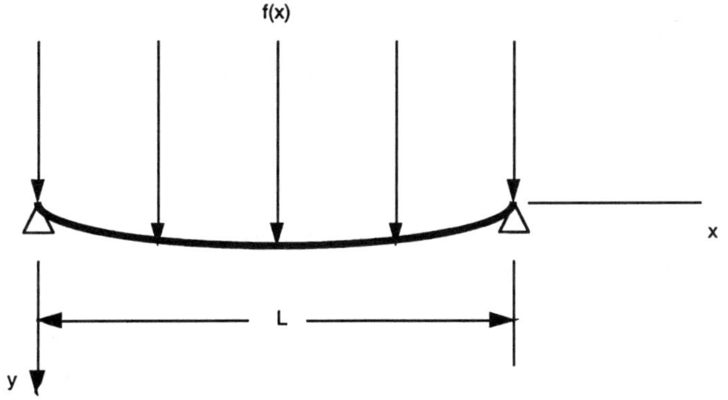

Figure 9-6 Configuration of a simply supported beam with a distributed transverse loading.

where τ_{xx} and ε_{xx} are the normal stress and the normal strain, respectively, M is the bending moment in a cross section of the beam, E is the modulus of elasticity of the beam material, and I is the moment of inertia of the beam cross section and is defined by

$$I = \int_A y^2 \, \mathrm{d}A.$$

For the beam, the strain energy is defined by

$$U = \int_V \frac{1}{2} \tau_{xx} \varepsilon_{xx} \, \mathrm{d}x \, \mathrm{d}A,$$

where V is the volume of the body and A is the cross-sectional area. The potential energy of the beam then becomes

$$V = \int_0^L \left[\frac{1}{2} EI (y'')^2 - f(x) y \right] \mathrm{d}x. \tag{9-20}$$

Equation (9-20) is the functional that requires minimization. It is also an example of the higher-order functional expression that was discussed in Section (9-3). The Euler equation that results from applying the calculus of variation to Eq. (9-20) is

$$F_y - \frac{\mathrm{d}}{\mathrm{d}x} F_{y'} + \frac{\mathrm{d}^2}{\mathrm{d}x^2} F_{y''} = 0.$$

For this beam,

$$\frac{\partial F}{\partial y} = -f(x),$$

$$\frac{\partial F}{\partial y''} = EIy'',$$

$$\frac{\mathrm{d}^2}{\mathrm{d}x^2}(F_{y''}) = EIy^{IV},$$

so that the Euler equation becomes

$$EIy^{IV} - f(x) = 0,$$

which is the governing differential equation for the beam.

We have thus demonstrated, at least for the sample problems that have been chosen, that the minimization of the potential energy in a structure yields the governing differential equation. The solution of this differential equation will then yield the required solution to the problem.

The following question then presents itself:

> If the minimization process yields the governing differential equations for the structural analysis problem, are we not merely returning to the use of a finite-difference technique for their solution?

The answer to this question is

> Since the exact minimization process yields an exact solution, then an approximate minimization will yield an approximate solution.

Thus we will seek ways of obtaining a minimization of the appropriate functional to a known degree of approximation.

9-5 RAYLEIGH–RITZ METHOD

One of the more common methods of approximating the minimization process in structural mechanics is the Rayleigh–Ritz method. A formal declaration of the technique can be expressed by the following.

Let a functional of the form

$$I(y) = \int_{x_0}^{x_1} F(x, y, y')\,dx, \tag{9-21}$$

such that $y(x_0) = y_0$ and $y(x_1) = y_1$. Let the exact solution to $I(y) =$ min be $y = y^*(x)$ and let $I(y^*)$ be the minimum of $I(y)$ on this set.

Construct $\bar{y}(x)$ [the approximating function to $y(x)$],

$$\bar{y}(x) = \phi(x, a_1, a_2, \ldots, a_k), \tag{9-22}$$

such that the end conditions are satisfied. Inserting Eq. (9-22) into Eq. (9-21) and minimizing the approximation $I(a_1, a_2, \ldots, a_k)$, we arrive at the minimizing conditions for this approach, viz.,

$$\frac{\partial I}{\partial a_r} = 0 \qquad (r = 1, 2, \ldots, k). \tag{9-23}$$

Solving this set of k algebraic equations for the unknown coefficients $\bar{a}_i$, we obtain

$$\bar{y}(x) = \phi(x, \bar{a}_1, \bar{a}_2, \ldots, \bar{a}_k) \tag{9-24}$$

Equation (9-24) is the approximate solution to the problem. We have arrived at this solution without the derivation of any differential equations and without having to resort to the finite-difference techniques. The approximating function that is chosen should be a complete set in the mathematical sense. If an incomplete set of functions is chosen, the Rayleigh–Ritz method yields an upper bound of the solution.

To illustrate this technique, we choose a simple example. Figure 9-7 shows a cantilever beam with uniformly distributed transverse loading.

Above we have shown that, for a beam, the principle of virtual potential energy becomes

$$V = \int_0^L \left[\frac{1}{2} EI(y'')^2 - f(x)y\right] dx.$$

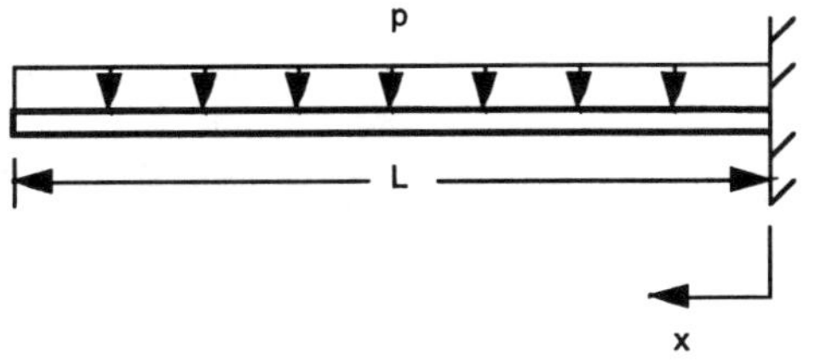

Figure 9-7 Configuration of a contilever beam with uniformly distributed loading.

In this case, $f(x) = p$ (a constant). We must now use an approximating interpolation function for the solution to the displacement y of the cantilever beam. As in all modeling, the choice that is made now for the function will determine the accuracy and the applicability of the solution that result from the minimization process. In other words, an expression that is complex and complete will yield, on minimization, a solution description that is highly accurate within the region. We must also remember that, in the Rayleigh–Ritz method, the interpolating function must satisfy the end condition for a complete and accurate solution. We now choose the expression

$$y(x) = ax^4 + bx^3 + cx^2 + dx + e \tag{9-25}$$

as the approximation to the solution within the cantilever beam.

The first step in the solution process is to ensure that Eq. (9-25) satisfies the boundary conditions

$$y(x)|_{x=0} = 0, \tag{9-26}$$

$$y'(x)|_{x=0} = 0. \tag{9-27}$$

If Eqs. (9-26) and (9-27) are applied to Eq. (9-25), we find that $e = 0$ and $d = 0$. Then the interpolating function (9-25), with the physical boundary conditions satisfied, becomes

$$y(x) = ax^4 + bx^3 + cx^2 \tag{9-28}$$

with

$$y''(x) = 12ax^2 + 6bx + 2c. \tag{9-29}$$

Inserting Eqs. (9-28) and (9-29) into the expression for the potential energy, we get

$$V = \int_0^L \left[\frac{1}{2}EI\{12ax^2 + 6bx + 2c\}^2 - p\{ax^4 + bx^3 + cx^2\}\right] \mathrm{d}x. \tag{9-30}$$

We now must carry out the indicated operations. Equation (9-30) then becomes the algebraic expression

$$V = a^2\left\{\frac{72EIL^5}{5}\right\} + a\left\{-\frac{pL^5}{5}\right\} + b\left\{-\frac{pL^4}{4}\right\} + b^2\{6EIL^3\} + c\left\{-\frac{pL^3}{3}\right\}$$
$$+ c^2\{2EIL\} + \frac{EI}{2}\{16acL^3 + 36abL^4 + 12bcL^2\}. \tag{9-31}$$

The application of the minimization procedure to Eq. (9-31), viz.,

$$\frac{\partial V}{\partial a} = 0,$$

$$\frac{\partial V}{\partial b} = 0,$$

$$\frac{\partial V}{\partial c} = 0,$$

results in three simultaneous algebraic equations for a, b, and c, i.e.,

$$\{144\,EIL^5\}a + \{90EIL^4\}b + \{40\,EIL^3\}c = pL^5, \tag{9-32}$$

$$\{72\,EIL^4\}a + \{48EIL^3\}b + \{24\,EIL^2\}c = pL^4, \tag{9-33}$$

$$\{24\,EIL^3\}a + \{18EIL^2\}b + \{12\,EIL\}c = pL^3. \tag{9-34}$$

The solution of Eqs. (9-32)–(9-34) yields

$$a = \frac{p}{24EI},$$

$$b = -\frac{pL}{6EI},$$

$$c = \frac{pL^2}{4EI}.$$

When the expressions for a, b, and c are substituted into Eq. (9-28), the deflection of the cantilever beam becomes

$$EIy(x) = \frac{p}{24}\{x^4 - 4Lx^3 + 6L^2x^2\}. \tag{9-35}$$

Equation (9-35) turns out to be the exact solution to the cantilever beam problem (as we can verify by consulting a textbook on the strength of materials).

In general, a minimization procedure will yield only an approximate solution to the physical problem. In this case, however, the exact solution was obtained because the physical boundary conditions were also satisfied at the free end of the beam, viz.,

$y''(L) = 0$ (the bending moment vanishes at $x = L$),

$y'''(L) = 0$ (the shear force vanishes at $x = L$).

9-6 HEAT TRANSFER FORMULATIONS

The three major fields in Mechanical Engineering are structures, heat transfer, and fluid flow. The energy formulation for structural analysis was discussed in Sections 9-4 and 9-5. This section addresses the minimization processes associated with heat transfer. Fluid flow, which tends to be nonlinear in its application, is reserved for Chapter 11 so that the formulation and the special nonlinear approaches in this area can be adequately explored.

In this work, heat transfer is confined to conduction, convection, and radiation. The variational approach in this field is presented along with the minimization processes that are required for deriving the governing equations. As for structural analysis, the development of a finite-element formulation for heat transfer is presented in Chapter 10. In heat transfer there is only one dependent variable, the temperature T, which can be a function of the coordinate directions x, y, and z. In contrast to the structural analysis field, there is no convenient energy functional (such as the potential energy) to form

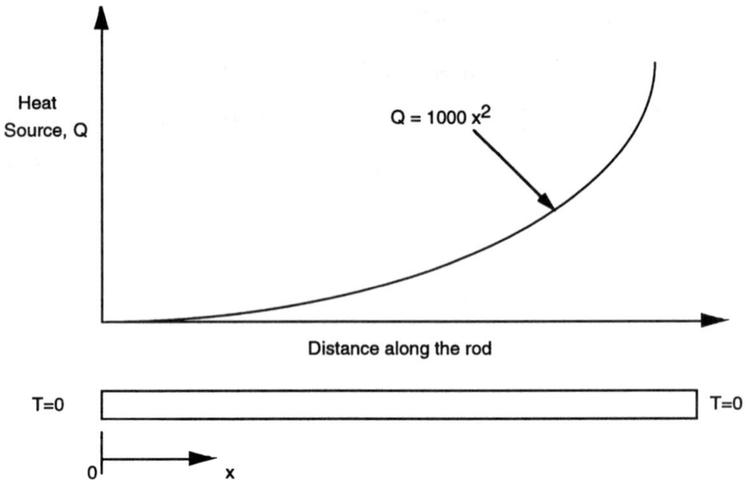

Figure 9-8 Heat conduction example.

the basis of the minimization process. We must therefore use the governing differential equations of the physical processes in order to derive (or infer) the correct functional.

Let us begin this process by considering a simple example of heat conduction. Figure 9-8 shows a circular cylinder whose ends are held at $T = 0$ and whose periphery is insulated. In addition, there is a heat source of the form of $Q = 1000x^2$ along the length of the cylinder. The governing differential equation for the steady-state heat conduction problem is

$$\frac{d^2T}{dx^2} + 1000x^2 = 0 \qquad 0 \le x \le 1, \tag{9-36}$$

with the physical boundary conditions

$$T(0) = 0,$$
$$T(1) = 0.$$

Since the values of the dependent variable are fixed at the ends of the geometry, only an Euler equation results from the minimization process. It is not necessary to satisfy natural boundary conditions for this problem. With reference to the discussions in Section 9-2, it would appear that Eq (9-11) would be the Euler equation that results from the minimization process. This equation is repeated here for the dependent variable T, viz.,

$$\frac{\partial F}{\partial T} - \frac{d}{dx}\left(\frac{\partial F}{\partial T_x}\right) = 0. \tag{9-37}$$

Comparing Eqs. (9-36) and (9-37) on a term-by-term basis, we can infer that

$$\frac{\partial F}{\partial T} = 1000x^2, \tag{9-38}$$

and substituting Eqs. (9-38) and (9-36) into Eq. (9-37), we get

$$-\frac{d}{dx}\left(\frac{\partial F}{\partial T_x}\right) = -\frac{d}{dx}\left(-\frac{\partial T}{\partial x}\right) = -\frac{d(-T_x)}{dx}, \tag{9-39}$$

so that

$$\frac{\partial F}{\partial T_x} = -T_x. \tag{9-40}$$

Integrating Eq. (9-38) yields

$$F = 1000x^2T + h(T_x). \tag{9-41}$$

Additionally, Eq. (9-40) provides

$$F = -\frac{1}{2}T_x^2 + g(T). \tag{9-42}$$

Comparing Eqs. (9-41) and (9-42) reveals that

$$h(T_x) = -\frac{1}{2}T_x^2,$$

$$g(T) = 1000x^2T,$$

so that

$$F = 1000x^2T - \frac{1}{2}T_x^2. \tag{9-43}$$

Equation (9-43) is the expression for F that forms the integrand of the functional I that is to be minimized over the region. If we now refer to the extremum

$$I(y) = F(x, y, y'),$$

which was presented in Section 9-2, we arrive at the functional that is to be minimized for this simple problem in heat conduction, i.e.,

$$I = \int_0^1 \left[1000x^2T - \frac{1}{2}\left(\frac{dT}{dx}\right)^2\right]dx. \tag{9-44}$$

Equation (9-44) represents the integral that, when minimized, yields a differential equation whose solution T represents the temperature distribution in the circular cylinder.

It should be noted here that we inferred the derivation of the functional by observing the governing differential equation for the process. This method of obtaining the minimizing functional is used in all subsequent heat transfer problems.

In order to complete this section with respect to heat transfer processes and in order to set the basis for the finite-element formulations of Chapter 10, we record some governing differential equations for heat transfer processes.

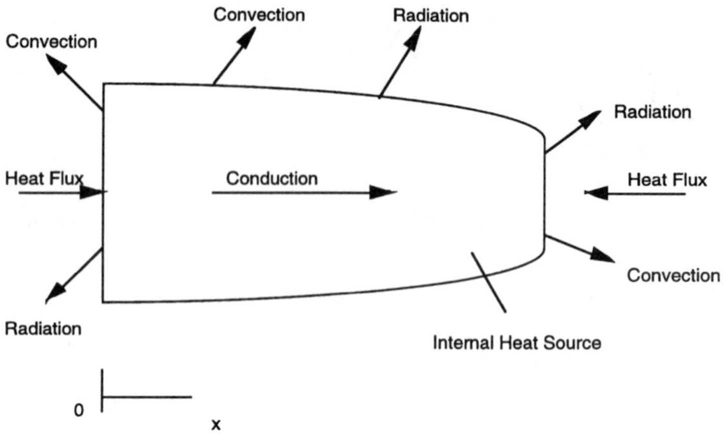

Figure 9-9 Nature of the one-dimensional thermal problem.

One-dimensional heat transfer Figure 9-9 shows the essential nature of the one-dimensional thermal problem. Heat may enter or leave the body by convection or by radiation. Additionally, an external heat source may be present in the problem along with an internal heat source. The heat transfer within the body is conduction. It can readily be shown that the governing differential equation is

$$\frac{\mathrm{d}}{\mathrm{d}x}\left(kA\frac{\mathrm{d}T}{\mathrm{d}x}\right) - hS(T - T_a) - \varepsilon\sigma S\left(T^4 - T_r^4\right) + QA = 0, \tag{9-45}$$

where k is the thermal conductivity within the body, h is the coefficient of convective heat transfer, ε is the emissivity of the surface area of the body, S is the surface perimeter, A is the cross-sectional area, T_a is the ambient temperature of the surrounding fluid, T_r is the radiation receiver temperature, Q is the rate of heat generation, and σ is the Stefan–Boltzmann constant. Additionally, all properties may be functions of temperature and vary within the body.

Two-dimensional heat transfer For two-dimensional steady-state heat conduction with a known heat source, the governing differential equation is

$$\frac{\partial}{\partial x}\left(k_x\frac{\partial T}{\partial x}\right) + \frac{\partial}{\partial y}\left(k_y\frac{\partial T}{\partial y}\right) + \frac{\partial}{\partial z}\left(k_z\frac{\partial T}{\partial z}\right) + q^B = 0, \tag{9-46}$$

where k_i are the conductivities of the material in the coordinate directions and q^B is the rate of heat generation per unit volume. On the surface of the body,

$$T|_{S_1} = T_e, \tag{9-47}$$

$$k_n\frac{\partial T}{\partial n}\bigg|_{S_2} = q^S, \tag{9-48}$$

where T_e is the environmental temperature, k_n is the thermal conductivity in the nth direction, q^S is the heat flow input on the surface of the body, and S are points on the surface of the body.

The convective boundary conditions are

$$q^S = h(T_e - T_S). \tag{9-49}$$

The radiation boundary conditions are

$$q^S = \kappa(T_r - T_S), \tag{9-50}$$

where

$$\kappa = h_r\left(T_r^2 + T_S^2\right)(T_r + T_S). \tag{9-51}$$

Then, by the methods that are discussed in this section, the functional that is to be minimized can be shown to be

$$\begin{aligned} I = &\int_V \frac{1}{2}\left\{k_x\left(\frac{\partial T}{\partial x}\right)^2 + k_y\left(\frac{\partial T}{\partial y}\right)^2 + k_z\left(\frac{\partial T}{\partial z}\right)^2\right\}\mathrm{d}V \\ &- \int_V Tq^B\,\mathrm{d}V - \int_{S_2} T^S q^S\,\mathrm{d}S - \sum_i T^i Q^i, \end{aligned} \tag{9-52}$$

where Q^i are the concentrated heat flow inputs.

Any further compilation of the governing equations in heat transfer is unnecessary here. A complete presentation is made in Chapter 10 in the sections in which the finite-element method for these problems is derived.

9-7 CONCLUSION

In this chapter, energy techniques were introduced as an alternative way of arriving at the solution of problems in applied physics and engineering. It began with the introduction of the mathematical technique of the calculus of variations. Then the concept of the virtual energy function was introduced as the important expression in structural mechanics. We subsequently applied these ideas to simple problems in structural mechanics to show that the minimization process leads to the derivation of the exact governing differential equations. Then the Rayleigh–Ritz method was introduced as a way to perform the minimization process in an approximate way rather than in an exact manner. This technique provided us with a technique that can be applied to any complex problem in engineering.

For problems that involve heat transfer in a body, the functional that is to be optimized was inferred from the governing differential equation. This process is generally straightforward. A full discussion is made in Chapter 10 for the generalized heat transfer problem.

The road that must be followed is now clearly marked. In a complex engineering problem, we must identify the relevant functional that, when minimized, would lead to the exact solution of the problem. Second, we must choose an interpolating (approximating) function that best describes the nature of the solution in the body. For complex problems, of course, it will be impossible to select an expression that is representative of the solution in the entire body. Therefore we are forced to apply the techniques of this chapter to small subregions of the body. We minimize these expressions over each subregion to obtain a valid approximation of the variables within that region, and then we connect each subregion by physical concepts. This sewing of the regions ultimately restores the entire body and yields a total solution to the problem. The technique that apples to the above ideas is called the finite-element method.

FURTHER READING

Anderson, D. A., Tannehill, J. C., and Pletcher, R. H. (1984). *Computational Fluid Mechanics and Heat Transfer*, Hemisphere, New York.

Burgreen, D. (1971). *Elements of Thermal Stress Analysis*, C.P. Press, New York.

Chou, P. C. and Pagano, N. J. (1967). *Elasticity—Tensor, Dyadic, and Engineering Approaches*, Van Nostrand, Princeton, New Jersey.

Collatz, L. (1966). *Functional Analysis and Numerical Mathematics*, Academic, New York.

Crandall, S. H. (1956). *Engineering Analysis—A Survey of Numerical Procedures*, McGraw-Hill, New York.

Desai, C. S. and Abel, J. F. (1972). *Introduction to the Finite Element Method—A Numerical Method for Engineering Analysis*, Van Nostrand Reinhold, New York.

Fung, Y. C. (1965). *Foundations of Solid Mechanics*, Prentice-Hall, Englewood Cliffs, New Jersey.

Langhaar, H. L. (1962). *Energy Methods in Applied Mechanics*, Wiley, New York.

Little, R. W. (1973). *Elasticity*, Prentice-Hall, Englewood Cliffs, New Jersey.

Meirovitch, L. (1980). *Computational Methods in Structural Dynamics*, Sijthoff & Noordhoff, Maryland.

Patankar, S. V. (1980). *Numerical Heat Transfer Analysis and Fluid Flow*, Hemisphere, New York.

Reddy, J. N. (1993). *An Introduction to the Finite Element Method*, McGraw-Hill, New York.

Sokolnikoff, I. S. and Redheffer, R. M. (1958). *Mathematics of Physics and Modern Engineering*, McGraw-Hill, New York.

Sokolnikoff, I. S. (1956). *Mathematical Theory of Elasticity*, McGraw-Hill, New York.

Tong, P. and Rossettos, J. N. (1977). *Finite-Element Method—Basic Technique and Implementation*, MIT, Cambridge, Massachusetts.

PROBLEMS

Heat Transfer Problems

9-1H For the heat conduction problem given by Eq. (9-36), use the Rayleigh–Ritz method to find the temperature distribution in the cylinder T'. Use the following approximating functions:

(a) $T' = a_1 x(1 - x^2)$.
Answer: $a_1 = 5000/48$.
(b) $T' = a_1 x(1 - x^4)$.
(c) $T' = a_1 x(1 - x^3)$.
(d) $T' = a_1 \sin \pi x$.

The exact solution for this problem is $T(x) = (1000/12)x(1 - x^3)$.

9-2H Figure 9-10 shows a tapered bar that is held at a temperature T_b at one end, insulated at the other end, and is allowed convective heat transfer along its surface. By performing a heat balance within the

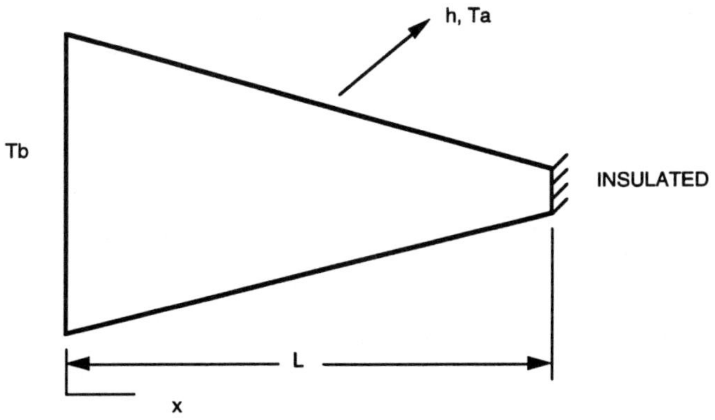

Figure 9-10 Tapered bar.

body, show that the governing differential equation becomes

$$\frac{\mathrm{d}}{\mathrm{d}x}\left(kA\frac{\mathrm{d}T}{\mathrm{d}x}\right) - hS(T - T_a) = 0,$$

where A is the cross-sectional area of the body, S is the surface area, k is the thermal conductivity, h is the convective coefficient, T_a is the ambient temperature, and T is the temperature distribution in the body.

For this equation, show that the functional becomes

$$I = \int_0^L \left\{ hST_aT - \frac{1}{2}hST^2 - \frac{1}{2}kA\left(\frac{\mathrm{d}T}{\mathrm{d}x}\right)^2 \right\}\mathrm{d}x - q_x AT\Big|_0^L$$

where

$$q_x = -k\frac{\mathrm{d}T}{\mathrm{d}x}$$

is the heat conduction in the body.

Figure 9-11 Square plate undergoing heat conduction.

9-3H Figure 9-11 shows a square plate undergoing heat conduction. The governing equation for this process is

$$-k\nabla^2 T = -k\left\{\frac{\partial^2 T}{\partial x^2} + \frac{\partial^2 T}{\partial y^2}\right\} = q_0,$$

subject to the boundary conditions

$$T = 0 \quad \text{on} \quad x = 1, \quad y = 1,$$

$$\frac{\partial T}{\partial n} = 0 \quad \text{at} \quad x = 0 \quad \text{and} \quad y = 0,$$

where n is the outward normal to the surface. Using the Rayleigh–Ritz method and an approximating function

$$T(x, y) = C_1(1 - x^2)(1 - y^2),$$

find the constant C_1.

$$\text{Answer:} \quad C_1 = \frac{5q_0}{16k}.$$

Variational Problems

9-1V For the differential equation

$$\frac{d^2 y}{dx^2} + 6y = 10x \qquad 0 \le x \le 2,$$

subject to the boundary conditions $y(0) = 1$ and $y(2) = 0$, show that the functional to be extremized is given by

$$I = \int_0^2 \left[3y^2 - 10xy - \frac{1}{2}\left(\frac{dy}{dx}\right)^2\right] dx.$$

9-2V For the differential equation

$$\frac{d^2 T}{dx^2} + 9T^2 = 5x^3 \qquad 0 \le x \le 5,$$

which is nonlinear and subject to the boundary conditions

$$T(0) = 9, \qquad T(5) = 0,$$

show that

$$I = \int_0^5 \left[3T^3 - 5x^3 T - \frac{1}{2}\left(\frac{dT}{dx}\right)^2\right] dx.$$

9-3V For the differential equation

$$\frac{d^4 y}{dx^4} + 8\frac{d^2 y}{dx^2} + 4y = 10 \qquad 0 \le x \le 5,$$

subject to the boundary conditions

$$y(0) = 0, \qquad y(5) = 100,$$

$$\frac{dy}{dx}(0) = 1, \qquad \frac{dy}{dx}(5) = 0,$$

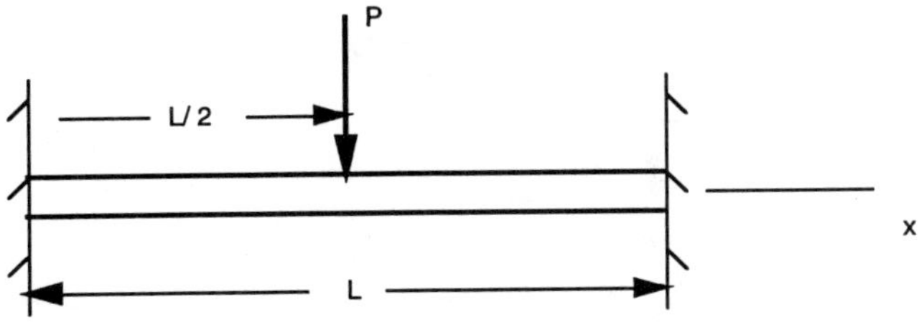

Figure 9-12 Fixed beam with length L subjected to a concentrated central load P.

show that

$$I = \int_0^5 \left[\frac{1}{2}\left(\frac{d^2y}{dx^2}\right)^2 - 4\left(\frac{dy}{dx}\right)^2 + 2y^2 - 10y \right] dx.$$

9-4V Show that

$$V = \int_V F(x, y, z, w, w_x, w_y, w_{xx}, w_{yy}, w_{zz}, w_{yz}, w_{zx}, w_{xy})\, dV$$

results (on minimization) in the Euler equation

$$\frac{\partial F}{\partial w} - \frac{\partial}{\partial x}\left(\frac{\partial F}{\partial w_x}\right) - \frac{\partial}{\partial y}\left(\frac{\partial F}{\partial w_y}\right) - \frac{\partial}{\partial z}\left(\frac{\partial F}{\partial w_z}\right) + \frac{\partial^2}{\partial x^2}\left(\frac{\partial F}{\partial w_{xx}}\right) + \cdots$$

$$+ \frac{\partial^2}{\partial y^2}\left(\frac{\partial F}{\partial w_{yy}}\right) + \frac{\partial^2}{\partial z^2}\left(\frac{\partial F}{\partial w_{zz}}\right) + \frac{\partial^2}{\partial y \partial z}\left(\frac{\partial F}{\partial w_{yz}}\right) + \frac{\partial^2}{\partial z \partial x}\left(\frac{\partial F}{\partial w_{zx}}\right)$$

$$+ \frac{\partial^2}{\partial x \partial y}\left(\frac{\partial F}{\partial w_{xy}}\right) = 0,$$

where integration by parts is carried out by use of the divergence theorem.

Rayleigh–Ritz Problems

9-1R Figure 9-12 shows a fixed beam of length L that is subjected to a concentrated central load P.

(a) Express the potential energy function that must be minimized for this beam.

(b) Show that the series

$$y = \sum_{n=2}^{\infty} A_n \left\{ \cos\frac{n\pi x}{L} - 1 \right\} \qquad n = 2, 4, 6, \ldots,$$

satisfies the boundary conditions at

$$x = 0, \qquad y = 0, \qquad \frac{dy}{dx} = 0,$$

$$x = L, \qquad y = 0, \qquad \frac{dy}{dx} = 0.$$

Take one term from the series and determine A_2 by the Rayleigh-Ritz method.

9-2R For the cantilever beam that is shown in Fig. 9-13.

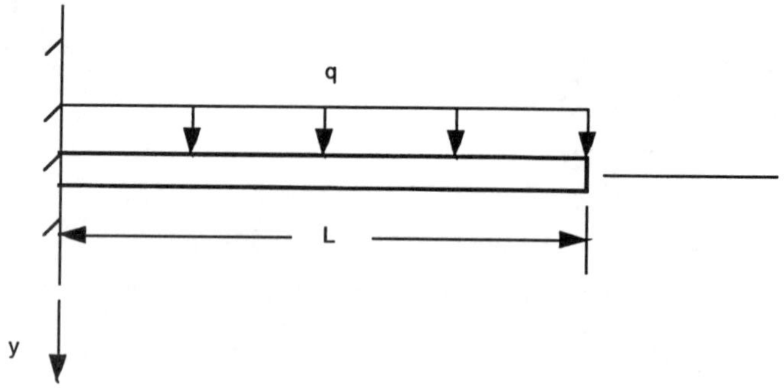

Figure 9-13 Cantilever beam.

(a) show that the infinite series

$$y(x) = \sum_{n=0}^{\infty} C_n \left\{ \left[\frac{2L}{(2n+1)\pi} \right]^3 \sin \frac{(2n+1)\pi x}{2L} + \frac{(-1)^n L x^2}{(2n+1)\pi} - \left[\frac{2L}{(2n+1)\pi} \right]^2 x \right\}$$

satisfies the physical boundary conditions

$$y = \frac{dy}{dx} = 0 \quad \text{at} \quad x = 0,$$

$$\frac{d^2 y}{dx^2} = \frac{d^3 y}{dx^3} = 0 \quad \text{at} \quad x = L.$$

(b) Determine C_0 by the Rayleigh–Ritz method using only one term of the series.
(c) Compare this solution with the exact solution

$$w = \frac{q}{24EI}(x^4 - 4Lx^3 + 6L^2x^2).$$

9-3R The simply supported beam that is shown in Fig. 9-14 is uniformly loaded. Using the Rayleigh-Ritz method, find the deflection $y(x)$ by considering

$$y = \sum_{0}^{\infty} A_n \sin \frac{n\pi x}{L}$$

for a one-term series, a two-term series, and a three-term series.
Why does the coefficient of $\sin[(2\pi x)/L]$ vanish?

$$\text{Answer:} \quad y = \frac{4pL^4}{\pi^5 EI} \left\{ \sin \frac{\pi x}{L} + \frac{1}{243} \sin \frac{3\pi x}{L} + \cdots + \right\}.$$

9-4R Figure 9-15 shows a cantlilever beam of length L that is subjected to a concentrated load P, at its end.

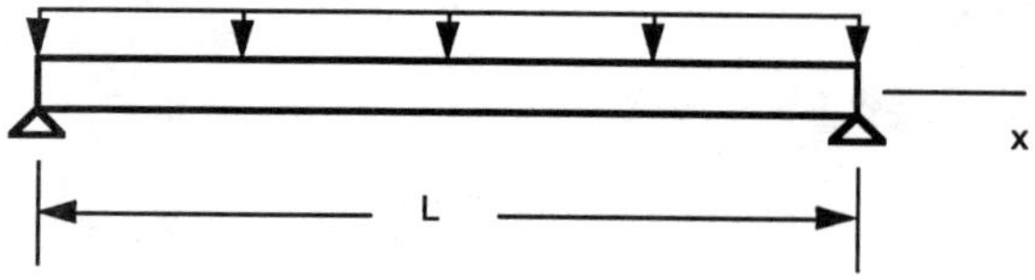

Figure 9-14 Simply supported beam.

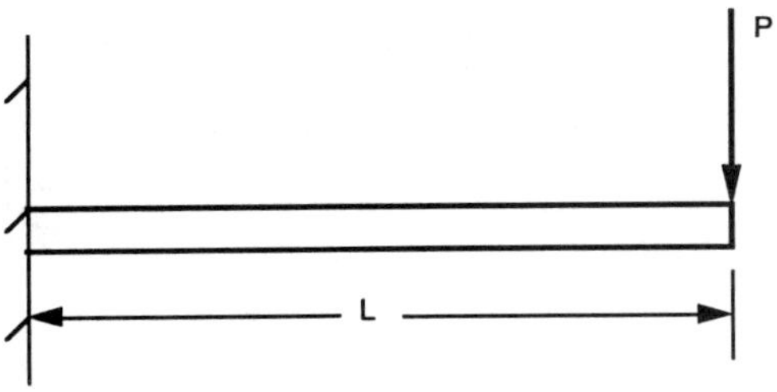

Figure 9-15 Cantilever beam of length L subjected to a concentrated load P.

(a) Derive an expression for the potential energy in the beam.
(b) Considering the approximating function

$$y(x) = a\left(1 - \cos\frac{\pi x}{2L}\right),$$

find the coefficient a by the Rayleigh–Ritz method. Compare your result with the exact solution.

$$\text{Answer: } a = \frac{32PL^3}{\pi^4 EI}.$$

9-5R For the simply supported beam shown in Fig. 9-16, which is loaded by a uniform load and subjected to the boundary conditions

$$y = \frac{d^2y}{dx^2} = 0 \quad \text{at} \quad x = 0, \quad x = L,$$

(a) find an approximation to the deflection of the beam $y(x)$ by using the Rayleigh–Ritz method and the approximating function

$$y(x) = C_1x + C_2x^2.$$

Compare your answer with the exact solution from the strength of materials.

$$\text{Answer: } C_1 = -7\alpha L, \qquad C_2 = -2\alpha, \quad \text{where} \quad \alpha = \frac{1}{180} f_0 L.$$

(b) Repeat part (a) for the approximating function

$$y = f_0 \sin\frac{\pi x}{L}.$$

9-6R The structure shown in Figure 9-17 is hanging under the influence of its own weight W and a concentrated load P. By doing a force balance on a section of the body, show that the equilibrium condition can be expressed as

$$\frac{d(\sigma A)}{dx} + bA = 0,$$

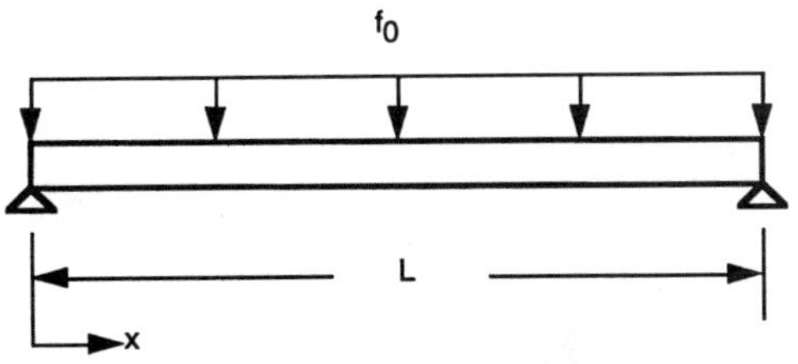

Figure 9-16 Simply supported beam.

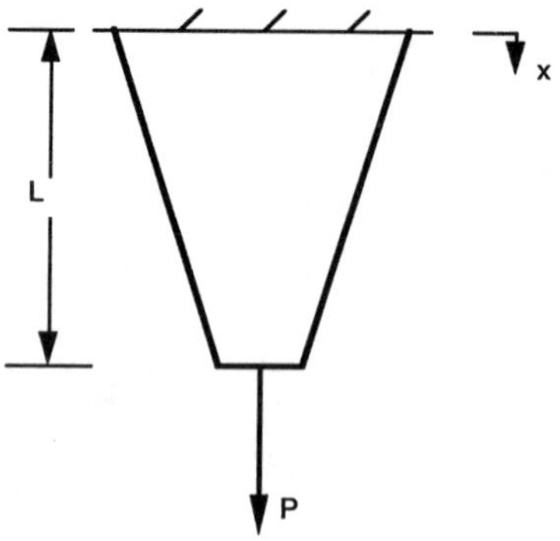

Figure 9-17

where b is the body force per unit volume, W is the weight such that $W = bA_{avg}L$, A is the cross-sectional area of the body, and σ is the stress in the body. Using the kinematic and constitutive conditions

$$\sigma = E\varepsilon = E\frac{du}{dx},$$

show that the equilibrium equation becomes

$$\frac{d}{dx}\left(AE\frac{du}{dx}\right) + bA = 0.$$

Finally, show that the functional for this problem is

$$I = \int_0^L \left\{ ubA - \frac{1}{2}AE\left(\frac{du}{dx}\right)^2 \right\} dx + Pu\Big|_0^L.$$

CHAPTER

TEN

FINITE-ELEMENT METHODS

NOMENCLATURE

ε	emissivity
ν	Poisson's ratio
Π	potential energy within a structure
ε_{ij}	strain component in a structure
σ_{ij}	stress component in a structure
A	cross-sectional area
E	modulus of elasticity of a material
h	coefficient of convective heat transfer
k_i	thermal conductivity in the ith coordinate direction
P	surface perimeter of a body
Q	heat source
q^s	surface heat transfer
T	temperature in a body
T_a	ambient temperature
T_r	receiver temperature for radiation heat transfer
u	dependent variable
W	work done by the external forces on a body
x	coordinate direction
y	coordinate direction and deflection of a structure

10-1 INTRODUCTION

We now turn our attention to a powerful numerical analysis technique known as the finite-element method (FEM). As noted in Chapter 9, the basis for this technique is the calculus of variations. This mathematical technique, when applied to physical problems, yields a procedure that has allowed engineers to solve problems that were intractable by the numerical methods that existed at the time that the FEM was first developed.[1]

The finite-difference techniques (described in Chapter 5) were used by scientists to solve the differential equations that resulted from the application of the physical principles of continuum mechanics to engineering problems. These equations, which are vectorial in nature, require the use of numerical approximations to describe the directional derivatives accurately. As a result, there is considerable difficulty in applying these techniques when the approximating grid is not chosen in the coordinate directions or when the boundaries of the body under examination do not coincide with the coordinate axes.

Because of the limitations of the vectorial techniques, it is important that energy techniques be examined to determine whether the limitations of finite-difference Techniques can be overcome. It was shown in Chapter 9 that, with the aid of the calculus of variations, we can develop energy techniques for both structural mechanics problems and for heat transfer problems that can be used to solve for the physical variables that constitute the solution to engineering problems. These techniques have the following advantages over finite-difference methods:

1. Irregular boundaries can be accurately described since the geometry is not dependent on rectangular grid approximations.
2. Boundary conditions can be described by the actual physical quantities rather than by derivatives.
3. The FEM, being scalar in nature, is not dependent on the directionality of the defining grid or discretization.

This chapter develops, in full, the finite-element procedures for both a structural analysis problem and for a problem in heat transfer. In addition, several examples of the proper usage of these techniques are demonstrated. In this textbook, all FEM problems are solved with a commercially available code, MARC.[2] It must be pointed out here that this code is used for demonstration purposes only. Any robust code such as ANSYS, ABAQUIS, etc., may be used with similarly accurate results.

10-2 PHILOSOPHY OF ELEMENT SELECTION

Before we begin to develop the mathematics that will create a typical element, it is useful to discuss the methods by which an element is chosen for use in a particular finite element analysis. An element in a FEM procedure is a region of the body in which the user makes an approximation as to the displacement field therein. As in the Rayleigh–Ritz method of Chapter 9, an approximating function is chosen for the displacements within the region. This function is applicable only within that particular element. Thus a different approximation can be selected for each element within the body. Since the displacement field is approximated within a given region, the size of the region is important and depends on how well the function approaches the actual (although unknown) physical displacements.

When one approaches the use of any robust finite-element package, the enormity of the number of different element types that are available to the user is intimidating. Therefore, a philosophical discussion regarding how one approaches the element choice is now given.

One must always remember that since the purpose of modeling is the prediction of actual physical phenomena, the finite-element simulation must give results that can be verified by experimentation. One must initially choose the element, then, by considering the nature of the geometry of the problem. In structural analysis, one encounters many geometric shapes such as beams, plates, three-dimensional structures, etc. Although all structures in nature are three dimensional, the theoretical development of the engineering formulation separates these structures into the above-mentioned categories. As an example, one need only consider the distinction between a solid three-dimensional structure (like a brick), a flat plate, and a beamlike structure (such as a fishing pole). The single parameter that differentiates these three structural types is known as the aspect ratio. The aspect ratio is the quotient between the largest dimension in the structure and the smallest. In a beam, for example, the aspect ratio between the length of the beam and any transverse dimension is large. Thus, for reasons that are explained below in this chapter, the element choice (for a finite-element simulation) must be one of the beam element types that are available in the computer package. Any other choice will lead to inaccurate results, regardless of the approximating function that is selected. In a plate, the ratio of a surface dimension to the plate thickness is large. Thus, for this type of structure, a plate element must be selected. Finally, in a bricklike structure, the aspect ratio tends to approach 1. In this case, a complete three-dimensional element (rather than a beam or a plate) should be chosen.

Having reduced the element choice possibilities by means of the above criteria, the user can now reduce the choices by considering the approximating function that will be used to formulate the element. In other words, each element uses a different algebraic function to approximate the dependent variables (usually the displacements in a structural member and the temperature in a member that is undergoing heat transfer) within the element, similar to the approximating function that was chosen in Section 9-5 for the Rayleigh–Ritz method. If one anticipates that the displacement, stress, and strain gradients are large (say near a crack), then either a high-order polynomial approximating function is required or small elements (with a low-order approximating polynomial) will be required. Depending on the choice of the user with respect to geometry, size, and approximation, the particular element choice may now be made. In Section 10-3 these trade-offs are explored, in depth, with the presentation of the construction of a sample finite element.

10-3 STRUCTURAL FINITE-ELEMENT ANALYSIS

Let us begin the presentation of element selection and construction by considering the arbitrary two-dimensional body shown in Fig. 10-1. By a two-dimensional structure is meant that one of the physical dimensions (the thickness in this case) is small relative to the other two dimensions. We now desire to choose an element type that will, at first,

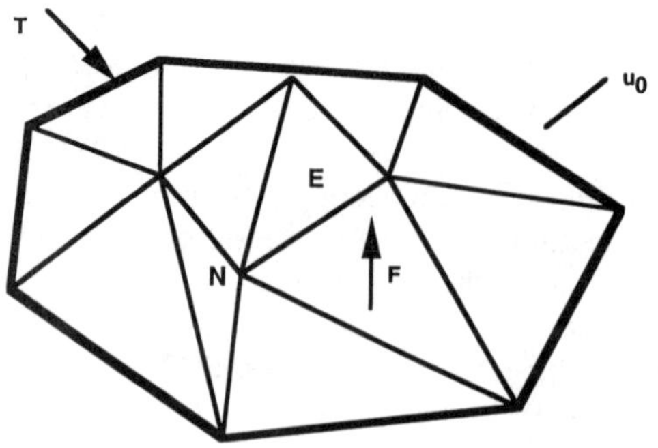

Figure 10-1 Element, nodes, and boundary conditions.

represent the geometric characteristics of the body. If one were to choose a bricklike element for this case, the aspect ratio would be large due to the surface area to thickness ratio. Instead, we choose an element type that already regards the thickness, not as a dimension that must be calculated, but as a parameter to be specified in the formulation. In other words, we specify that there are no displacement variations within the thickness. The problem is then reduced to 2 degrees of freedom, i.e., the displacement in the x and the y directions. The boundaries of the body are arbitrarily constructed and the structure is subjected to the following boundary conditions:

1. Tractions T are applied to specified portions of the boundary of the structure.
2. Displacements u_0 are applied to the remaining portions of the boundary.
3. Internal body forces F (such as centrifugal forces and gravitational forces) are distributed within the structure.

When this body is loaded, as shown, it will deform. Each point within the body will have a horizontal displacement $u(x, y)$, a vertical displacement $v(x, y)$, as well as an internal stress state and strain state, σ_{ij} and ε_{ij}, respectively. We want to solve for these state variables within the body.

Let us now imagine that this structure is subdivided into small areas, called elements, that are connected to each other at points that are called nodes. It is now our job to find a function (or a set of functions) that can be used to approximate the deformation field (u, v) in each of these elements. We use the methods of Chapter 9 to accomplish our task.

Before proceeding directly to the task of creating a finite-element methodology for the above problem, let us once again demonstrate that by minimizing the appropriate functional, the exact solution to the structural problem is attained. To review, the governing energy function is the potential energy defined as

$$\text{potential energy} = \text{strain energy—work done by the Forces.}$$

If we choose $u_i(u_1 = u, u_2 = v)$ as the displacements in the body in the x and the y directions, respectively, the potential energy can be written as

$$I(u_i) = \int_A W(u_i)\, dA - \int_A F_i u_i\, dA - \oint_C T_i u_i\, dS, \tag{10-1}$$

where W is the strain energy density, T_i are the boundary forces, and F_i are the body forces.

If we apply the theory of elasticity to Eq. (10-1), the potential energy becomes

$$I(u_i) = \int_A \left[\frac{1}{2}\lambda \left(\frac{\partial u}{\partial x} + \frac{\partial v}{\partial y} \right)^2 + \mu \left\{ \left(\frac{\partial u}{\partial x} \right)^2 + \left(\frac{\partial v}{\partial y} \right)^2 + \frac{1}{2} \left(\frac{\partial u}{\partial y} + \frac{\partial v}{\partial x} \right)^2 \right\} - F_x u - F_y v \right] \mathrm{d}A - \oint_C (T_x u + T_y v)\, \mathrm{d}S, \tag{10-2}$$

where λ and μ are elastic material constants.

If we now apply the exact minimization process to this functional (as discussed in Chapter 9), we achieve the Euler equations:

$$\mu \nabla^2 u + (\lambda + \mu) \frac{\partial}{\partial x} \vartheta + F_x = 0, \tag{10-3}$$

$$\mu \nabla^2 v + (\lambda + \mu) \frac{\partial}{\partial y} \vartheta + F_y = 0, \tag{10-4}$$

where the dilatation $\vartheta = \varepsilon_{xx} + \varepsilon_{yy}$. Equations (10-3) and (10-4) are known to be the exact equations for the solution of structural analysis problems and are known as the Navier equations. Thus the functional relationship given by Eq. (10-2) is the energy equation that now forms the basis for the development of a FEM for a structural member.

The choice of the approximating function for the displacements is not an easy one. It is worthwhile to discuss the trade-offs that must be made in the selection. In Chapter 4 we discussed the methodology for the approximation of functions that accurately represent a given algebraic curve. Some of the same ideas are relevant here. Basically, there is a definite correlation between

(a) The size of the discretization (elements)
(b) The number of elements
(c) The number of nodes
(d) The engineering development time to create a FEM formulation.

Since the nature of the displacement functions in the body is unknown *a priori*, there is usually no physical criterion to guide the element selection based on the quantities (a)–(d). If a simple approximating function, such as a linear equation, is chosen for u and v, then the size of the elements will have to be small so that a linear approximation will accurately represent the actual displacements within the region of the body that is covered by the element. This choice will therefore require a large number of elements and nodes for completely describing the geometry. The engineering effort that is necessary to develop the formulation of a linear element will be small. However, because the large number of elements that would be required, the size of the resulting matrix will be very large. The time of matrix inversion and subsequent solution, as discussed in Chapter 1, will be increased. This will result in large computational times for a given problem. Since engineering cost is directly related to the computational time, the choice of this simple element will be borne each time that the analysis (with this element) is conducted. On the other hand, if a more complex functional approximation is chosen for the displacements, the engineering costs in element development time will increase. This, however, will be repaid during each analysis. Because of the ability of the engineer to choose a large grid size, a decreased number of elements will be required, and as a result, reduced computational time will be needed for matrix inversion and solution.

For a general structural problem, there is usually no guide to element selection or to discretization size. In some special cases, however, the nature of the problem will act as a guide to the selection of the finite-element methodology. For example, in problems that involve cracks that are embedded within the structure, the physics will require the choice of small elements near the crack for determining the resulting large stress gradients. The element size may then be allowed to grow as one moves away from the crack. Thus an irregular grid size would be recommended for crack problems. In areas where sharp corners exist in the structure, small element sizing (near the irregularity) would also be advised.

We now see that the nature of the discretization and element selection is an art form to which prior knowledge of the particular problem and other user experiences are brought to bear. In most engineering applications, a nonuniform grid is usually chosen.

With these thoughts in mind, let us now turn our attention, once again, to the two-dimensional structural problem shown in Fig. 10-1. We now develop the formulation for a typical element in the discretized body. We imagine that each element is independent of the others and that the formulation of a particular element may be made without regard to any other element. When we are finished with the development process, we will discuss the procedure of recombining the elements to recreate the original structure.

For demonstration purposes, let us choose the least complex approximating function for a two-dimensional structural element, i.e., a "linear" function to simulate the displacement distribution. Figure 10-2 shows a triangular element that has been separated from the body. In the process of choosing the mathematical form for the element

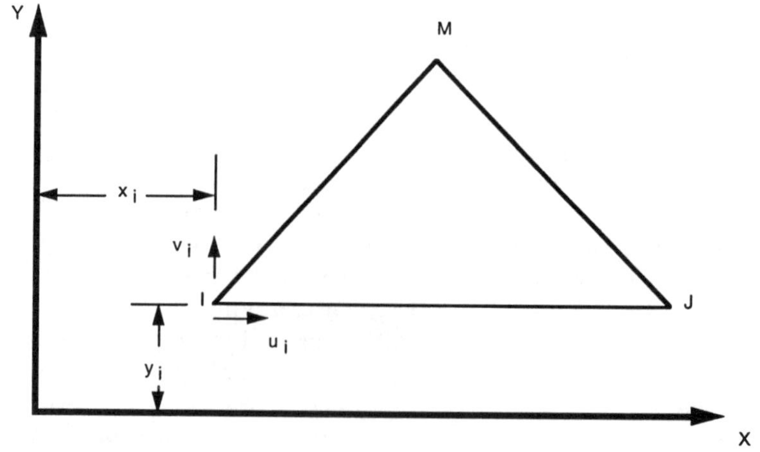

Figure 10-2 Two-dimentional element.

approximation, we must consider the number of degrees of freedom at a node and the number of nodes in the chosen element. In this case, there are 2 degrees of freedom (u, v) at each node and there are three nodes in the triangular element. Thus there are six unknown values (also known as degrees of freedom) that must be determined by the approximating function. Therefore, for this case, we choose

$$u = \alpha_1 + \alpha_2 x + \alpha_3 y, \tag{10-5}$$

$$v = \alpha_4 + \alpha_5 x + \alpha_6 y, \tag{10-6}$$

as the approximation of the displacements within the element.

Equations (10-5) and (10-6) represent linear approximating functions for the displacements within this triangle only. There are six unknown constants $\alpha_1, \ldots, \alpha_6$. As noted above, because of the simple nature of the approximating function, a large number of small elements will have to be chosen within the entire body to describe the displacement field accurately. Equation (10-5) can now be evaluated at the nodal points I, J, and M (of Fig. 10-2), viz.,

$$u_I = \alpha_1 + \alpha_2 x_I + \alpha_3 y_I, \tag{10-7}$$

$$u_J = \alpha_1 + \alpha_2 x_J + \alpha_3 y_J, \tag{10-8}$$

$$u_M = \alpha_1 + \alpha_2 x_M + \alpha_3 y_M, \tag{10-9}$$

or

$$\begin{vmatrix} 1 & x_I & y_I \\ 1 & x_J & y_J \\ 1 & x_M & y_M \end{vmatrix} \begin{vmatrix} \alpha_1 \\ \alpha_2 \\ \alpha_3 \end{vmatrix} = \begin{vmatrix} u_I \\ u_J \\ u_M \end{vmatrix},$$

which can be expressed in matrix notation as

$$[\mathbf{A}][\alpha] = [\mathbf{b}]. \tag{10-10}$$

Solving for the unknown constants α and noting that the area of the triangle Δ is related to the determinant of matrix $\mathbf{A}$, we get

$$\alpha = A^{-1}b,$$

$$\det A = 2\Delta = \{x_J y_M - x_M y_J\} - \{x_I y_M - x_M y_I\}$$
$$+ \{x_I y_J - x_J y_I\} = 2 * (\text{area of triangle}).$$

The unknown coefficients α_1, α_2, and α_3 are thereby expressed in terms of the nodal coordinates and the nodal displacements. By inserting the expressions for α_1, α_2, and α_3 into Eq. (10-5), we can now express the horizontal displacement u at any point within the element in terms of the displacements at the nodal points, viz.,

$$u = \frac{1}{2\Delta}\{(a_I + b_I x + c_I y)u_I + (a_J + b_J x + c_J y)u_J$$
$$+ (a_M + b_M x + c_M y)u_M\}, \tag{10-11}$$

where

$$
\begin{aligned}
a_I &= x_J y_M - x_M y_J,\\
a_J &= x_M y_I - x_I y_M,\\
a_M &= x_I y_J - x_J y_I,\\
b_I &= y_J - y_M,\\
b_J &= y_M - y_I,\\
b_M &= y_I - y_J,\\
c_I &= x_M - x_J,\\
c_J &= x_I - x_M,\\
c_M &= x_J - x_I.
\end{aligned}
$$

If we perform a similar analysis on the vertical displacement v, we get

$$
\begin{aligned}
v = \frac{1}{2\Delta}\{&(a_I + b_I x + c_I y)v_I + (a_J + b_J x + c_J y)v_J\\
&+ (a_M + b_M x + c_M y)v_M\}.
\end{aligned}
\tag{10-12}
$$

Using matrix notation, we can express Eqs. (10-11) and (10-12) as

$$
|u| = \left\{ \begin{matrix} u \\ v \end{matrix} \right\} = [\mathbf{D}]|q_n|,
\tag{10-13}
$$

where

$$
|q_n^T| = \{u_I \quad v_I \quad u_J \quad v_J \quad u_M \quad v_M\},
$$

$$
[\mathbf{D}] = \begin{bmatrix} f_1 & 0 & f_2 & 0 & f_3 & 0 \\ 0 & f_1 & 0 & f_2 & 0 & f_3 \end{bmatrix},
$$

$$
f_i(x, y) = \frac{a_i + b_i x + c_i y}{2\Delta} \qquad (i = 1, 2, 3).
$$

The subscripts (1, 2, 3) are directly related to the nodal points as $1 = I$, $2 = J$, and $3 = M$.

Since the displacements are now defined within the entire element in terms of the nodal displacements, we may calculate the strain within that element by using the strain-displacement relations from the theory of elasticity, i.e.,

$$
|\varepsilon| = \left\{ \begin{matrix} \varepsilon_x \\ \varepsilon_y \\ \gamma_{xy} \end{matrix} \right\} = \begin{bmatrix} \dfrac{\partial}{\partial x} & 0 \\ 0 & \dfrac{\partial}{\partial y} \\ \dfrac{\partial}{\partial y} & \dfrac{\partial}{\partial x} \end{bmatrix} \left\{ \begin{matrix} u \\ v \end{matrix} \right\} = [\mathbf{E}]|q_n|,
\tag{10-14}
$$

where

$$
[\mathbf{E}] = \frac{1}{2\Delta} \begin{bmatrix} b_1 & 0 & b_2 & 0 & b_3 & 0 \\ 0 & c_1 & 0 & c_2 & 0 & c_3 \\ c_1 & b_1 & c_2 & b_2 & c_3 & b_3 \end{bmatrix}.
$$

It may now be pointed out that since the **E** matrix and the q_n vector are constant (since they are in terms of geometric parameters that will not vary), the choice of the linear approximating function yields a finite element that has a constant strain within the entire element. This type of element is called a constant stress–constant strain element and has the tetrahedron as its three-dimensional counterpart. This result emphasizes, once again, the need for selecting small element sizes within the original structural member when this type of element is chosen. Since, in general, the stress and the strain vary with position, this element describes the actual physical behavior only when it covers a very small region of the body.

With the application of a constitutive behavior law, the constant stress within the element becomes

$$[\sigma] = \begin{Bmatrix} \sigma_x \\ \sigma_y \\ \tau_{xy} \end{Bmatrix} = [\mathbf{C}][\varepsilon]. \tag{10-15}$$

In an isotropic material in plane stress,

$$[\mathbf{C}] = \frac{E}{1-\nu^2}\begin{bmatrix} 1 & \nu & 0 \\ \nu & 1 & 0 \\ 0 & 0 & \frac{1-\nu}{2} \end{bmatrix}.$$

Note that this set of geometric and material properties applies only to this element. Another element (in the same body) can possess its own unique properties. It should be apparent that, in using the FEM, nonhomogeneous, anisotropic materials do not complicate the element formulation as would occur in the use of the finite-difference procedures.

The potential energy, which was derived in Eq. (10-2), becomes, for this approximating function,

$$\Pi = \int_V \left[\frac{1}{2}\sigma_{ij}\varepsilon_{ij} - F_i u_i\right] \mathrm{d}V - \int_S T_i u_i \,\mathrm{d}s \tag{10-16}$$

or

$$\Pi = \int_V \left[\frac{1}{2}[\varepsilon]^T[\sigma] - |u|^T|F|\right] \mathrm{d}V - \int_S |u|^T T \,\mathrm{d}S. \tag{10-17}$$

Substituting Eqs. (10-13)–(10-15) into Eq. (10-17), we get

$$\Pi_n = \frac{1}{2}q_n^T\left\{\int_V [\mathbf{E}]^T[\mathbf{C}][\mathbf{E}]\,\mathrm{d}V\right\}q_n - q_n^T\left\{\int_V [\mathbf{D}]^T|F|\,\mathrm{d}V + \int_S [\mathbf{D}]^T|T|\,\mathrm{d}S\right\} \tag{10-18}$$

or

$$\Pi_n = \frac{1}{2}q_n^T K_n q_n - q_n^T Q_n, \tag{10-19}$$

where

$$K_n = \int_V [\mathbf{E}]^T[\mathbf{C}][\mathbf{E}]\,\mathrm{d}V \tag{10-20}$$

is the stiffness matrix of the element, and

$$Q_n = \int_V [\mathbf{D}]^T [\mathbf{F}]\,\mathrm{d}V + \int_S [\mathbf{D}]^T [\mathbf{T}]\,\mathrm{d}S \tag{10-21}$$

is the total force acting on the element.

The theoretical development for a typical element is now complete. The steps that were used in the elemental development process for this element can be followed regardless of the mathematical nature of the approximating functions. When all the elements in the structure are summed, the potential energy in the entire body becomes

$$\Pi = \sum_{\text{all elements}} \left[\frac{1}{2} q_n^T K_n q_n - q_n^T Q_n\right]. \tag{10-22}$$

If, for the triangular element, the thickness t is constant, then

$$K_n = \int_V [\mathbf{E}]^T [\mathbf{C}][\mathbf{E}] t\,\mathrm{d}x\,\mathrm{d}y.$$

If, in addition, we allow the individual triangle to be homogeneous, then the matrices **E** and **C** are constant over the entire area. The stiffness reduces to

$$K_n = tA[\mathbf{E}]^T [\mathbf{C}][\mathbf{E}],$$

where A is the area of the triangle.

It is important to point out that Eq. (10-20) is unique, in that all the terms under the integral are constant. In general, an actual integration of the matrix structure that defines the stiffness matrix would be necessary. In those cases, the use of the Gauss quadrature approach, which was presented in Chapter 4, is an excellent choice for performing the indicted integration in an exact manner.

If we now minimize the functional given by Eq. (10-19), we get

$$\mathrm{d}\Pi = 0 \rightarrow \left(\frac{1}{2} q_n^T K_n\right)^T + \frac{1}{2} k_n q_n - Q_n = 0. \tag{10-23}$$

Since

$$(K_n q_n)^T = q_n^T K_n^T = q_n^T K_n,$$
$$(q_n^T K_n)^T = K_n^T q_n = K_n q_n,$$

Eq. (10-23) becomes

$$K_n q_n = Q_n. \tag{10-24}$$

Equation (10-24) represents the matrix equation. K_n is the stiffness matrix of the body within the element domain, q_n is the vector that represents the displacements at the nodes, and Q_n is the applied force in the element. We note that Eq. (10-24) has the form of a spring with an applied force. The finite-element formulation has the effect of transforming a solid body into a spring of stiffness K_n.

A methodology must now be developed that recombines all the elements into the original body so that the overall matrix for the body can be derived. Consider the two

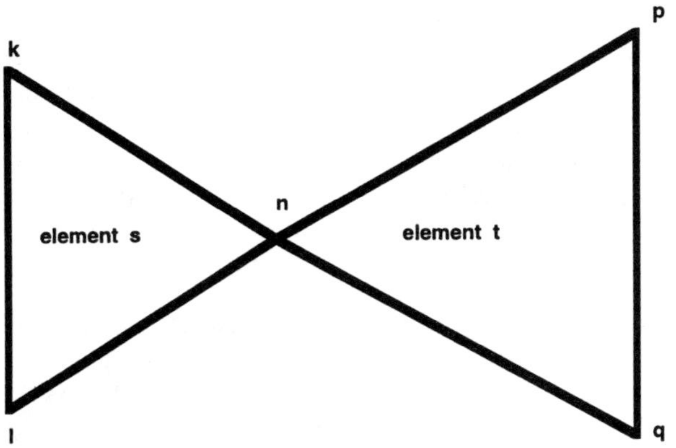

Figure 10-3 Combination of two elements.

triangular elements that are shown in Fig. 10-3. Element s and element t share a common node n at which point they will be attached.

For element $s(k, l, n)$ the finite-element construction will produce

$$|F|_s = [\mathbf{K}]_s |\delta|_s. \tag{10-25}$$

Similarly, for element $t(n, p, q)$ the methodology yields

$$|F|_t = [\mathbf{K}]_t |\delta|_t. \tag{10-26}$$

We have arbitrarily chosen node n to be the third node of element s and the first node of element t. Since there are two degrees of displacement freedom in each element, rows 5 and 6 of Eq. (10-25) relate to node n. Rows 1 and 2 of Eq. (10-26) relate to node n. The two elements will be joined physically by satisfaction of the continuity of the displacements at node n and the equilibrium condition at node n. These can be expressed as

$$F_x^n = F_1^t + F_5^s, \tag{10-27}$$

$$F_y^n = F_2^t + F_6^s, \tag{10-28}$$

$$\delta_5^s = \delta_1^t, \tag{10-29}$$

$$\delta_6^s = \delta_2^t, \tag{10-30}$$

where the superscripts indicate the element and the subscripts represent the vector positions.

Using Eqs. (10-25) and (10-26), we get

$$F_5^s = K_{51}^s\delta_1^s + K_{52}^s\delta_2^s + K_{53}^s\delta_3^s + K_{54}^s\delta_4^s + K_{55}^s\delta_5^s + K_{56}^s\delta_6^s, \tag{10-31}$$

$$F_1^t = K_{11}^t\delta_1^t + K_{12}^t\delta_2^t + K_{13}^t\delta_3^t + K_{14}^t\delta_4^t + K_{15}^t\delta_5^t + K_{16}^t\delta_6^t, \tag{10-32}$$

and substituting Eqs. (10-29)–(10-32) into Eq. (10-27), we get

$$\begin{aligned} F_x^n = {} & K_{51}^s\delta_1^s + K_{52}^s\delta_2^s + K_{53}^s\delta_3^s + K_{54}^s\delta_{54}^s + \left(K_{55}^s + K_{11}^t\right)\delta_1^t \\ & + \left(K_{56}^s + K_{12}^t\right)\delta_2^t + K_{13}^t\delta_3^t + K_{14}^t\delta_4^t + K_{15}^t\delta_5^t + K_{16}^t\delta_6^t. \end{aligned} \tag{10-33}$$

In the same manner, we can derive an expression for F_y^n. When this is done, elements t and s have been rejoined in an attempt to recreate the entire structure. Following this process of applying continuity and equilibrium to all of the common nodes within the body, we rebuild the entire structure. When the process is completed, only the boundary nodes are available for the application of displacement or force boundary conditions. The entire force/displacement relationship for the body can now be expressed by

$$F_{\text{boundary}} = K_{\text{entire body}}\, \delta_{\text{boundary}}. \tag{10-34}$$

Equation (10-34) represents the matrix formulation of the structural problem. At any of the boundary nodes, either the displacement or the load is known. The remaining unknown quantity (load or displacement) can be calculated at the boundary nodes by use of the matrix inversion techniques of Chapter 1. The displacements at the interior nodes, the stresses in the elements, and the strains in the elements are calculated with equations that have previously been derived [e.g., Eqs. (10-11), (10-12), (10-14), (10-15), etc.].

This formulation has proceeded with a two-dimensional triangular element and a linear displacement function to formulate the two-dimensional elasticity problem. Of course, had the problem been three dimensional, the equivalent element would have been the tetrahedron (as noted above). The approximating function could still be linear, except the displacements (u, v, w) would now be functions of (x, y, z).

10-4 OTHER STRUCTURAL FINITE-ELEMENT METHOD APPROXIMATIONS

Before we consider a structural example in which the body is discretized, analyzed, and evaluated by the methods just explained, let us now return to the discussion of element choice and element type that are included in a finite-element package.

The consideration of aspect ratio was given in Section 10-2 as one of the criteria for the selection of a particular type of element. Referring to the example in Section 10-3, we can see that equations (10-11), (10-12), (10-14), etc., which are part of the formation of the stiffness matrix **K**, are functions of the distances between nodal points in the element. Thus, if the aspect ratio of an element were large, there would be large discrepancies in the nodal distances. The inclusion of these values into the stiffness matrix would then result in an illconditioned stiffness matrix that would have to be eventually inverted in order to solve for the unknown values of the displacements. In Chapter 1 the consequences of inverting a set of equations whose matrix is ill conditioned are discussed. The solution results are generally extremely poor. Thus, in the above example, had we chosen a three-dimensional brick element to solve the structural problem of the thin plate, our element would have either had a large aspect ratio, or the other dimensions of the element would have been chosen as very small (in order to be of the same order as the thickness). In the latter case, an excessive amount of elements would have been required for creating the entire body. This would have resulted in a huge matrix that must be inverted. Considering that the size of the final matrix is given by the formula

$$\text{N size} = (\text{number of nodes}) \times (\text{degrees of freedom per node}),$$

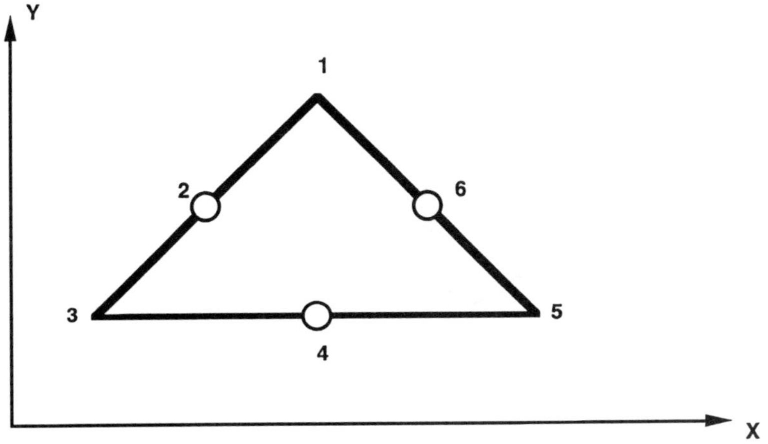

Figure 10-4 Higher-order two-dimensional element.

then by choosing an element type (as in Section 10-3) that already had the small thickness in its construction, we could ignore any displacement variation in the transverse direction and reduce the number of degrees of freedom at each node to two instead of the three that would have resulted if a brick were chosen. We have thus reduced the size of the resultant matrix and have eliminated the need to consider the aspect ratio problem.

In order to complete the discussion of other element types, we will now present a sample of other element choices and their application.

1. Figure 10-4 shows a two-dimensional plane triangular element that has six nodal points, each possessing 2 degrees of freedom, (u, v). Here again, the thickness is considered small relative to the other dimensions. Since we have 12 degrees of freedom, rather than 6 in the above example, we must use higher-order approximating functions to describe the displacements within the element, viz.,

$$u = a_1 + a_2x + a_3y + a_4x^2 + a_5y^2 + a_6xy, \tag{10-35}$$

$$v = b_1 + b_2x + b_3y + b_4x^2 + b_5y^2 + b_6xy. \tag{10-36}$$

   Since Eqs. (10-35) and (10-36) are quadratic polynomial functions, they approximate the displacement functions over a larger region than the above-used linear functions. Thus, larger (and thereby a smaller number of) elements can be used to construct a body with the same accuracy that resulted from a finer mesh in the linear formulation. The derivation of the element formulation will, in this case, require a bit more mathematical effort than the above-shown linear element, but the savings in computational time will soon make the choice of this element the more desirable.
2. Figure 10-5 shows a tetrahedral element that is the counterpart to the triangular element in Section 10-3. Since we have four nodes, each having 3 degrees of freedom (u, v, w), there are 12 degrees of freedom that must be evaluated. Choosing the linear

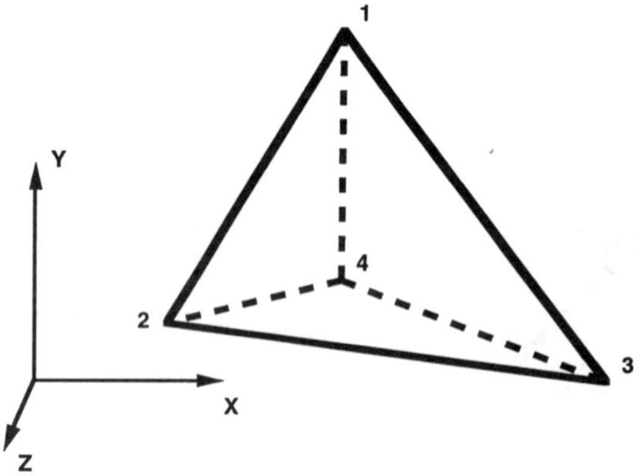

Figure 10-5 Tetrahedral finite element.

functions for the displacement approximations, viz.,

$$u = a_1 + a_2x + a_3y + a_4z, \tag{10-37}$$

$$v = b_1 + b_2x + b_3y + b_4z, \tag{10-38}$$

$$w = c_1 + c_2x + c_3y + c_4z \tag{10-39}$$

we can construct a constant stress–constant strain element by using the techniques that are described above.

3. Figure 10-6 shows a beam element that can be used to construct all or part of a structure that has the geometric characteristics of a beam. By this is meant that the region under consideration has a significant dimension that is much larger than the two remaining transverse directions. Choosing a bricklike element for this geometry would create aspect ratio problems. Instead, the choice of an element that ignores displacement variations in the transverse directions will lead to a candidate for finite-element simulation. The only independent variable in this case is the distance along the beam, which is called the arc length ζ. Since the element has three nodes, each

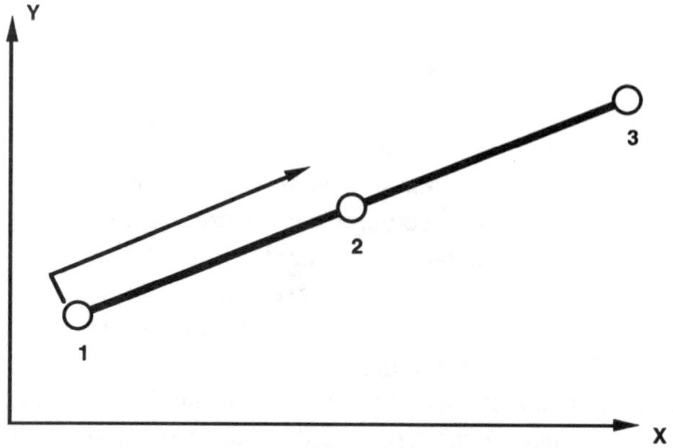

Figure 10-6 Beam element with three nodes.

having 3 degrees of freedom for its displacements, the approximating functions are

$$u = a_0 + a_1\zeta + a_2\zeta^2, \tag{10-40}$$

$$v = b_0 + b_1\zeta + b_2\zeta^2, \tag{10-41}$$

$$w = c_0 + c_1\zeta + c_2\zeta^2. \tag{10-42}$$

These quadratic functions will create a beam element that can be used in regions where the structure has a rodlike geometry and has only small curvatures.

Four examples (the original model and the subsequent three element types) have been shown that form only part of a complete finite-element analysis package. Many more beam elements, with different approximating functions, are usually available. Similarly, many plate, two-dimensional, and three-dimensional elements are contained within a good finite-element code. Finally, in Section 10.6, the element types that support heat transfer analysis are presented. A discussion of the various heat elements is reserved until that presentation.

10-5 FINITE-ELEMENT STRUCTURAL EXAMPLE

Figure 10-7 shows a tensile bar that is subjected to a free-end displacement of 0.015 in. The material properties of the steel bar are modulus of elasticity $E = 30(10^6)$ psi, density $\rho = 0.278$ lb/in., and Poisson's ratio $\nu = 0.3$.

With the MARC finite-element code, the data for this problem called **tdisp.dat**, were constructed and are shown in Appendix 10.1. As discussed above, any analysis must contain the following basic ingredients.

1. A geometric description section that contains the physical locations.
2. A material description.
3. Boundary conditions.

In the **tdisp.dat** file, the geometric descriptions are contained under the categories of Connectivity and Coordinates. The first provides the nodal connections that constitute each element for a three-dimensional bricklike element (called element 7, in MARC). The Coordinates section contains the physical locations of the nodal points in a global coordinate system. The material properties section called Isotropic defines the material

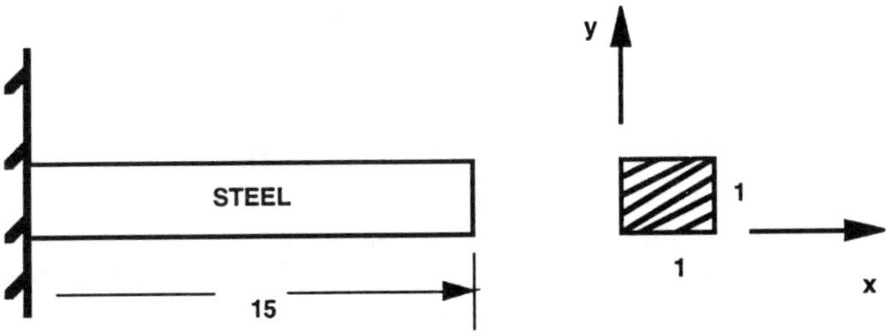

Figure 10-7 Tensile bar with an end displacement.

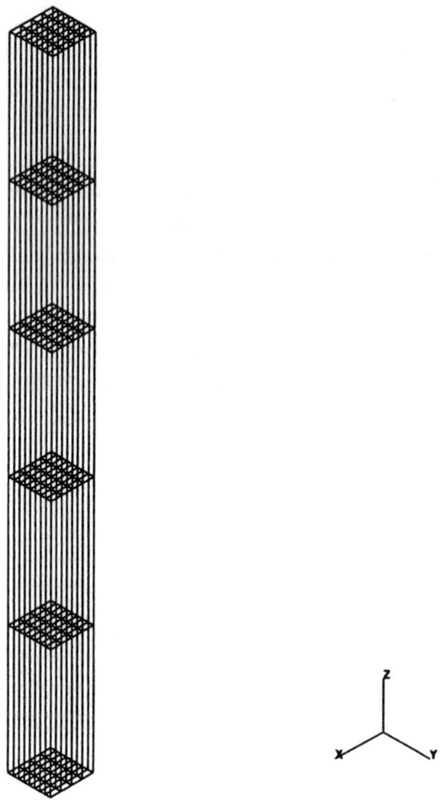

Figure 10-8 Basic finite-element model of a tensile bar.

behavior for the resulting 160 elements in the model. The boundary conditions are given under the section called Fixed Disp. The left end of the rod is restrained in the axial direction while the right end is given the displacement 0.015. Care is taken, however, not to overconstrain the rod in the transverse direction so as to allow for the Poisson effect in the bar. All other data in Appendix 10.1 are for input/output purposes or are required by MARC and unique only to this code.

Figure 10-8 shows the basic finite-element model that was used for this analysis. An 8×4 mesh was used for the cross-sectional area, while five sets of solid elements were used along the axial direction. 160 elements were thus created. In this problem, the stress and the strain are uniform throughout the bar while the displacements vary linearly. Thus, for this problem, almost any discretization, regardless of its crudity, would have yielded an accurate solution. Figure 10-9 verifies that the transverse displacement in the x direction is uniform in the axial direction and is a function of the transverse position only. The displacement in the y direction (not shown) has similar characteristics. Figure 10-10 shows that the displacement in the z direction is a function of the axial position but does not vary in the transverse direction. All these results are to be anticipated when one

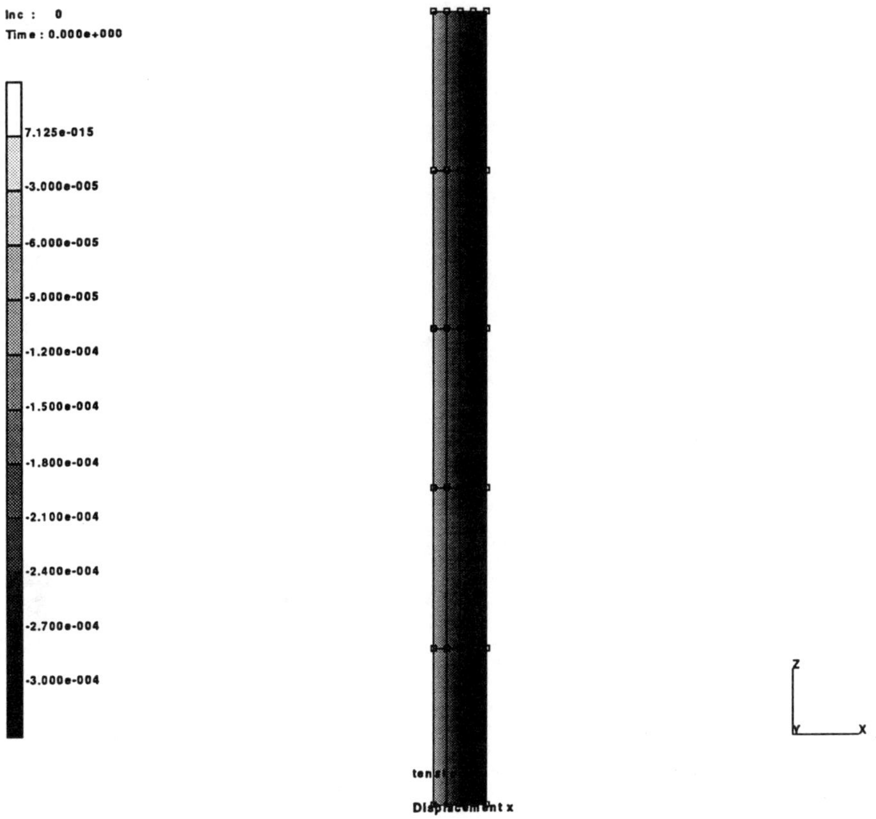

Figure 10-9 X-displacement in the tensile bar.

considers the physics of this problem. Additionally, Fig. 10-11 verifies that the axial stress in the entire rod is uniform.

The problem that was chosen in this section was for demonstration purposes only. Problems that will challenge readers are included in the Problems section of this chapter.

10-6 THERMAL FINITE-ELEMENT ANALYSIS

We now turn our attention to the finite-element modeling of problems that involve heat transfer. We confine our discussion to the transfer of heat due to conduction, convection, and radiation. Any effects on the heat transfer process due to fluid flow are not within the scope of this section.

The differential equation that governs the steady-state conduction of heat in two dimensions is

$$\frac{\partial}{\partial x}\left(k\frac{\partial T}{\partial x}\right) + \frac{\partial}{\partial y}\left(k\frac{\partial T}{\partial y}\right) + Q = 0, \qquad (10\text{-}43)$$

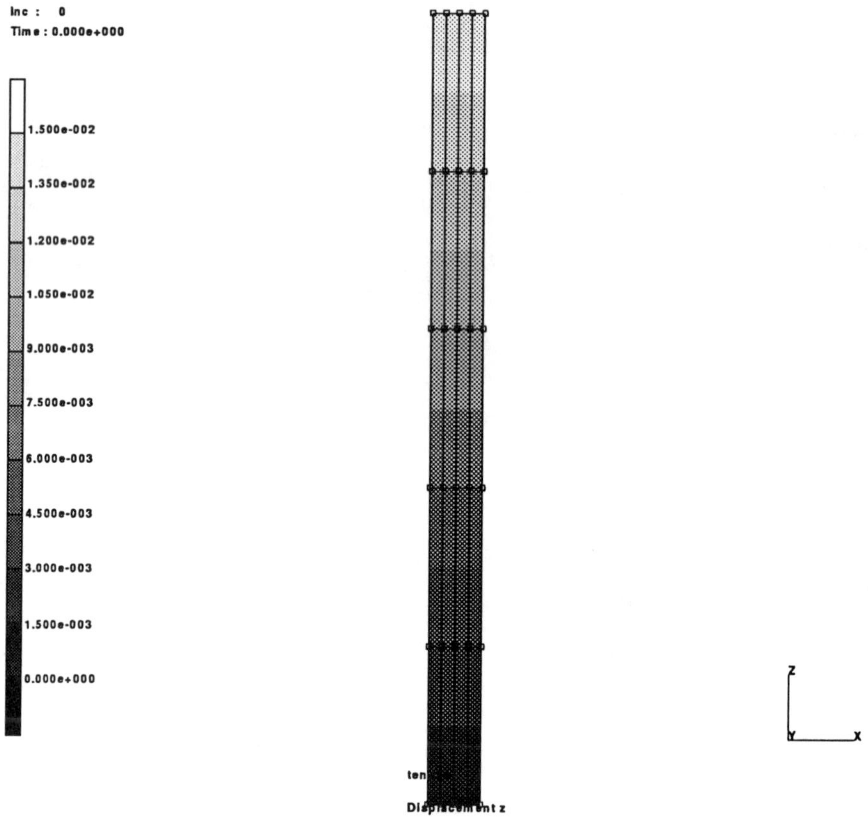

Figure 10-10 Z-displacement in the tensile bar.

where k is the thermal conductivity, T is the temperature, x and y are coordinate directions, and Q is the external heat source. As shown in Chapter 9, the governing functional equation that must be minimized for this problem becomes

$$I = \int_A \left[-\frac{1}{2} \left\{ k_x \left(\frac{\partial T}{\partial x} \right)^2 + k_y \left(\frac{\partial T}{\partial y} \right)^2 \right\} + QT \right] \mathrm{d}A - \int_S q_{sb} T \, \mathrm{d}S, \qquad (10\text{-}44)$$

where k_x is the thermal conductivity component in the x direction, k_x is the thermal conductivity component in the y direction, A is the cross-sectional area of the region, S is the bounding curve of the region, and q_{sb} is the heat that flows across the boundary of the region.

In this problem the temperature is the only dependent variable. If we assume that the physical region is, once again, shown in Figs. 10-1 and 10-2, the development of the finite-element construction that was presented in Section 10-3 for a structure is directly

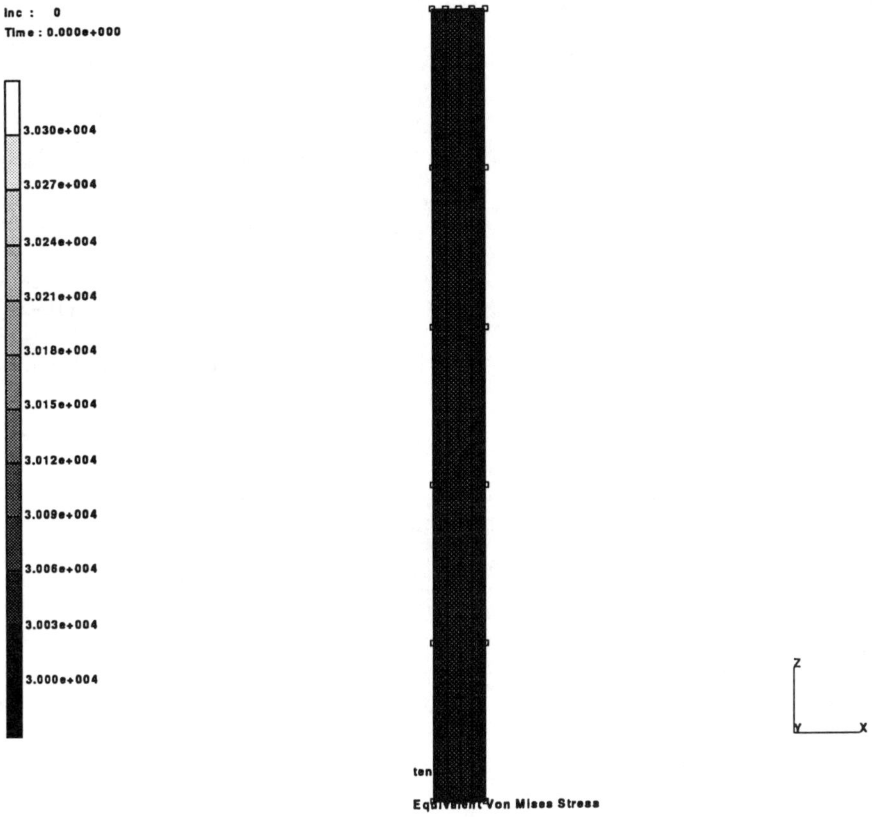

Figure 10-11 Axial stress in a tensile bar.

applicable here. The temperature approximating function is chosen to be

$$T = c_1 + c_2x + c_3y = [1 \quad x \quad y]\begin{bmatrix} c_1 \\ c_2 \\ c_3 \end{bmatrix}. \tag{10-45}$$

Equation (10-45) is the one-dimensional counterpart of Eqs. (10-5) and (10-6). Using the triangular element of Fig. (10-2) for this development, the temperature becomes, at nodes I, J, and M,

$$T_I = c_1 + c_2x_I + c_3y_I, \tag{10-46}$$

$$T_J = c_1 + c_2x_J + c_3y_J, \tag{10-47}$$

$$T_M = c_1 + c_2x_M + c_3y_M, \tag{10-48}$$

or

$$a^e = \begin{bmatrix} T_I \\ T_J \\ T_M \end{bmatrix} = \begin{bmatrix} 1 & x_I & y_I \\ 1 & x_J & y_J \\ 1 & x_M & y_M \end{bmatrix}\begin{bmatrix} c_1 \\ c_2 \\ c_3 \end{bmatrix}. \tag{10-49}$$

Inverting Eq. (10-49) and inserting the results into Eq. (10-45), we get

$$T = N_I(x, y)T_I + N_J(x, y)T_J + N_M(x, y)T_M, \tag{10-50}$$

where

$$N_I(x, y) = \frac{1}{2A}\{(x_J y_M - x_M y_J) + (y_J - y_M)x + (x_M - x_J)y\}, \tag{10-51}$$

$$N_J(x, y) = \frac{1}{2A}\{(x_M y_I - x_I y_M) + (y_M - y_I)x + (x_I - x_M)y\}, \tag{10-52}$$

$$N_M(x, y) = \frac{1}{2A}\{(x_I y_J - x_J y_I) + (y_I - y_J)x + (x_J - x_I)y\}, \tag{10-53}$$

and, as above, A is the area of the triangular element. The temperature distribution in the element may now be expressed as

$$[\mathbf{T}] = [\mathbf{N}][a^e]. \tag{10-54}$$

With the application of the strategies that have been delineated for the structural member, the minimization of the functional for this problem becomes

$$\frac{\mathrm{d}I}{\mathrm{d}a^e} = \int_A \left[QN^T - k\frac{\partial N^T}{\partial x}\frac{\partial N}{\partial x}a^e - k\frac{\partial N^T}{\partial y}\frac{\partial N}{\partial y}a^e\right]\mathrm{d}A + \int_S q_{sb}N^T\,\mathrm{d}S = 0. \tag{10-55}$$

If we choose

$$k^e_{xx} = \int_A \frac{\partial N^T}{\partial x}k\frac{\partial N}{\partial x}\,\mathrm{d}A, \tag{10-56}$$

$$k^e_{yy} = \int_A \frac{\partial N^T}{\partial y}k\frac{\partial N}{\partial y}\,\mathrm{d}A, \tag{10-57}$$

$$f^e_Q = \int_A N^T Q\,\mathrm{d}x\,\mathrm{d}y, \tag{10-58}$$

$$f^e_{sb} = \int_S N^T q_{sb}\,\mathrm{d}x\,\mathrm{d}y, \tag{10-59}$$

Eq. (10-55) becomes, on minimization,

$$K^e a^e = f^e, \tag{10-60}$$

where

$$K^e = k^e_{xx} + k^e_{yy} \tag{10-61}$$

is the elemental stiffness matrix and

$$f^e = f^e_Q + f^e_{sb} \tag{10-62}$$

is the thermal load that is applied to the region.

The problem of heat transfer in a medium is thereby reduced to solving a matrix equation similar to the structural problem. The superscript e in Eqs. (10-56)–(10-62) is, of course, used to denote element quantities only. The assembly of the stiffness matrix for the entire body proceeds as in the structural example. In this case, temperature compatibility at the connecting nodes is enforced.

Other levels of element approximation are similar to those described in Section 10-4 for beam, plate, and solid geometries. In fact, a complete finite-element package has many approximating heat transfer elements, similar to those for structural analysis. Once again, the elements are distinguished by their approximation functions (linear, quadratic, etc.) and by their geometric nature (beam, plate, solid, etc.).

10-7 EXAMPLE IN HEAT TRANSFER

The same structure that was used in the above example and shown in Fig. 10-8 is now subjected to fixed temperatures at its ends. One face is held at 0° while the opposite face along the axial direction is held at 1000°. The lateral surfaces are adiabatic so that heat can flow along only the axis of the rod. Appendix 10.2 contains the MARC finite-element code, called **heat.dat**, that was used to solve this problem. Only steady-state heat conduction was considered. Figure 10-12 shows the results of this analysis.

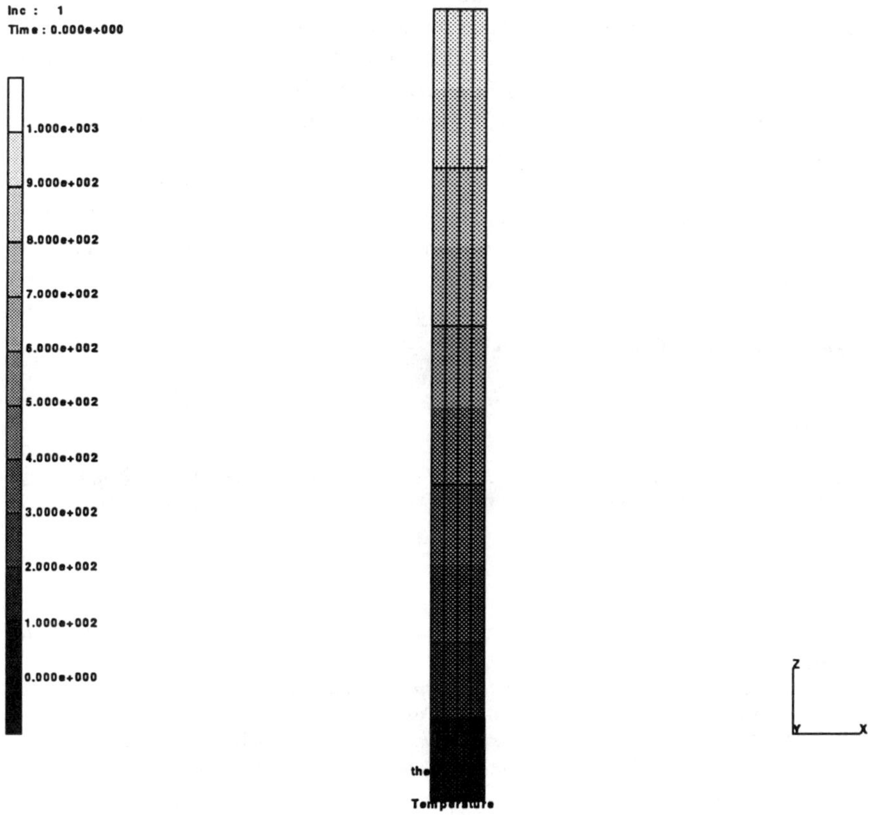

Figure 10-12 Temperature distribution in a rod.

The temperature distribution is linear along the axial direction and does not vary in the transverse direction. MARC element type 43, a three-dimensional solid heat transfer element, was chosen here. The coefficient of thermal conductivity was 1.0*e*-06. In the event that readers desire to conduct a thermal-stress analysis on this body, a data set that combines the two examples in this chapter would be constructed. Readers are asked to construct such a combination in the Problems section of this chapter.

10-8 CONCLUSION

In this chapter an attempt has been made to present the methodology that is required for building a finite-element formulation that would be used to solve engineering problems. We have discussed the trade-offs and the considerations that must be made as the development process proceeds. Parameters such as element type, element size, aspect ratio, and matrix inversion strategies were among the criteria that were explored. Some of these ideas were illustrated with reference to a commercially available finite-element code, called MARC. It must be pointed out, once again, that any robust code can be used with the same ideas that are delineated herein. Since there are so many fine FEM codes available, this author strongly urges users not to attempt to develop codes of their own. If, however, a special element formulation is required (such as a large strain model or a crack model) then users should develop an element that exhibits the required behavior and connect that element to an existing FEM code. MARC is a code that allows that attachment of a user-supplied element to its overall structure. The user should choose a commercial code that has the same capability.

Readers should now able to formulate their own engineering analyses, using the FEM of their choice. The problems that are contained in this chapter introduce the users to finite-element simulations by solving problems for which an exact solution is known. Appendixes 10.1–10.4 contain the formulations for a selected number of these problems. The solutions are developed for MARC, but are adequately annotated so that they can be transformed to the syntax of any other code.

REFERENCES

1. Zienkiewicz, O. C. (1971). *The Finite Element Method in Engineering Science*, McGraw-Hill, London.
2. MARC Analysis Research Corporation, 260 Sheridan Avenue, Suite 309, Palo Alto, California.

FURTHER READING

Anderson, D. A., Tannehill, J. C., and Pletcher, R. H. (1984). *Computational Fluid Mechanics and Heat Transfer*, Hemisphere, New York.

Desai, C. S. and Abel, J. F. (1972). *Introduction to the Finite Element Method—A Numerical Method for Engineering Analysis*, Van Nostrand Reinhold, New York.

Patankar, S. V. (1980). *Numerical Heat Transfer Analysis and Fluid Flow*, Hemisphere, New York.

Tong, P. and Rossettos, J. N. (1977). *Finite-Element Method—Basic Technique and Implementation*, MIT, Cambridge, Massachusetts.

Zienkiewicz, O. C. (1971). *The Finite Element Method in Engineering Science*, McGraw-Hill, London.

PROBLEMS

10-1 A prismatic bar whose length is 15 in. and whose cross section is a unit square is shown in Fig. 10-7. This bar is loaded as a tensile bar so that all stresses vanish except for the normal stress in the z direction. The bar is made of steel whose material properties are modulus of elasticity $E = 30 \times 10^6$ psi, density $\rho = 0.278$ lb/in., Poisson's ratio $\nu = 0.293$, and thermal conductivity $\alpha = 6.7 \times 10^{-6}$ in./(in./deg F).

The bar is loaded by application of a uniform displacement of 0.015 in. to the right end. The left end is restrained so that only a normal tensile stress is produced in the z direction. All other stress components vanish.

This problem is to be solved with the finite-element technique. Since the solution to this problem has been determined in a course of strength of materials, you have full knowledge of its exact nature. Therefore, when solving this problem, you can compare your solution with the exact solution. A complete solution for this problem involves the description of the stresses, the strains, and the displacements in the entire body.

This problem was actually solved in the body of this chapter. Students should attempt to show how, if at all, the solution would vary, if other mesh discretizations were used.

10-2 A prismatic bar whose length is 15 in. and whose cross section is a unit square is shown in Fig. 10-7. This bar is loaded as a tensile bar so that all stresses vanish except for the normal stress in the z direction. The bar is made of steel whose material properties are modulus of elasticity $E = 30 \times 10^6$ psi, density $\rho = 0.278$ lb/in., Poisson's ratio $\nu = 0.293$, and thermal conductivity $\alpha = 6.7 \times 10^{-6}$ in./(in./deg F).

The bar is loaded by application of a concentrated load of 30,000 lb to the right end. The left end is restrained so that only a normal tensile stress is produced in the z direction. All other stress components vanish.

This problem is to be solved with the finite-element technique. Since the solution to this problem has been determined in a course of strength of materials, you have full knowledge of its exact nature. Therefore, when solving this problem, you can compare your solution with the exact solution. A complete solution for this problem involves the description of the stresses, the strains, and the displacements in the entire body.

A sample FEM data file, called **tload.dat**, which uses the MARC finite-element code, is shown in Appendix 10.3. Student should construct their own data files and mesh distributions and comparethem with the code in the appendix.

10-3 A prismatic bar whose length is 15 in. and whose cross section is a unit square is shown in Fig. 10-13. This bar is loaded as a fixed beam with a central concentrated load. The bar is made of

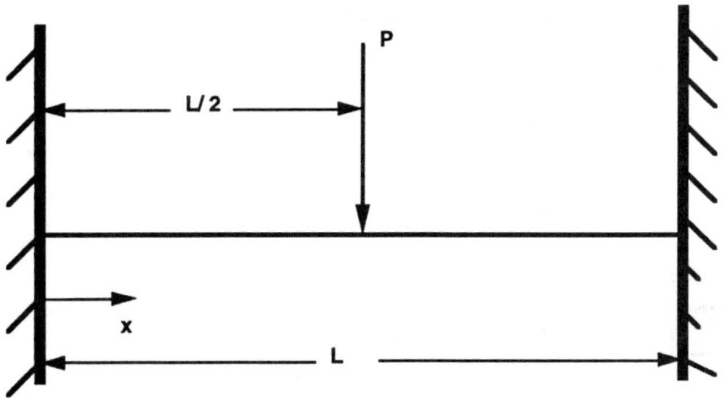

Figure 10-13 Fixed beam with a center load.

steel whose material properties are modulus of elasticity $E = 30 \times 10^6$ psi, density $\rho = 0.278$ lb/in., Poisson's ratio $\nu = 0.298$, and thermal conductivity $\alpha = 6.7 \times 10^{-6}$ in./(in./deg F).

The bar is loaded by application of a force of 10,000 lb at the midspan. Both ends of the beam are completely clamped to the walls.

This problem is to be solved with the finite-element technique. Students must choose a beam element from their FEM package. Since the solution to this problem has been determined in a course of strength of materials, you have full knowledge of its exact nature. Therefore, when solving this problem, you can compare your solution with the exact solution. A complete solution for this problem involves the description of the stresses, the strains, and the displacements in the entire body. You must choose the element type and the element size so that an accurate description of the physical parameters (i.e., stresses, strains, and displacements) may be calculated.

A sample MARC data file, called **beam.dat**, is shown in Appendix 10.4.

Exact solution (from strength of materials)

$$y(z) = \left\{\frac{Pz^2}{192\,EI}\right\}[16z - 12L] \qquad 0 \le z \le \frac{L}{2} \qquad \text{displacement,}$$

$$\frac{dy}{dz} = \theta = \left[\frac{Pz}{8\,EI}\right]\{2z - L\} \qquad 0 \le z \le \frac{L}{2} \qquad \text{slope.}$$

The cross section is in the x–y plane while the span of the beam is in the z direction.

10-4 Re-solve Problem 10-3 with a brick element rather than a beam element. Discuss the difficulties that you will encounter and the mesh discretation strategies that will have to be used in order to obtain an accurate result. Consider, the aspect ratio discussion that was given in the body of this chapter.

10-5 A prismatic bar whose length is 15 in. and whose cross section is a unit square is shown in Fig. 10-14. This bar will be subjected to the following thermal environment. The left end of the bar is to be held at 70° F. The right end of the bar is to be held at 1000° F. The upper and the lower surfaces are adiabatic (i.e., no heat transfer occurs across these boundaries). Only conduction heat transfer within the bar need be considered. The bar is a bimetallic strip composed of steel and aluminum, as shown.

For Steel, the thermal properties are

$$\lambda = 9.4\,\text{Btu/h ft.}\,^\circ\text{F} = 2.176 \times 10^{-4}\,\text{Btu/s in.}^\circ\text{F},$$

$$c = 0.11\text{Btu/lb}_m{}^\circ\text{F},$$

$$\rho = 488\,\text{lb}_m/\text{ft}^3 = 0.282\,\text{lb}_m/\text{in.}^3 = 7.3057 \times 10^{-4}\,\text{lb}_f/\text{in.}^3,$$

$$a = k/c = 0.15\,\text{ft.}^2/\text{h}.$$

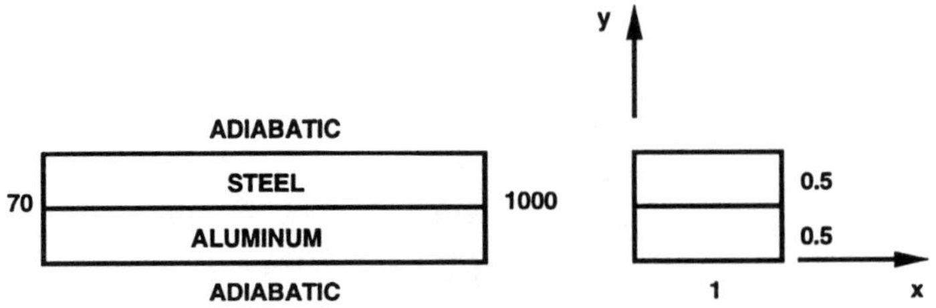

Figure 10-14 Bimetallic strip heat conduction and thermal stress.

For aluminum, the thermal properties are

$$\lambda = 119\,\text{Btu/h ft.}\,^\circ\text{F} = 2.754 \times 10^{-3}\,\text{Btu/s in.}^\circ\text{F},$$
$$c = 0.208\,\text{Btu/lb}_m\,^\circ\text{F},$$
$$\rho = 169\,\text{lb}_m/\text{ft.}^3 = 9.78 \times 10^{-2}\,\text{lb}_m/\text{in.}^3 = 2.534 \times 10^{-4}\,\text{lb}_f/\text{in.}^3,$$
$$a = k/c = 3.33\,\text{ft.}^2/\text{h}.$$

This problem is to be solved with the finite-element technique. Only the steady-state temperature field is required within the bar. Care should be taken when determining the temperature field since these values will form the basis of the next problem. The output temperature should be formatted in a manner that will be consistent with the technique that you will use in the determination of the thermal stresses in the next problem.

A complete solution for this problem involves the description of the temperatures in the entire body.

10-6 When the thermal field of Problem 10-5 is imposed on the bimetallic strip, stresses are induced in the structure. It is now required that these stresses, strains, and displacements be evaluated. The temperatures that were calculated in Problem 10-5 must be used as the driver for this problem.

The elastic material properties for steel are modulus of elasticity $E = 30 \times 10^6$ psi and Poisson's ratio $\nu = 0.293$; for aluminum they are modulus of elasticity $E = 10 \times 10^6$ psi and Poisson's ratio $\nu = 0.300$. Show the deflected shape of the strip.

CHAPTER

ELEVEN

DYNAMIC METHODS IN APPLIED MECHANICS

NOMENCLATURE

β	factor for exact solution of modal frequencies
ρ	mass density
ω	natural frequency
ν	Poisson's ratio
A	cross-sectional area
E	modulus of elasticity of a material
I	moment of inertia
L	length of beam
x	coordinate direction
y	coordinate direction and deflection of a structure

11-1 INTRODUCTION

This chapter is written to introduce readers to the special computational methods that are available when they are solving dynamics problems in structures. In Chapter 6 the finite-difference procedures were applied to the differential equations that represented both static and dynamic processes. These methods are still applicable if one wants to obtain the entire temporal behavior of the structure from initial conditions to an arbitrarily specified time. If the structure is undergoing both transient and steady-state behaviors, the above techniques will allow the user to determine the values of all the dependent variables at all the independent variable locations. If, however, only the steady-state behavior is required, then the methods that have been delineated above are both inefficient and time consuming. This is particularly important when a long time interval is required for reaching the steady state or when the body is geometrically so complex that a great amount of computational time is required for each time step in the simulation process. The calculation of vibration modes is another example of a case in which the solution of the entire differential equation will not achieve the desired results when the techniques of the preceding chapters are used.

If, on the other hand, the finite-element techniques are used to solve dynamics problems, special methodologies that have not been exposed would have to be used. For both modal extraction and for transient behavior analysis, the finite-element theory must be modified from the formulation that was presented in Chapter 10.

The theory of this chapter, coupled with the computational tools of Chapter 10, should equip readers with all the background that will be necessary to conduct robust engineering calculations in the dynamics domain of physics and engineering.

11-2 LUMPED-PARAMETER SYSTEMS

When one analyzes a structure or continuous system, a differential equation of motion is usually derived. This equation, for an undamped system, is of the form

$$\mathbf{M}\ddot{X} + \mathbf{K}X = F(t), \tag{11-1}$$

where $\mathbf{M}$ is the matrix that represents the dynamic quantities of the system (i.e., the masses), $\mathbf{K}$ is the matrix that represents the elasticity of the system (i.e., the springs), and X represents the displacements of the system. In a continuous system, say a beam with 6 degrees of freedom, a complex set of partial differential equations is derived. These equations, coupled with the constitutive equations of the material generally constitute a set of 12 simultaneous partial differential equations that must be solved for the unknown variables. This is a difficult task for the experienced computational researcher and is an insurmountable task for the novice. Therefore, some simplifying methods are required for converting this task to one that becomes more tractable. One of these methods is known as the *lumped-parameter method.*

Basically, any continuous structure is converted, through the process of computational separation, into a series of lumped masses in which the dynamic quantities are unknown and a series of massless structural members in which the flexibility of the system is determined. In other words, for computational simplicity, the dynamic portion of the structural model is separated from the constitutive portion of the model. We then apply our dynamic equations to these individual masses. We apply the material constitutive relations directly to these massless entities.

In order to demonstrate the use of the lumped-parameter technique and to explore its computational strengths, let us consider the elastic cantilever beam shown in Fig. 11-1, which is undergoing uncoupled flexural vibrations while being subjected to a centrifugal loading due to the rotation of the beam about a transverse axis. This model can serve to simulate a blade in either the turbine or the compressor of a jet engine. The engine speed will be assumed to be Ω, as shown in the figure. We begin the simulation process by subdividing the beam into any desired number of segments N. The mass of the beam

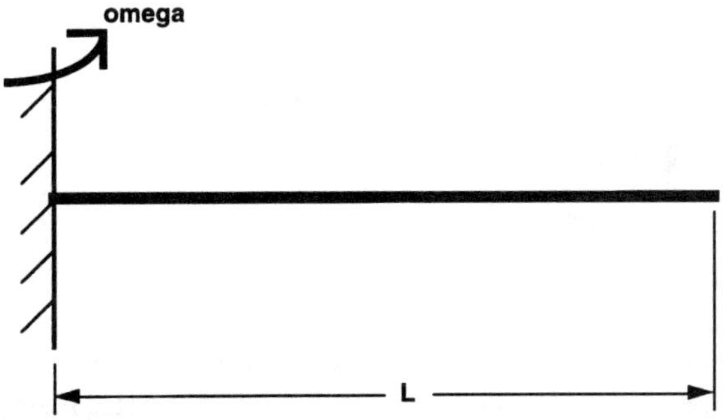

Figure 11-1 Rotating cantilever beam.

is then lumped into point masses for each segment. Figures 11-2 and 11-3 show this discretization along with the internal forces that act on the masses and the connecting segments, which are massless and represent the flexibility in the beam. The internal forces are labeled with respect to the nodal mass on which they act. Thus, on the left-hand side of mass n, the forces are V_n^L and M_n^L while on the right-hand side of the mass are forces M_n^R and V_n^R. If we assume a sinusoidal vibration analysis (modal analysis) of the beam, then the translatory inertia force that acts on the mass n is given by $\omega^2 m_n y_n$. In this case, ω (not the rotational frequency Ω) is the frequency of steady-state vibrations while y_n is the magnitude of the displacement of the nth node in the beam. The free-body diagram in Fig. 11-4 shows the effect of the centrifugal force on node n in the deformed beam, which is due to the engine rotation. This completes the force and dynamic simulation.

We now turn our attention to the constitutive relations that relate the forces and moments at one end of the massless segment to the force system at the other end. Figure 11-5 shows the elastic deformation parameters of the beam under consideration. The angles θ_{n-1} and θ_n represent the slopes of the beam at either end of the segment l_n.

To derive the equations that form the lumped-parameter simulation for this example, let us temporarily ignore the centrifugal force that is due to engine rotation. This effect will be subsequently added to the analysis. Since we are ignoring rotatory inertia of the mass in this example, a moment summation across mass n in Fig. 11-3, yields

$$M_n^L = M_n^R. \tag{11-2}$$

The force summation across the node is

$$V_n^L = V_n^R - \omega^2 m_n y_n, \tag{11-3}$$

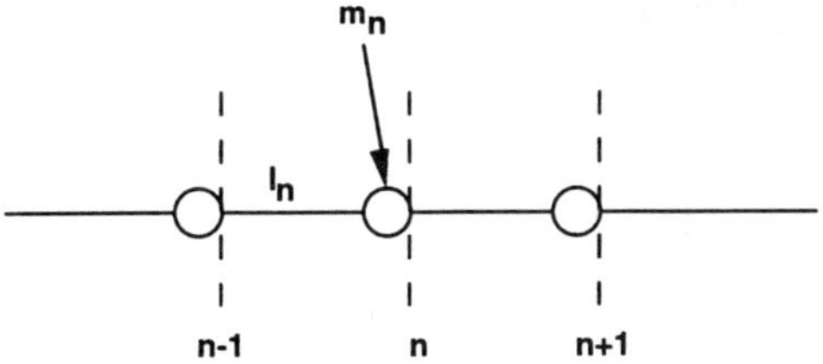

Figure 11-2 Lumped-parameter representation of a beam.

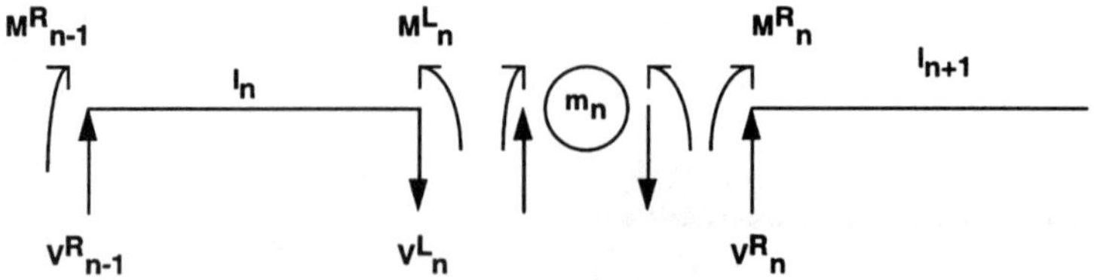

Figure 11-3 Internal forces for the lumped-parameter simulation.

where the last term represents the translatory inertia in the mass. Keeping in mind that segment n is massless, we find that the relationship between the forces on the left-hand side of segment n and the forces on the right-hand side may be expressed as

$$V_{n-1}^R = V_n^L, \tag{11-4}$$

$$M_{n-1}^R = M_n^L - V_n^L l_n. \tag{11-5}$$

Figure 11-5 allows us to form geometric relationships that govern the deflection and the slope of segment n, viz.,

$$y_n = y_{n-1} + l_n\theta_{n-1} + M_n^L \frac{l_n^2}{2(EI)_n} - V_n^L \frac{l_n^3}{3(EI)_n}, \tag{11-6}$$

$$\theta_n = \theta_{n-1} + \frac{M_n^L l_n}{(EI)_n} - \frac{V_n^L l_n^2}{2(EI)_n}. \tag{11-7}$$

If the centrifugal loading is ignored, then the relationship between variables in segment n can be expressed as

$$\begin{bmatrix} y \\ \theta \\ M \\ V \end{bmatrix}_n^L = \begin{bmatrix} 1 & l & \frac{l^2}{2EI} & \frac{l^3}{6EI} \\ 0 & 1 & \frac{l}{EI} & \frac{l^2}{2EI} \\ 0 & 0 & 1 & l \\ 0 & 0 & 0 & 1 \end{bmatrix}_n \begin{bmatrix} y \\ \theta \\ M \\ V \end{bmatrix}_{n-1}^R, \tag{11-8}$$

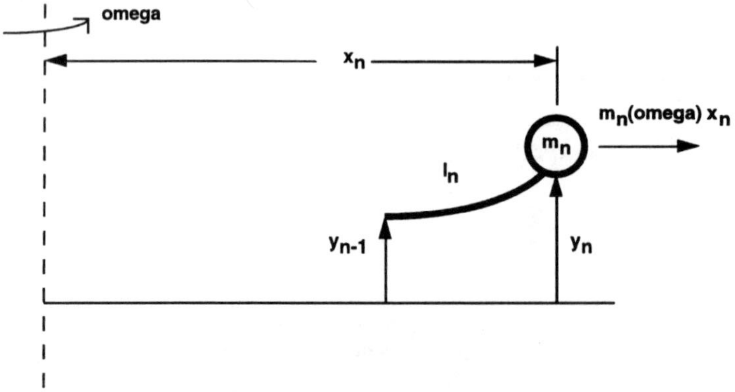

Figure 11-4 Centrifugal force on a mass in a deformed beam.

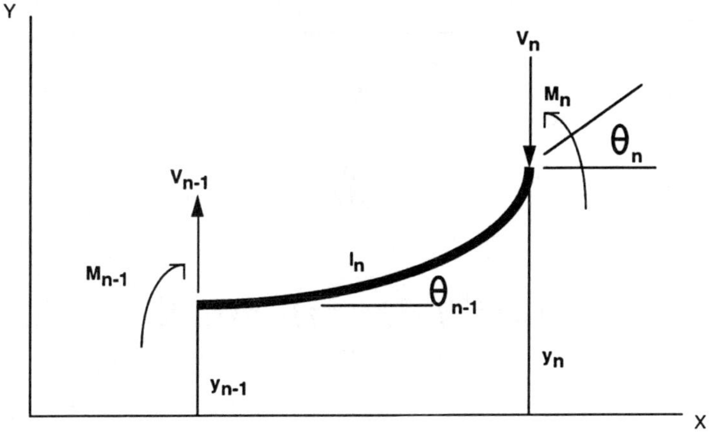

Figure 11-5 Eleastic configuration of a beam section.

and the point mass equations yield

$$\begin{bmatrix} y \\ \theta \\ M \\ V \end{bmatrix}_n^R = \begin{bmatrix} 1 & 0 & 0 & 0 \\ 0 & 1 & 0 & 0 \\ 0 & 0 & 1 & 0 \\ \omega^2 m & 0 & 0 & 1 \end{bmatrix}_n \begin{bmatrix} y \\ \theta \\ M \\ V \end{bmatrix}_n^L . \tag{11-9}$$

If we substitute Eqs. (11-9) into Eqs. (11-8), we can express the relationship between the variables on the right-hand side of mass n and the variables at the left end of segment n, viz.,

$$\begin{bmatrix} y \\ \theta \\ M \\ V \end{bmatrix}_n^R = \begin{bmatrix} 1 & l & \frac{l^2}{2EI} & \frac{l^3}{6EI} \\ 0 & 1 & \frac{l}{EI} & \frac{l^2}{2EI} \\ 0 & 0 & 1 & l \\ \omega^2 m & \omega^2 ml & \frac{\omega^2 ml^2}{2EI} & 1+\omega^2 m\frac{l^3}{6EI} \end{bmatrix}_n \begin{bmatrix} y \\ \theta \\ M \\ V \end{bmatrix}_{n-1}^R . \tag{11-10}$$

We now introduce the centrifugal force on mass n, $m_n\Omega^2 x_n$, so that the axial force becomes

$$F_n^L = F_n^R + m_n\Omega^2 x_n, \tag{11-11}$$

where

$$F_n^L = \sum_{i=n}^{N} m_i\Omega^2 x_i . \tag{11-12}$$

Because of the introduction of the force from Eq. (11-11), Eqs. (11-5)–(11-7) for the moment, deflection, and slope must be modified as

$$M_{n-1}^R = M_n^L - V_n^L l_n - F_n^L\left(y_n^L - y_{n-1}^R\right). \tag{11-13}$$

$$y_n^L = y_{n-1}^R + l_n\theta_{n-1}^R - \theta_n^L F_n^L \frac{l^3}{3(EI)_n} + M_n^L \frac{l_n^2}{2(EI)_n} - V_n^L \frac{l_n^3}{3(EI)_n}, \tag{11-14}$$

$$\theta_n^L = \theta_{n-1}^R + \frac{M_n^L l_n}{(EI)_n} - \frac{V_n^L l_n^2}{2(EI)_n} - \frac{\theta_n^L F_n^L l^2}{2(EI)_n}. \tag{11-15}$$

Using Eqs. (11-13)–(11-15) to modify Eqs. (11-10), we arrive at our final matrix relationship for the nth segment of the beam, i.e.,

$$\begin{bmatrix} y \\ \theta \\ M \\ V \end{bmatrix}_{n-1}^{R} = \begin{bmatrix} \left(1+\frac{m\omega^2 l^3}{6EI}\right) & -\left(l+\frac{Fl^3}{6EI}\right) & \frac{l^2}{2EI} & -\frac{l^3}{6EI} \\ -\frac{m\omega^2 l^2}{2EI} & \left(1+\frac{Fl^2}{2EI}\right) & -\frac{l}{EI} & \frac{l^2}{2EI} \\ m\omega^2\left(l+\frac{Fl^3}{6EI}\right) & -F\left(l+\frac{Fl^3}{6EI}\right) & \left(1+\frac{Fl^2}{2EI}\right) & -\left(l+\frac{Fl^3}{6EI}\right) \\ -m\omega^2 & 0 & 0 & 1 \end{bmatrix}_n \begin{bmatrix} y \\ \theta \\ M \\ V \end{bmatrix}_n^R. \tag{11-16}$$

If Eqs. (11-16) are now applied successively to all the N segments of the beam, we obtain a functional relationship between the force and the geometric parameters at each end of the entire beam, viz.,

$$\begin{bmatrix} y \\ \theta \\ M \\ V \end{bmatrix}_N = \begin{bmatrix} a_{11} & a_{12} & a_{13} & a_{14} \\ a_{21} & a_{22} & a_{23} & a_{24} \\ a_{31} & a_{32} & a_{33} & a_{34} \\ a_{41} & a_{42} & a_{43} & a_{44} \end{bmatrix} \begin{bmatrix} y \\ \theta \\ M \\ V \end{bmatrix}_0. \tag{11-17}$$

Equations (11-17) represent the matrix that transfers the force and the deflection from the left end of the beam to the right end. Regardless of the number of segments into which the original beam is discretized, the final matrix size is only 4×4. This small matrix size allows for easy matrix inversion and does not carry the overhead of a finite-element analysis. This advantage is reemphasized in the sections in which the finite-element analysis is introduced for dynamic problems.

The formulation of this section can be used in one of two ways. First, with the application of proper boundary conditions at each end of the beam, the forces, moments, deflections, and slopes of a beam that is being rotated, can be determined. Second, with the above formulation, we can extract the natural frequencies of a rotating member and thereby estimate the effect of vibrations on the performance of a turbine or compressor blade.

In this section we use the lumped-parameter formulation to calculate the static deflection of a cantilever beam (i.e., $\omega = 0$). In Section 11-3, a discussion of modal analysis (by lumped parameter techniques) is presented.

Let us now consider a cantilever beam 10 in. long that will be divided into 10 segments, so that there are 11 nodal points in the simulation. The modulus of elasticity is $E = 30 \times 10^6$ psi while the moment of inertia is $I = 0.08333$ in.^{4}. The density of the beam will be chosen as 0.278 lb/in.^{3}. Since a cantilever beam is being simulated, we will choose $[y, \theta]_0 = [0, 0]$ and $[\mathbf{F}, \mathbf{M}]_N = [1000, 0]$. Appendix 11.1 contains a computer code, called **cantilever.f**, that solves this problem by the lumped-parameter methods of this chapter. Figures 11-6–11-8 show the deflection, the slope, and the bending moment, respectively, in the entire cantilever beam.

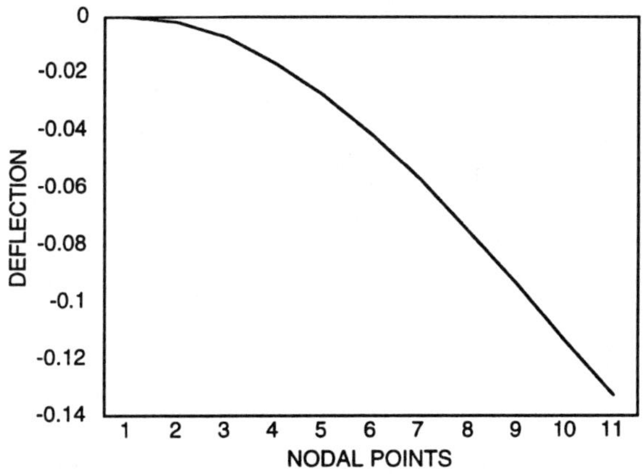

Figure 11-6 Deflection of a cantilever beam.

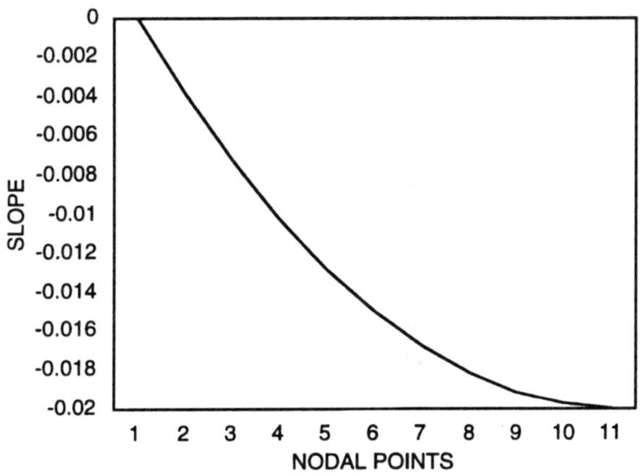

Figure 11-7 Angular deflection in a cantilever beam.

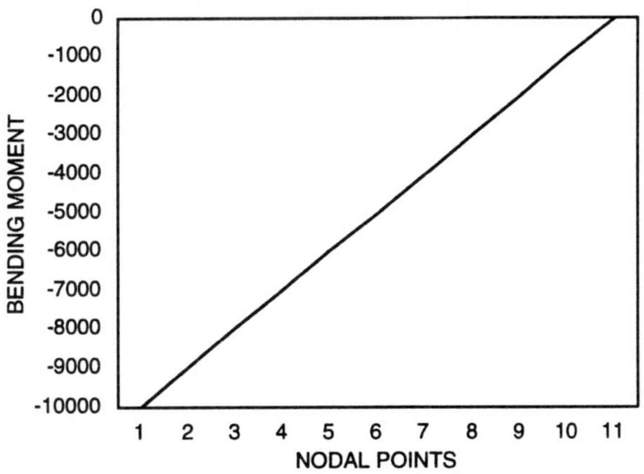

Figure 11-8 Bending moment in a cantilever beam.

11-3 MODAL EXTRACTION BY LUMPED PARAMETERS

The matrix that is presented in Eqs. (11-16) contains the natural frequencies ω that, when coupled with the mode shapes, form the normal modes of vibration for a beam. The search for these frequencies depends on finding the values of ω that will make the determinant of the matrix shown in Eqs. (11-17) vanish. The search algorithm that is used is a trial- and-error procedure that begins at $\omega = 0$ proceeds until a zero of the determinant of the matrix is achieved by either the bisection method or the false-position method of Chapter 2. All the preceding considerations for these search processes, which were described in Chapter 2, are relevant here.

If we apply this search methodology to the cantilever beam of Section 11-2, we can calculate the first several natural frequencies of vibrations. Because of accuracy considerations during the search process, the cantilever beam was subdivided into 30 segments (31 mass points) for the modal extraction phase. After the natural frequencies of vibration are calculated, we require the determination of the mode shapes that are associated with each frequency. After inserting each of the calculated frequencies back into the matrix and using the boundary conditions of the beam, we can then calculate the mode shapes. Appendix 11.2 contains a computer code, **modal.f**, that uses the lumped-parameter method to calculate the system matrix and then uses the bisection method to extract the natural frequencies. The mode shapes are then calculated for each normal mode. With 31 grid points (lumped masses), the first three natural frequencies are calculated to be $\omega_1 = 1997.3$ rad/s (with an error of 3.7% from the exact solution), $\omega_2 = 12\,577$ rad/s (with an error of 4.7%), and $\omega_3 = 35\,236.8$ rad/s (with an error of 3.1%). The normal modes are calculated and displayed in Figs. 11-9, 11-10, and 11-11, respectively. Since the number of grid points and the number of normal modes are arbitrary in **modal.f**, we may use the lumped-parameter method to calculate all the required modes to any arbitrary accuracy.

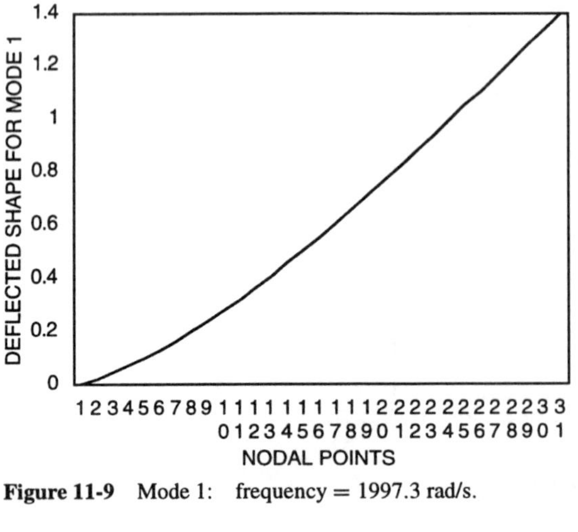

Figure 11-9 Mode 1: frequency = 1997.3 rad/s.

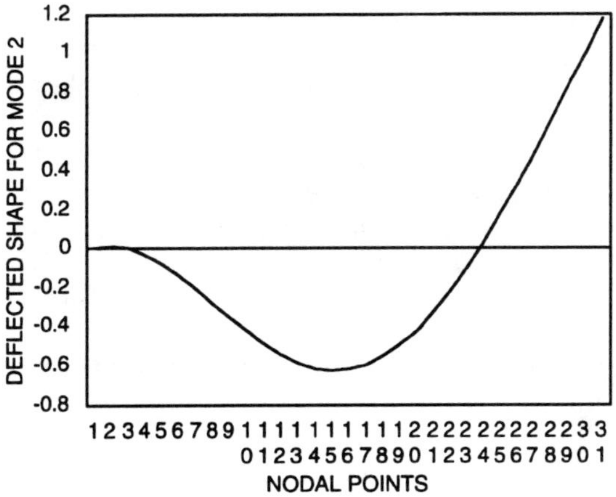

Figure 11-10 Mode 2: frequency = 12 577 rad/s.

11-4 MODAL EXTRACTION BY FINITE ELEMENTS

In Chapter 10, the finite-element method for problems in structural mechanics and heat transfer was presented. In static problems, the structure was reduced to an equivalent spring so that the entire body was governed by the matrix equation

$$[\mathbf{K}][\delta] = [\mathbf{F}], \tag{11-18}$$

where the displacement field is given by

$$\{f\} = [N]\{\delta\}^e = [N_i \quad N_j \quad N_m \quad \cdots] \left\{ \begin{array}{c} \delta_i \\ \delta_j \\ \delta_m \\ \cdots \end{array} \right\} \tag{11-19}$$

where $[N]$ is the interpolation function and $\{\delta\}^e$ are the nodal displacements in the elements. For dynamic problems then, we continue the process of reducing the continuous

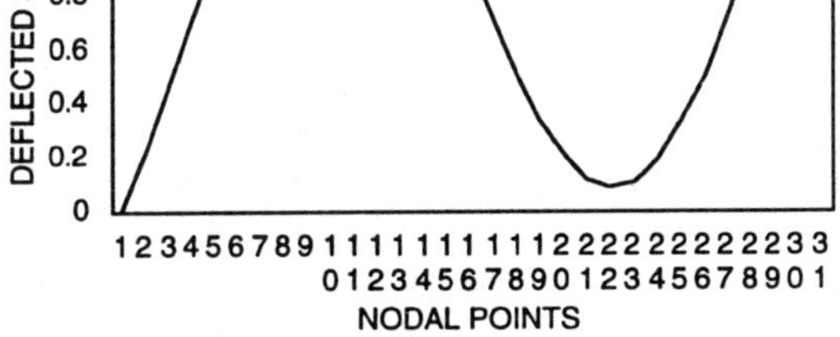

Figure 11-11 Mode 3: frequency = 35 236.8 rad/s.

(actual) structural system to a mass-spring-dashpot system of the form

$$[\mathbf{K}]\{\delta\} + [\mathbf{C}]\frac{\partial}{\partial t}\{\delta\} + [\mathbf{M}]\frac{\partial^2}{\partial t^2}\{\delta\} + \{F\} = 0, \tag{11-20}$$

where [**K**] is the stiffness matrix of the entire system, [**C**] is the damping matrix, [**M**] is the mass matrix, and $\{F\}$ represents the structural loading. The stiffness matrix may be assembled from the element topology with the above-mentioned rules (Chapter 10), i.e.,

$$[c_{ij}]^e = \iiint \{[N_i]^T \mu [N_j]\}\, \mathrm{d}V \tag{11-21}$$

as the element damping, where μ is the damping constant. The element mass matrix is calculated with

$$[m_{ij}]^e = \iiint \{[N_i]^T \rho [N_j]\}\, \mathrm{d}V, \tag{11-22}$$

where ρ is the mass density of the material of the structure. Equations (11-21) and (11-22) may be assembled by the rules given in Chapter 10 to create Eq. (11.20).

In order to configure Eqs. (11-20) for vibrations and eigenvalue analyses, we let the forcing function $F(t)$ and the displacements $\delta(t)$ be sinusoidal, viz.,

$$\{F(t)\} = \{F_0\}e^{i\omega t}, \tag{11-23}$$

$$\{\delta(t)\} = \{\delta_0\}e^{i\omega t}. \tag{11-24}$$

If we substitute Eqs. (11-23) and (11-24) into Eqs. (11-20), we get

$$([\mathbf{K}] + i\omega[\mathbf{C}] - \omega^2[\mathbf{M}])\{\delta_0\} + F_0 = 0. \tag{11-25}$$

For modal extraction, we use the usual technique of disregarding damping and considering the free vibrations configuration. Equation (11-25) then becomes

$$([\mathbf{K}] - \omega^2[\mathbf{M}])\{\delta_0\} = 0. \tag{11-26}$$

To extract the natural frequencies of the system, we desire to find the values of ω that cause the determinant of the system to vanish,[1] i.e.,

$$\det |[\mathbf{K}] - \omega^2[\mathbf{M}]| = 0. \tag{11-27}$$

Thus, when performing modal extraction using the FEM, one must construct the stiffness matrix and the mass matrix for each element before combining these matrices to form the overall system matrix by using Eqs. (11-26). The system matrix now contains the parameter ω, which is unknown. Using a searching technique, similar to one of the methods that was introduced in Chapter 2, one may extract the desired natural frequencies and mode shapes for the desired system.

Appendix 11.3 contains the data file, **cant.dat**, that performs modal extraction for a cantilever beam with the following geometric and mass parameters: modulus of elasticity $E = 30 \times 10^6$ psi, area $A = 1.0$ in.^{2}, moment of inertia $I = 0.08333$ in.^{4}, length $L = 10.0$ in., mass density $\rho = 0.278$ lb/in.^{3}.

Using the MARC[2] finite-element code with the above data file, the first three natural frequencies and their mode shapes are calculated. Since the chosen FEM model contains 6 degrees of freedom and the cross-sectional area is square, there is a symmetry that is associated with the bending modes of vibration. This means that the

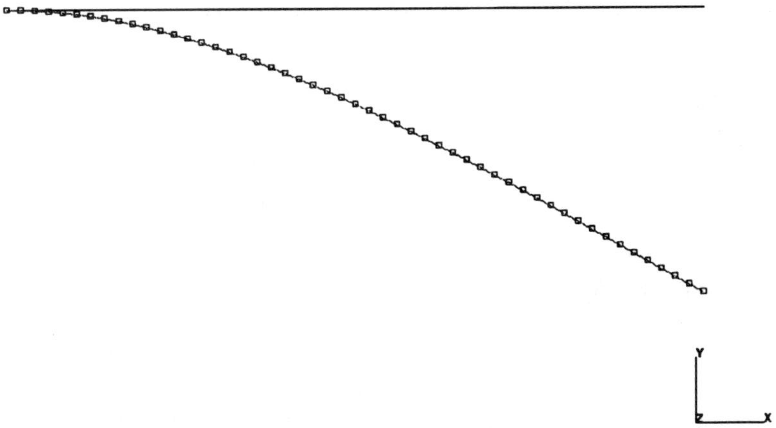

Figure 11-12 Mode 1 = mode 2; $f = 2073$ rad/s.

structure is unable to distinguish between vibrations in any of the transverse directions. Therefore, on extracting the frequencies of vibration, modes 1 and 2 (which represent the fundamental mode), modes 3 and 4 (which represent the second bending mode), and modes 7 and 8 (which represent the third bending mode) appear in equal pairs in the analysis. Mode 6, which represents an axial stretch mode, is unique since it is not associated with a symmetrical mode shape. The mode shapes that are associated with each of the above frequencies have been calculated and are shown in Figs. 11-12–11-16. The FEM code was, of course, capable of extracting higher-order modes, but

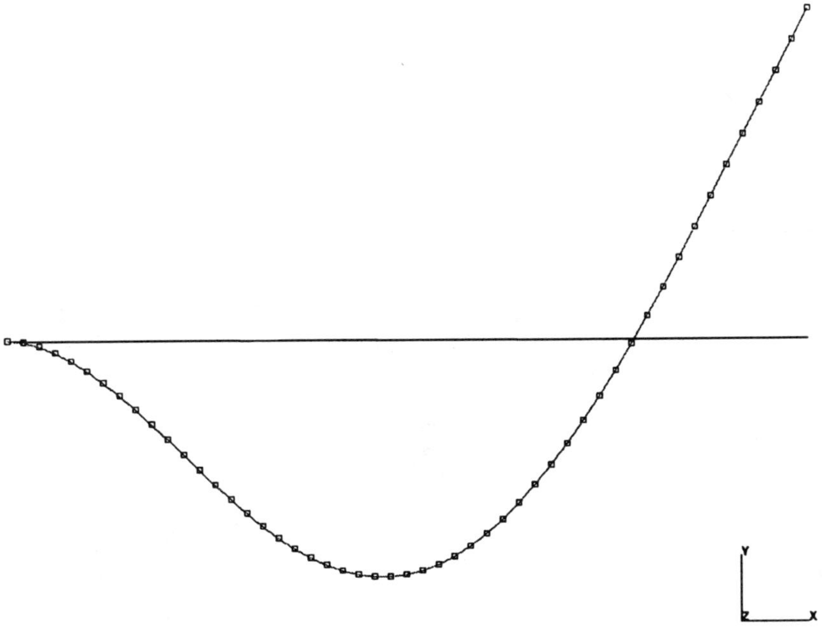

Figure 11-13 Mode 3 = mode 4; $f = 12\,990$ rad/s.

Figure 11-14 Mode 6; $f = 32\,080$ rad/s.

that task is left to the readers. As the number of desired modes increases, their frequencies tend to bunch and the task of modal extraction becomes increasingly more difficult.

11-5 DYNAMIC ANALYSIS BY FINITE-ELEMENT METHODS

As a final example of the use of dynamic methods in structural analysis problems, let us consider the cantilever beam of the preceding sections when it is loaded by a concentrated end load of 3000 lb, which increases, then decreases linearly. Figures 11-17–11-19 show the histories of the deflections, the applied forces, and the reaction forces, respectively, at the end nodes of the beam. Appendix 11.4 contains the data file, called **history.dat**, that was used to perform the analysis of this section. Because of the elastic nature of the material, the response of the beam to the applied load is also linear.

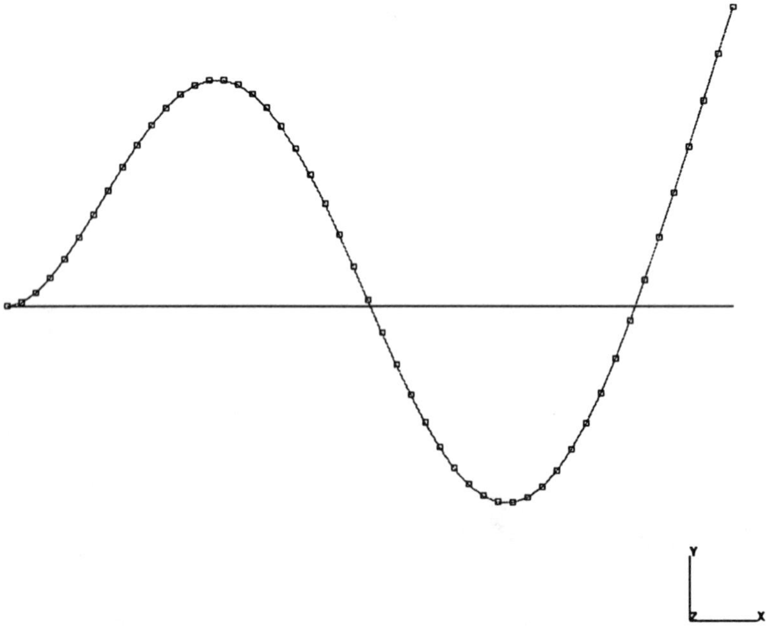

Figure 11-15 Mode 7 = mode 8; $f = 36\,370$ rad/s.

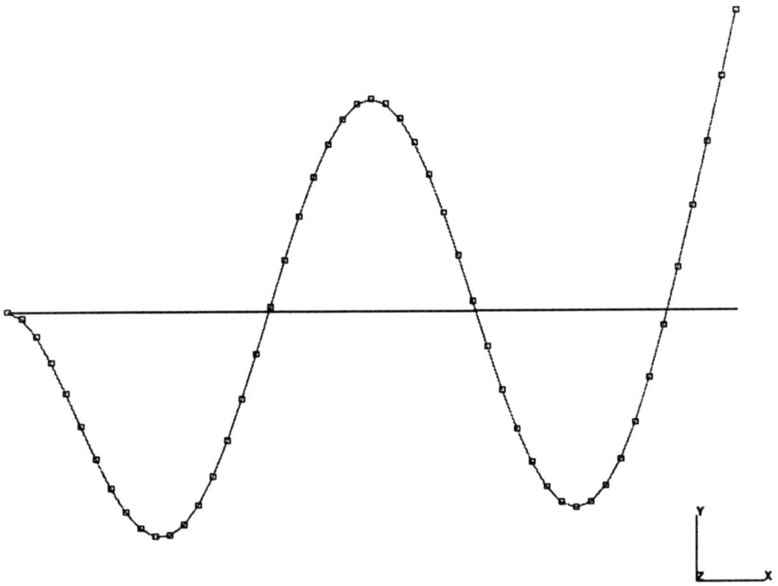

Figure 11-16 Model 10; $f = 71\,270$ rad/s.

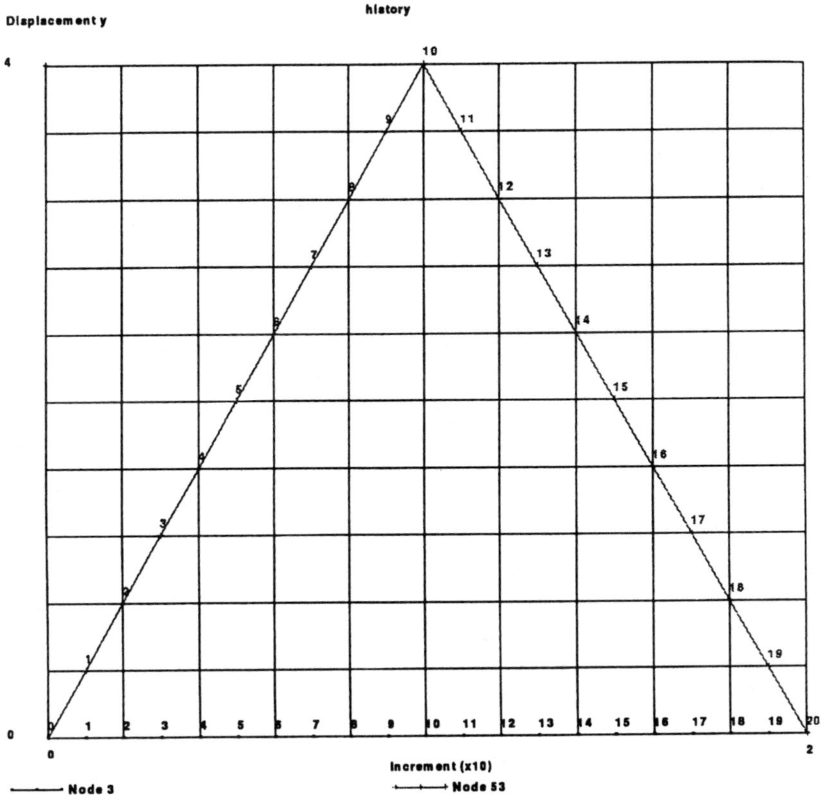

Figure 11-17 Displacement history of the end nodes.

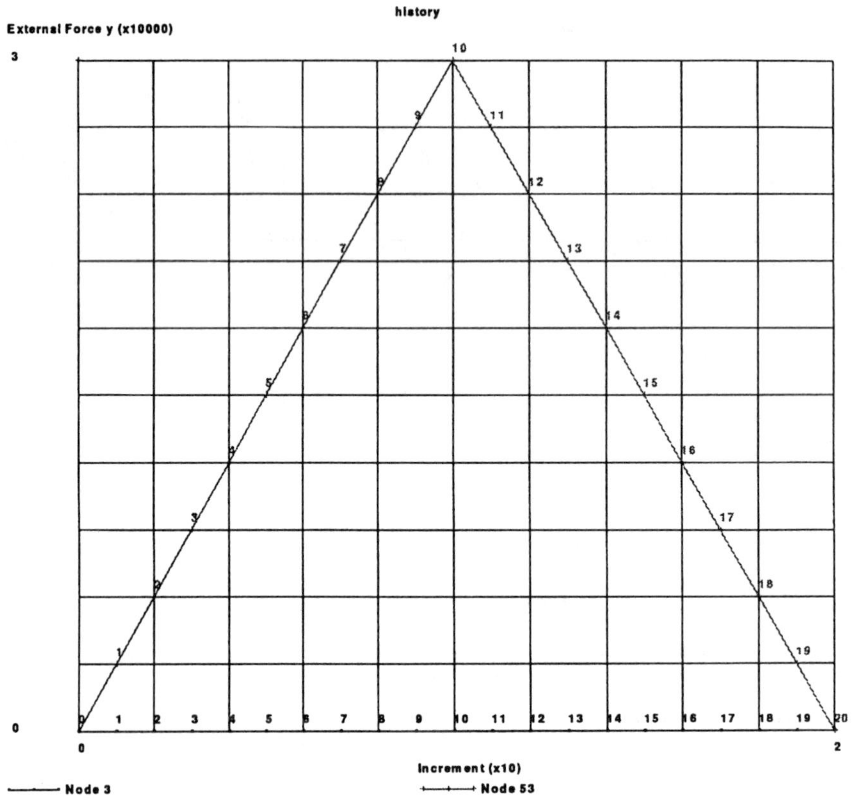

Figure 11-18 Applied load history of the end nodes.

11-6 CONCLUSION

This chapter has been included in this textbook to show the readers how one may solve dynamic problems in structures without having to resort to the finite-difference simulation of differential equations.

The lumped-parameter methods are an attempt to convert a continuous system that would be governed by a complex set of differential equations into a system in which the masses are lumped at discrete points. This approach results in an algebraic system of simultaneous equations that are more readily solvable than the differential equations. Normal mode extraction, in these cases, results in the use of the nonlinear searching techniques that were derived in Chapter 2. The lumped-parameter method can be carried over into the FEM's so that the modal analysis proceeds directly from a simple geometric model.

Finally, structural problems that involve spatial as well as temporal distributions, are solved by FEM methods that require only a history definition in addition to the basic static structural data that would be required if the problem were static.

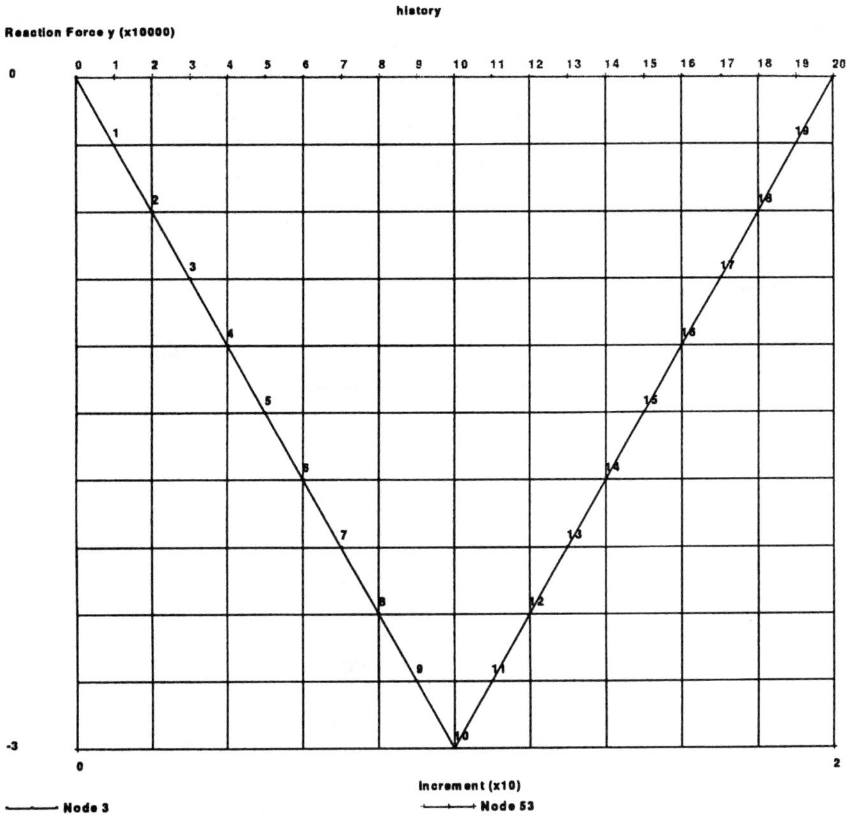

Figure 11-19 Reaction force history of the end nodes.

REFERENCES

1. Zienkiewicz, O. C. (1971). *The Finite Element Method in Engineering Science*, McGraw-Hill, London.
2. MARC Analysis Research Corporation, 260 Sheridan Avenue, Suite 309, Palo Alto, California.

FURTHER READING

Desai, C. S. and Abel, J. F. (1972). *Introduction to the Finite Element Method—A Numerical Method for Engineering Analysis*, Van Nostrand Reinhold, New York.

Rao, S. S. (1990). *Mechanical Vibrations*, Addison-Wesley, Reading Massachusetts.

Thomson, W. T. (1972). *Theory of Vibration with Applications*, Prentice-Hall, Englewoods Cliffs, New Jersey.

Tong, P. and Rossettos, J. N. (1977). *Finite-Element Method—Basic Technique and Implementation*, MIT, Cambridge, Massachusetts.

Zienkiewicz, O. C. (1971). *The Finite Element Method in Engineering Science*, McGraw-Hill, London.

PROBLEMS

The problems in this chapter require either the static deflected shape of a beam, the modal extraction for a beam, or the dynamic response of a beam to an applied load. For the static problems, readers should use the analytical lumped parameter procedure, as outlined in Section 11-1. For the modal extraction problems, students should use either the analytical procedures of Section 11-2 or the FEM modal extraction procedures that are described in Section 11-3, the specific problem.

For the problems in this chapter that require the calculation of the normal modes of vibration for various geometric configurations, the first three natural frequencies and their corresponding mode shapes are desired. For each beam, the following mass and geometric parameters apply: modulus of elasticity $E = 30 \times 10^6$ psi, density $\rho = 0.278$ lb/in.^{3}, Poission's ratio $\nu = 0.3$, length $L = 10.0$ in., and moment of inertia $I = 0.08333$ in.^{4}.

The exact solutions for the modes of vibration in a beam may be given by

$$\omega_n = \beta^2 \sqrt{\frac{gEI}{\rho}}.$$

The values of β for the first three natural frequencies are given for the following cases:

	$(\beta_1 l)^2$	$(\beta_2 l)^2$	$(\beta_3)^2$
Beam configuration	Mode 1	Mode 2	Mode 3
Simply supported	9.87	39.5	88.9
Cantilever	3.52	22.4	61.7
Free–free	22.4	61.7	121.0
Clamped–hinged	15.4	50.0	104.0
Clamped–clamped	22.4	61.7	121.0
Hinged–free	0	15.4	50.0

The exact solutions for the static beam problems can be found in any textbook that deals with the mechanics of materials.

11-1 For a cantilever beam, calculate the static deflected shape (using the lumped-parameter approach) for an end load of 1000 lb. Compare your results with the exact solution to determine the optimum nodal spacing.

11-2 For a cantilever beam, use the analytical lumped-parameter method of Section 11-2 to find the first three mode shapes. Compare your results with the exact solution to determine the optimum nodal spacing.

11-3 For a cantilever beam, use the FEM method of Section 11-3 to find the first three mode shapes. Compare your results with the exact solution to determine the optimum nodal spacing.

11-4 For a simply supported beam, calculate the static deflected shape (using the lumped-parameter approach) for a center load of 1000 lb. Compare your results with the exact solution to determine the optimum modal spacing.

11-5 For a simply supported beam, use the analytical lumped-parameter method of Section 11-2 to find the first three mode shapes. Compare your results with the exact solution to determine the optimum nodal spacing.

11-6 For a simply supported beam, use the FEM method of Section 11-3 to find the first three mode shapes. Compare your results with the exact solution to determine the optimum nodal spacing.

11-7 For a clamped–clamped beam, calculate the static deflected shape (using the lumped-parameter approach) for a center load of 1000 lb. Compare your results with the exact solution to determine the optimum nodal spacing.

11-8 For a clamped–clamped beam, use the analytical lumped-parameter method of Section 11-2 to find the first three mode shapes. Compare your results with the exact solution to determine the optimum nodal spacing.

11-9 For a clamped–clamped beam, use the FEM method of Section 11-3 to find the first three mode shapes. Compare your results with the exact solution to determine the optimum nodal spacing.

11-10 For a clamped–hinged beam, calculate the static deflected shape (using the lumped-parameter approach) for a center load of 1000 lb. Compare your results with the exact solution to determine the optimum nodal spacing.

11-11 For a clamped–hinged beam, use the analytical lumped–parameter method of Section 11-2 to find the first three mode shapes. Compare your results with the exact solution to determine the optimum nodal spacing.

11-12 For a clamped–hinged beam, use the FEM method of Section 11-3 to find the first three mode shapes. Compare your results with the exact solution to determine the optimum nodal spacing.

11-13 For a free–free beam, use the analytical lumped-parameter method of Section 11-2 to find the first three mode shapes. Compare your results with the exact solution to determine the optimum nodal spacing.

11-14 For a free–free beam, use the FEM method of Section 11-3 to find the first three mode shapes. Compare your results with the exact solution to determine the optimum nodal spacing.

11-15 For a hinged–free beam, use the analytical lumped-parameter method of Section 11-2 to find the first three mode shapes. Compare your results with the exact solution to determine the optimum nodal spacing.

11-16 For a hinged–free beam, use the FEM method of Section 11-3 to find the first three mode shapes. Compare your results with the exact solution to determine the optimum nodal spacing.

PART THREE

ADVANCED COMPUTATIONAL TOOLS

CHAPTER

TWELVE

METHOD OF CHARACTERISTICS

NOMENCLATURE

ϕ_t	angular velocity (rad/s)
ρ	density (lb/in.^{3})
σ_θ	hoop stress (psi)
w_t	linear velocity (in./s)
ε_x	normal strain (in./in.)
σ_x	normal stress in the x direction (psi)
ν	Poisson ratio
M_r	radial moment per unit length (lb in./in.)
σ_r	radial stress (psi)
k_2	shear correction factor
M_θ	tangential moment per unit length (lb in./in.)
Q_r	transverse shear force per unit length (lb/in.)
c	wave speed (in./s)
D	flexural rigidity (psi)
E	modulus of elasticity (psi)
G	shear modulus (psi)
h	plate thickness (in.)
r	radial direction (in.)
t	time (s)
u	displacement (in.)
v	linear velocity $= \partial u/\partial t$ (in./s)
x	coordinate direction (in.)
y	coordinate direction (in.)

12-1 INTRODUCTION

In Chap. 8 it was implied that partial differential equations may be grouped into three categories: elliptic, parabolic, and hyperbolic. These groupings are not merely mathematical in nature. They represent physical processes that differ greatly. In this chapter we see that the type of partial differential equation that occurs for problems involving wave propagation (deformable body dynamics) differ from those that represent elastic behavior and from those that represent heat transfer processes. In fact, it is those differences that direct the type of solution approach that is used to solve for the state variables that are unknown in those physical applications.

Hyperbolic differential equations differ greatly from the rest of the mathematical representations. Therefore in this chapter our attention is focused on only those problems that produce hyperbolic differential equations. The usual computational strategy that is used on these equations is called the method of characteristics, which forms the basis for this chapter.

Before 1964, the method of characteristics was used primarily to solve problems in fluid mechanics. These applications included shock waves in compressible fluids flows.[1] Chou *et al.*[2] used this technique to solve nonlinear fluid problems that resulted from hypervelocity impacts. In 1965, Chou and Koenig[3] demonstrated that the method of characteristics may be generalized to problems involving wave propagation in elastic solids. Since that time, researchers have been using the method of characteristics for all problems that have discontinuities in their physical behavior.

12-2 CLASSIFICATION OF PARTIAL DIFFERENTIAL EQUATIONS

Consider the generalized nonlinear second-order partial differential equation,

$$A\frac{\partial^2\phi}{\partial x^2} + 2B\frac{\partial^2\phi}{\partial x\partial y} + C\frac{\partial^2\phi}{\partial y^2} = D$$

or

$$A\phi_{xx} + 2B\phi_{xy} + C\phi_{yy} = D, \tag{12-1}$$

where ϕ_{xx}, ϕ_{xy}, and ϕ_{yy} are shorthand notations for the second derivatives of ϕ, while A, B, C, and D are functions of ϕ, ϕ_x, and ϕ_y. Since our initial solution strategy would be to use the methods of Chap. 5, we begin by constructing a Taylor series for ϕ, viz.,

$$\phi(x, y) = \phi_0 + \frac{\partial\phi}{\partial x}\,\mathrm{d}x + \frac{\partial\phi}{\partial y}\,\mathrm{d}y + \cdots + \frac{\partial^2\phi}{\partial x\partial y}\,\mathrm{d}x\,\mathrm{d}y + \cdots + . \tag{12-2}$$

If the finite-difference techniques, which depend on continuous derivatives, are to be used here, then we must ensure that all the derivatives in the Taylor series exist. We may calculate expressions for the second derivatives

$$\frac{\partial^2\phi}{\partial x^2}, \frac{\partial^2\phi}{\partial y^2}, \frac{\partial^2\phi}{\partial x\partial y}$$

and determine whether there are any conditions under which these derivatives may be discontinuous within the x–y plane and thereby cease to exist. Using Eqs. (12-1) along with the chain rule of differentiation for the first derivatives, we can form the following equations:

$$d\phi_x = \phi_{xx}\,dx + \phi_{xy}\,dy, \tag{12-3}$$

$$d\phi_y = \phi_{xy}\,dx + \phi_{yy}\,dy, \tag{12-4}$$

$$D = \phi_{xx}A + \phi_{xy}(2B) + \phi_{yy}C. \tag{12-5}$$

Equations (12-3)–(12-5) may be viewed as three simultaneous equations for the unknown second derivatives ϕ_{xx}, ϕ_{xy}, and ϕ_{yy}. If we arbitrarily choose to solve for ϕ_{xy} by Cramer's rule and seek to determine whether it can be made to be undefined, we get

$$\phi_{xy} = \frac{\begin{vmatrix} dx & d\phi_x & 0 \\ 0 & d\phi_y & dy \\ A & D & C \end{vmatrix}}{\begin{vmatrix} dx & dy & 0 \\ 0 & dx & dy \\ A & 2B & C \end{vmatrix}} = \frac{0}{0}. \tag{12-6}$$

The determinant of the denominator of Eq. (12-6) yields

$$C\,dx^2 + A\,dy^2 - 2B\,dx\,dy = 0$$

or

$$A\left(\frac{dy}{dx}\right)^2 - 2B\left(\frac{dy}{dx}\right) + C = 0. \tag{12-7}$$

Solving Eqs. (12-7) by the quadratic formula (for an algebraic polynomial), we get

$$\left(\frac{dy}{dx}\right)_{\text{characteristic } I^{\pm}} = \frac{B \pm \sqrt{B^2 - AC}}{A}. \tag{12-8}$$

The families of curves that are represented by Eq. (12-8) are called the characteristic curves, or characteristics. These are the curves along which ϕ_{xy} does not exist. We can now use Eq. (12-8) to categorize partial differential equations such as the one that is given in Eq. (12-1).

If $B^2 - AC > 0$, then two real characteristics exist and the differential equation is hyperbolic.

If $B^2 - AC = 0$, then only one real characteristic exists and the differential equation is parabolic.

If $B^2 - AC < 0$, then there are no real characteristics and the differential equation is elliptic.

From these criteria we can now evaluate the type of physical processes that are described by certain types of partial differential equations. Elliptic equations can exist for problems only in which there are no discontinuities. The theory of elasticity is an example of a process that produces elliptic partial differential equations. Parabolic equations allow only one real characteristic, which would allow a discontinuity in the second derivative. However, this condition does not qualify as being a physical entity that would carry a discontinuity. Therefore, only problems that allow two real characteristics are considered to be truly discontinuous physical problems. This is because, in engineering problems, the discontinuity in the dependent variable must be able to propagate in both the positive and the negative coordinate directions from its point of origin. For this to happen, we must have a characteristic that can carry information in the positive direction and one to carry the information in the negative direction, i.e., two real characteristics. In parabolic equations, the one real characteristic can propagate information in only one direction, which is physically impossible. Partial differential equations that describe the process of conduction heat transfer are parabolic. Thermal information can be carried in only one direction and cannot signal its information in the other direction. Therefore applications in conductive heat transfer are not amenable to the use of the method of characteristics.

The remainder of this chapter is devoted to the exploration of hyperbolic differential equations and the solution processes that must be used to solve for the physical variables. The author must remind the readers, once again, that the values of the dependent variables in hyperbolic differential equations may not be calculated from a direct application of finite differences since these techniques depend on the existence of all the derivatives. In order to be able to calculate the values of the dependent variables in hyperbolic differential equations, we turn to the evaluation of the numerator of Eq. (12-6). By expanding the determinant of the numerator and using Eq. (12-8), we get

$$C\,\mathrm{d}x\,\mathrm{d}\phi_y + \mathrm{d}\phi_x\,\mathrm{d}y A - D\,\mathrm{d}y\,\mathrm{d}x = 0$$

or

$$\left(\frac{\mathrm{d}\phi_y}{\mathrm{d}\phi_x}\right)_{I^+,\,I^-} = -\frac{B \pm \sqrt{B^2 - AC}}{C} + \frac{D}{C}\left(\frac{\mathrm{d}y}{\mathrm{d}\phi_x}\right). \tag{12-9}$$

These two equations describe how the variables ϕ_x and ϕ_y vary along the characteristics I^+ and I^-. Equations (12-9) are called the characteristics equations and are valid only along the physical characteristics that are represented by Eq. (12-8). The upper sign refers to the I^+ characteristics, while the lower sign refers to the I^- characteristic. Basically, then, the method of characteristics has transformed a partial differential equation in the x–y plane [Eq. (12-1)] to ordinary differential equations [Eqs. (12.9)] that are valid only along the characteristic curves.

Figure 12-1 shows a schematic representation of the physical characteristics. The dependent variables are unknown at point 3. The characteristic curves that were derived

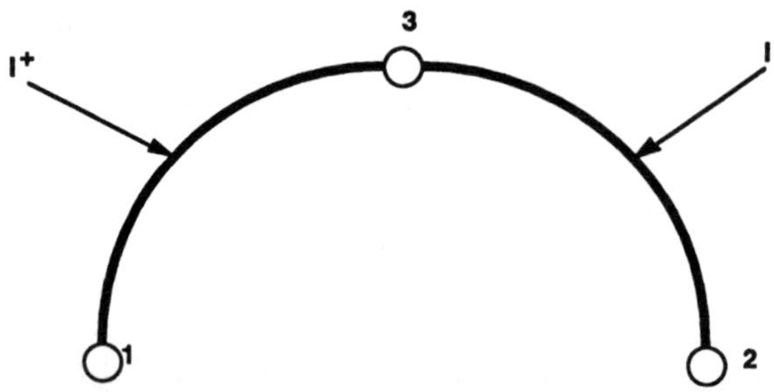

Figure 12-1 Typical nonlinear characteristic curve.

above are nonlinear, since the coefficients A, B, C, and D are functions of the dependent variables. Thus, emanating from points 1 and 2 are nonlinear curves that intersect at point 3.

As a final note, if one had chosen to solve for either ϕ_{xx} or ϕ_{yy} by using Eqs. (12-3)–(12-5), the resulting physical characteristics and characteristic equations would have been identical to those that have already been derived. The proof is left to the readers.

12-3 INVESTIGATIONS IN ENGINEERING

Before delineating the process by which a solution to a physical application may be achieved by using the method of characteristics, let us use the above techniques to demonstrate the use of the method of characteristics in categorizing problems in solid mechanics and in fluid mechanics. In the following sections of this chapter, complete solution equations are developed for two representative problems.

I. Theory of elasticity. The differential equation that governs a two-dimensional problem in the theory of elasticity in terms of the displacement is the Laplace equation:

$$\frac{\partial^2 u}{\partial x^2} + \frac{\partial^2 u}{\partial y^2} = 0. \tag{12-10}$$

When we add the chain-rule equations for the changes in the first derivatives, we get

$$\mathrm{d}u_x = \frac{\partial^2 u}{\partial x^2}\,\mathrm{d}x + \frac{\partial^2 u}{\partial x\,\partial t}\,\mathrm{d}t, \tag{12-11}$$

$$\mathrm{d}u_t = \frac{\partial^2 u}{\partial x\,\partial t}\,\mathrm{d}x + \frac{\partial^2 u}{\partial t^2}\,\mathrm{d}t, \tag{12-12}$$

where u_x and u_t are the first partial derivatives with respect to x and t, respectively. The physical characteristics then become

$$\left(\frac{\mathrm{d}x}{\mathrm{d}t}\right)^2 = -1$$

or

$$\frac{\mathrm{d}x}{\mathrm{d}t} = \text{imaginary}.$$

Thus, for a formulation in the theory of elasticity, there are no characteristics. Discontinuities therefore are not allowed in elastic formulations.

II. Heat conduction. For a body undergoing an unsteady one-dimensional heat conduction process, the governing differential equation becomes

$$\frac{\partial^2 T}{\partial x^2} - k\frac{\partial T}{\partial t} = 0. \tag{12-13}$$

The chain-rule equations for the first derivatives are

$$\mathrm{d}x\left(\frac{\partial^2 T}{\partial x^2}\right) + \mathrm{d}t\left(\frac{\partial^2 T}{\partial x \partial t}\right) = \mathrm{d}u_x \tag{12-14}$$

$$\mathrm{d}x\left(\frac{\partial^2 T}{\partial x \partial t}\right) + \mathrm{d}t\left(\frac{\partial^2 T}{\partial t^2}\right) = \mathrm{d}u_t \tag{12-15}$$

Applying the techniques of Section 12-2, we find that the characteristics for a heat conduction problem become

$$\mathrm{d}t = 0$$

or

$$\frac{\mathrm{d}x}{\mathrm{d}t} \to \infty. \tag{12-16}$$

Expression (12-16) implies that a thermal impulse that is applied at one end of a bar is sensed immediately at the other end of the bar from which it is generated (since the speed of propagation approaches infinity). This result essentially regards a heat conduction problem as having no physically real characteristics in the true sense. This is because a discontinuity in a physical quantity cannot be sensed simultaneously at both ends of the bar.

III. Dynamic wave propagation. For the physical problem in which a one-dimensional tensile bar is struck by an impulsive load at one end, the governing differential equation is

$$\frac{\partial^2 u}{\partial x^2} - \frac{1}{c^2}\frac{\partial^2 u}{\partial t^2} = 0. \tag{12-17}$$

Adding the chain-rule equations, as in the preceding two examples, and using the methods of this chapter, we find that the characteristic curves become

$$\left(\frac{\mathrm{d}x}{\mathrm{d}t}\right)_{I^+,I^-} = \pm c. \tag{12-18}$$

Equation (12-18) shows that the speed of wave propagation in the bar is c. Additionally, there is one characteristic I^+ that carries an impulse in the positive x direction and a second characteristic I^- that carries a signal in the negative x direction.

We have thus shown by the above examples that, in solid mechanics, only dynamic problems in materials will yield differential equations that have real characteristics and are applicable to the methods of this chapter.

12-4 APPLICATIONS IN STRUCTURAL ANALYSIS

12-4.1 Continuous Functions

We now turn our attention to the actual computational methods that may be used in conjunction with the method of characteristics for the solution of problems in structural

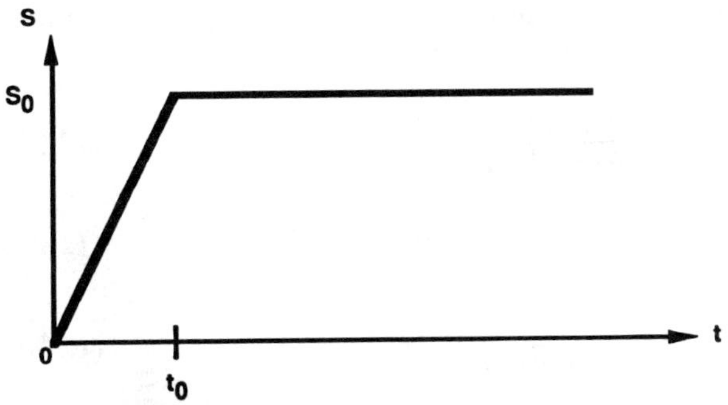

Figure 12-2 Ramp loading on a tensile bar.

mechanics. The simplest example that can be presented is the one-dimensional tensile bar that is subjected to a dynamic end load. The mathematical nature of the load is arbitrary. However, in this chapter we confine our discussions to a load that may be a ramp function, as shown in Fig. 12-2, or a step function, as shown in Fig. 12-3. The differential equation that governs the dynamic inertia of the bar may be expressed in terms of the stress in the bar σ_x and the displacement in the bar u, viz.,

$$\frac{\partial \sigma_x}{\partial x} = \rho \frac{\partial^2 u}{\partial t^2} \tag{12-19}$$

where ρ is the mass density of the material of the bar. The elastic constitutive behavior of the material may be expressed as

$$\varepsilon_x = \frac{\partial u}{\partial x} = \frac{1}{E}\sigma_x. \tag{12-20}$$

Let us assume that the ramp loading has been applied to the bar. In this case, then, the stress is continuous and the particle velocity is continuous. However, the derivatives of these variables are discontinuous. Since the stress is continuous, problems dealing with a step stress application cannot be solved with the following formulations. This restriction is addressed in Subsection 12-4.2.

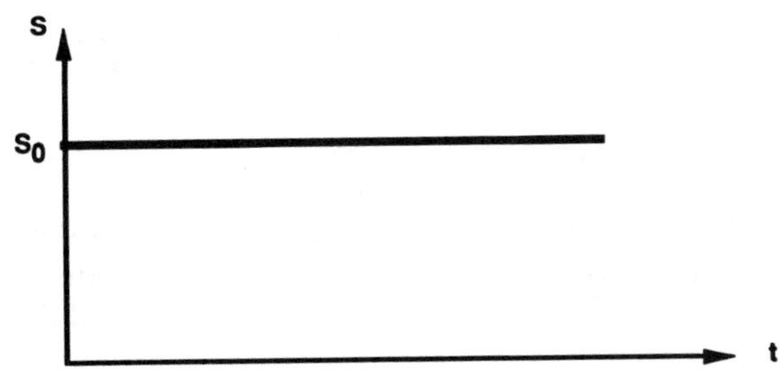

Figure 12-3 Step loading on a tensile bar.

If we now apply the techniques of this chapter, we get the following four equations for the four unknown first derivatives, $\partial\sigma_x/\partial x$, $\partial\sigma_x/\partial t$, $\partial u/\partial x$, and $\partial u/\partial t$:

$$\frac{\partial\sigma_x}{\partial x} - \rho\frac{\partial v}{\partial t} = 0, \tag{12-21}$$

$$-\frac{1}{E}\frac{\partial\sigma_x}{\partial t} + \frac{\partial v}{\partial x} = 0, \tag{12-22}$$

$$\mathrm{d}t\left(\frac{\partial\sigma_x}{\partial t}\right) + \mathrm{d}x\left(\frac{\partial\sigma_x}{\partial x}\right) = \mathrm{d}\sigma_x, \tag{12-23}$$

$$\mathrm{d}t\left(\frac{\partial v}{\partial t}\right) + \mathrm{d}x\left(\frac{\partial v}{\partial x}\right) = \mathrm{d}v, \tag{12-24}$$

where $v = \partial u/\partial t$ is the particle velocity in the bar.

Using Cramer's rule to solve for a typical first derivative, say $\frac{\partial\sigma_x}{\partial t}$, we get

$$\frac{\partial\sigma_x}{\partial t} = \frac{\begin{vmatrix} 0 & 0 & -\rho & 0 \\ -\frac{1}{E} & 0 & 0 & 1 \\ \mathrm{d}t & \mathrm{d}\sigma_x & 0 & 0 \\ 0 & \mathrm{d}v & \mathrm{d}t & \mathrm{d}x \end{vmatrix}}{\begin{vmatrix} 0 & 1 & -\rho & 0 \\ -\frac{1}{E} & 0 & 0 & 1 \\ \mathrm{d}t & \mathrm{d}x & 0 & 0 \\ 0 & 0 & \mathrm{d}t & \mathrm{d}x \end{vmatrix}} = \frac{0}{0}. \tag{12-25}$$

Evaluating the determinant of the denominator, we get

$$\left(\frac{\mathrm{d}x}{\mathrm{d}t}\right)_{I^+,I^-} = \pm\sqrt{\frac{E}{\rho}} = \pm c. \tag{12-26}$$

Equation (12-26), which indicates the physical characteristics, specifies the speed of have propagation in the bar. Because the constant material properties, the wave speed is constant. Figure 12-4 shows a representation of these characteristic curves, which are linear in this case.

In order to calculate how the dependent variables (i.e., the stress and the velocity) change along these characteristics (and therefore within the body), we evaluate the determinant of the numerator in Eq. (12-25), viz.,

$$\mp c\,\mathrm{d}\sigma_x = -E\,\mathrm{d}v, \tag{12-27}$$

where the upper sign refers to the I^+ characteristic and the lower sign refers to the I^- characteristic. Equations (12-26) and (12-27) form the basis for a computational methodology that can be used to solve for the stresses, the particle velocities, and the

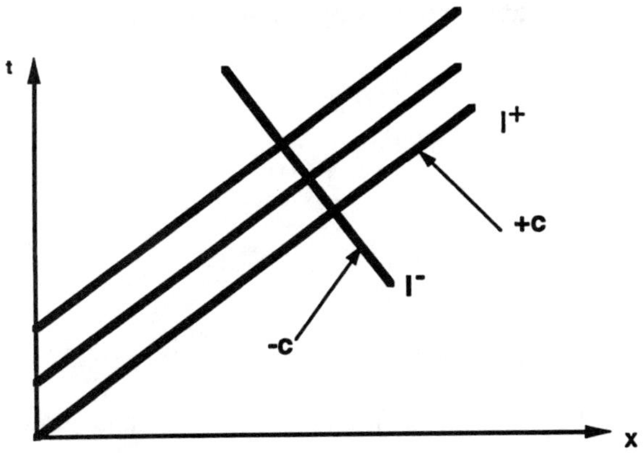

Figure 12-4 Characteristic curves for one-dimensional wave propagation.

displacements of points in the rod. If we refer to Fig. 12-4 and form a grid of I^+ and I^- characteristics within the entire x–t plane, then the variables σ_x and v may systematically be evaluated. Figure 12-5 shows a typical grid of four points, $(I-1, J-1)$, $(I-1, J)$, $(I, J-1)$, and (I, J), within the x–t plane. Let us assume that the variables have been evaluated at three of the points and must be determined at point (I, J). Writing the characteristic Eq. (12-27) from point $(I-1, J)$ to (I, J) and from $(I, J-1)$ to (I, J), we get

$$-c\{\sigma_x(I, J) - \sigma_x(I-1, J)\} = -E\{v(I, J) - v(I-1, J)\} \tag{12-28}$$

along the I^+ characteristic and

$$c\{\sigma_x(I, J) - \sigma_x(I, J-1)\} = -E\{v(I, J) - v(I, J-1)\} \tag{12-29}$$

along the I^- characteristic or,

$$\begin{bmatrix} -c & E \\ c & E \end{bmatrix} \begin{bmatrix} \sigma_x(I, J) \\ v(I, J) \end{bmatrix} = \begin{bmatrix} -c\sigma_x(I-1, J) + Ev(I-1, J) \\ c\sigma_x(I, J-1) + Ev(I, J-1) \end{bmatrix}, \tag{12-30}$$

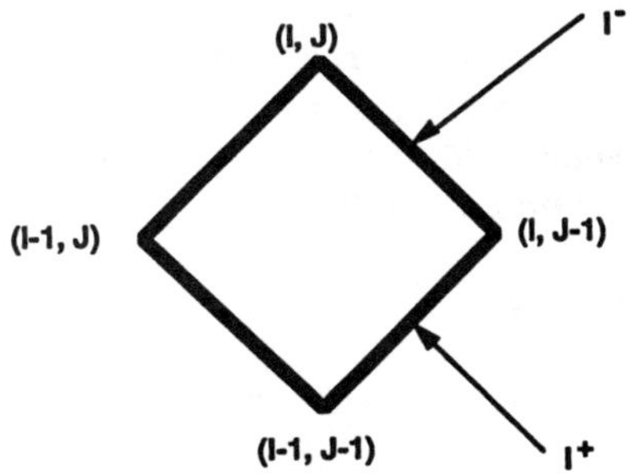

Figure 12-5 Typical grid for computations.

from which we can solve for the unknown stress and velocity at point (I, J), i.e.,

$$\sigma_x(I, J) = \frac{1}{2}\{\sigma_x(I-1, J) + \sigma_x(I, J-1)\} + \frac{E}{2c}\{v(I, J-1) - v(I-1, J)\}, \tag{12-31}$$

$$v(I, J) = \frac{c}{2E}\{\sigma_x(I, J-1) - \sigma_x(I-1, J)\} + \frac{1}{2}\{v(I-1, J) + v(I, J-1)\}. \tag{12-32}$$

Continuous application of Eqs. (12-31) and (12-32) in the entire x–t plane allows us to calculate the entire stress and velocity field within the bar. At the end of the bar where the initial impulse was applied (i.e., $x = 0$), the stress is known at all time intervals $\mathrm{d}t = \mathrm{d}x/c$. The particle velocity at these points may be calculated with the I^- characteristic only (since there is no material to the left of $x = 0$):

$$v(I, J) = v(I, J-1) - \frac{c}{E}\{\sigma_x(I, J) - \sigma_x(I, J-1)\}. \tag{12-33}$$

One must remember that characteristics I^+ and I^- have physical significance since they represent the speeds of the propagating waves in the material. Therefore, if the initial impulse occurs at $x = 0$, then the region $x \geq c\,\mathrm{d}t$ has no stress, strain, velocity, or displacement since the initial wave has not reached that position in the rod.

The above procedures may be used for any stress or velocity function that is continuous at $x = 0$ or at the far end of the rod at $x = L$. If, however, the initial impulse is discontinuous (such as a step function), then the above procedure must be modified.

12-4.2 Discontinuities in the Variables

The methodology of Subsection 12-4.1 cannot be directly applied to the numerical simulation of the bar that is subjected to a suddenly applied load (or velocity) at one end of the rod. Equation (12-27) shows that if a discontinuity occurs in one of the dependent variables, then a discontinuity in the other variable is also present. Thus, if the bar is struck by a blow that results in a suddenly applied stress σ_0, the particle velocity v also changes suddenly. Figure 12-6 shows two characteristics, I_1^+ and I_2^+, each of which carry a discontinuity in the stress and the velocity at points A and B. This discontinuity will always occur when the initial impulse is discontinuous, such as an explosive shock.

In order to calculate the conditions that must be imposed on the theory of Subsection 12-4.1 when a jump occurs, we write Eq. (12-27) from $A \to B$ along I^-, viz.,

$$\sigma_{x_A} - \sigma_{x_B} + \frac{E}{c}(v_A - v_B) = 0.$$

If the discontinuity is stress at point A is called $\delta\sigma_x$ and the discontinuity in velocity δv, then, as A approaches B,

$$\delta\sigma_x + \frac{E}{c}\delta v = 0$$

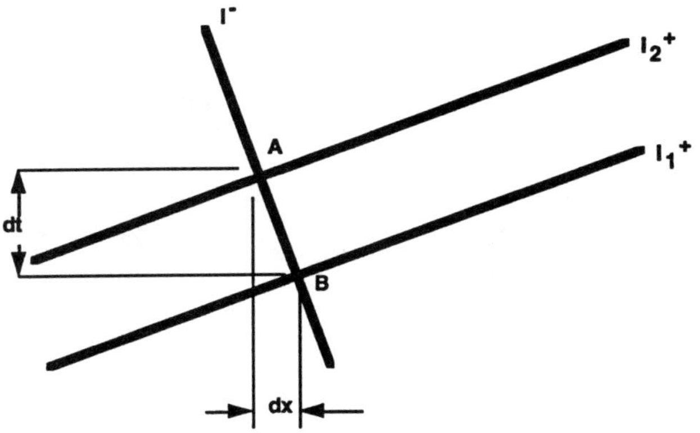

Figure 12-6 Jump conditions.

or

$$\delta\sigma_x = -\frac{E}{c}\delta v. \tag{12-34}$$

Equation (12-34) is called a jump condition and shows that a tensile discontinuity in the stress will produce a negative jump in the particle velocity if the stress is applied to the left end of the rod. Equation (12-34) describes the relationship between the stress and the velocity at any point where a discontinuity is encountered in one of the physical variables in the rod. However, this equation does not allow the researcher to evaluate how these discontinuities vary as they propagate from point to point inside the rod. Equation (13-34) also does not allow the calculation of the discontinuity when it is reflected from the other end of the bar. Before we can use the above result, therefore, we must derive a formula for the manner by which this jump condition propagates along the rod. For example, if a jump stress σ_0 is applied at $x = 0$, then Eq. (12-34) describes how the velocity changes at $x = 0$. It does not describe how the jump propagates into the rod. To answer this question, consider the propagation of the discontinuities along I_1^+ from point A and along I_2^+ from point B. Thus, from Eq. (12-27),

$$-\mathrm{d}\sigma_{x_A} + \frac{E}{c}\,\mathrm{d}v_A = 0, \tag{12-35}$$

$$-\mathrm{d}\sigma_{x_B} + \frac{E}{c}\,\mathrm{d}v_B = 0, \tag{12-36}$$

as $I_1^+ \to I_2^+$,

$$-\mathrm{d}(\delta\sigma_x) + \frac{E}{c}\,\mathrm{d}(\delta v) = 0$$

or, by substitution of the jump condition that is given by Eq. (12-34),

$$+\mathrm{d}\left(\frac{E}{c}\delta v\right) + \frac{E}{c}\,\mathrm{d}(\delta v) = 0$$

or

$$\mathrm{d}(\delta v) = 0. \tag{12-37}$$

Equation (12-37) may be integrated to yield

$$\delta v = K_1 = \text{const.}, \tag{12-38}$$

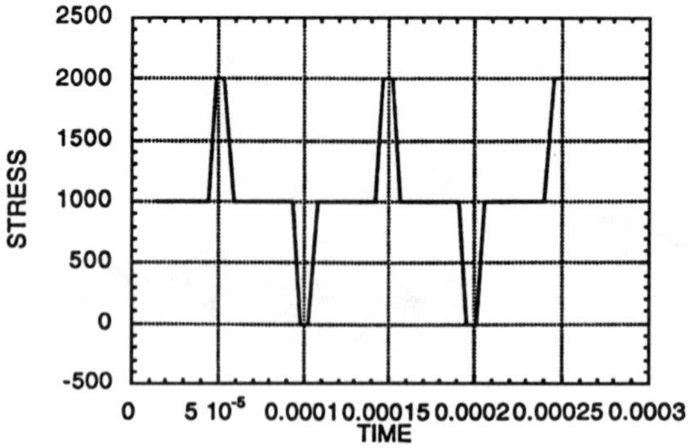

Figure 12-7 Stress history at $x = 3.0$.

and, by Eq. (12-34),

$$\delta\sigma_x = -\frac{E}{c}K_1. \tag{12-39}$$

Equations (12-38) and (12-39) indicate that any discontinuity in the rod propagates without magnification or attenuation. For example, if the stress jumps to σ_0 at $x = 0$, then this value will propagate unchanged throughout the rod until it encounters another boundary. The velocity will jump to a value of $-\frac{c}{E}\sigma_0$ at $x = 0$ and will propagate in a constant manner throughout the rod, until it hits another boundary.

Appendix 12.1 contains a computer code, called **moc.f**, that was written to solve for the stresses and velocities in a rod that is hit at one end by a suddenly applied stress load ($\sigma = 1000$ psi) on a bar whose length is 10 in. The modulus of elasticity of the material is $E = 30 \times 10^6$ psi, and the density is $\rho = 0.278$ lb/in.^{3} The bar is restrained from motion at the far end. Figures 12-7–12-9 show the stress histories in the bar at $x = 3.0$, 5.0, and 7.0 in., respectively. Because of the fixed boundary condition, the stress doubles in magnitude when it reflects from that end. The stress wave front then propagates toward the impacted end. Since the problem is elastic, there is no mechanism to remove energy from the bar. Therefore the wave propagation continues, in a repeating pattern, indefinitely. Figures 12-10–12-12 show the velocities at the above points. Again, the

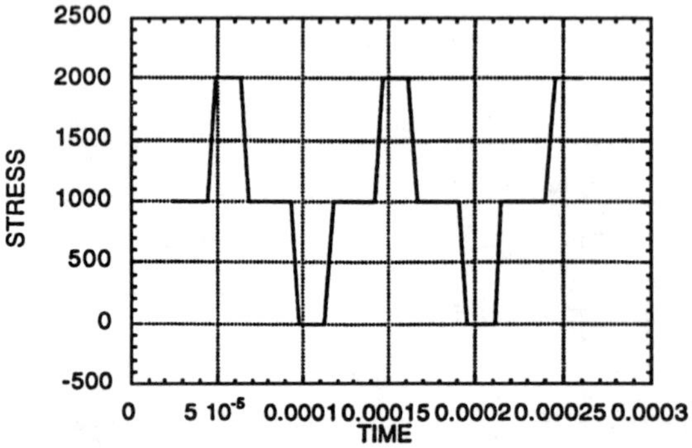

Figure 12-8 Stress history at $x = 5.0$.

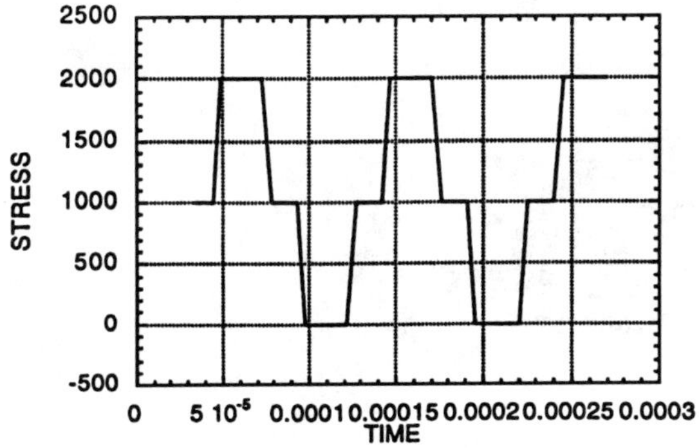

Figure 12-9 Stress history at $x = 7.0$.

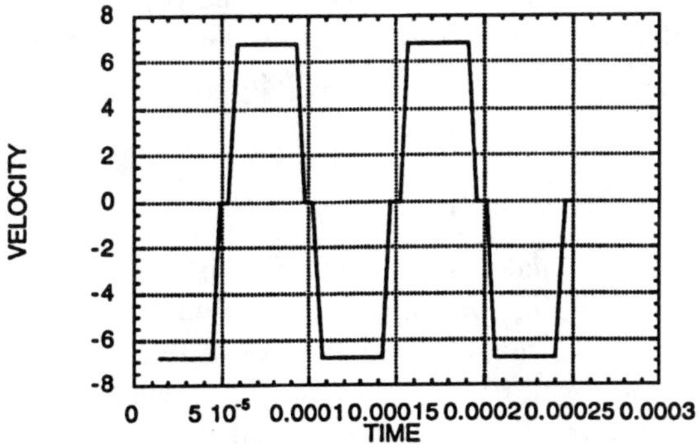

Figure 12-10 Velocity history at $x = 3.0$.

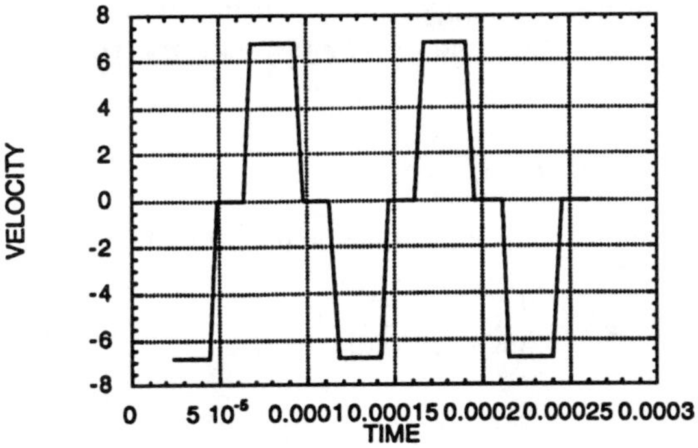

Figure 12-11 Velocity history at $x = 5.0$.

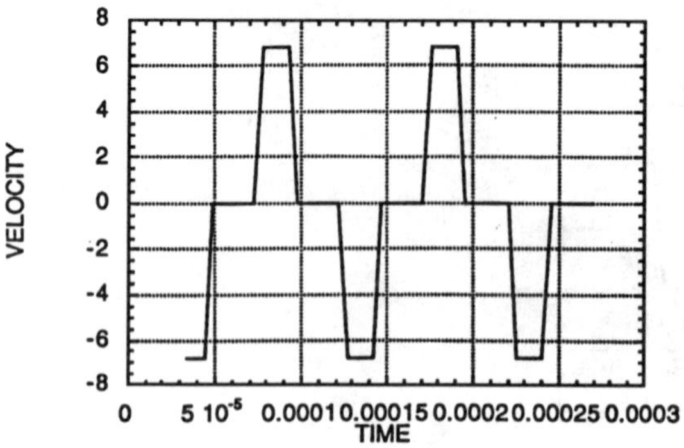

Figure 12-12 Velocity history at $x = 7.0$.

wave pattern repeats. The zero-velocity boundary condition causes the particle velocity in the bar to vanish when the reflected wave propagates back into the medium.

The Problems at the end of this chapter contain examples of other geometric configurations, some of which have jump conditions that propagate with a constant value in the medium and some that have jump conditions that vary as the wave propagates into the medium. The solutions of those problems will add to the reader's ability to use the method of characteristics in the solution of structural problems in engineering.

12-5 APPLICATION IN FLUID MECHANICS

In the analysis of engineering problems that model fluid mechanical processes, the method of characteristics can contribute greatly since the physics of the problem usually allows for discontinuities in the flow field (e.g., shocks). The physics of fluid flow is so complicated that its application is usually well beyond the scope of an elementary textbook. The fluid flow example that is given herein is a simplified version of the complete set of flow equations that are necessary to model this physics.

In general, fluid flow processes must satisfy the equations of continuity, momentum, and energy. To these equations are usually added an equation of state for the fluid. The results of this type of formulation produce five state variables, i.e., density ρ, three components of the fluid velocity u, v, w, and the temperature T in the fluid. If we consider a two-dimensional simplification of these equations and relate the velocity vector $\bar{V}$ to a potential ϕ such that

$$\bar{V} = \nabla\phi = \left(\frac{\partial\phi}{\partial x}\right)\bar{i} + \left(\frac{\partial\phi}{\partial y}\right)\bar{j},$$

we can reduce the two-dimensional fluid flow equations to the potential equation

$$\left(1 - \frac{\phi_x^2}{c^2}\right)\phi_{xx} - 2\frac{\phi_x\phi_y}{c^2}\phi_{xy} + \left(1 - \frac{\phi_y^2}{c^2}\right)\phi_{yy} = 0 \tag{12-40}$$

or

$$A\phi_{xx} + B\phi_{xy} + C\phi_{yy} = 0.$$

Using the methods that have been developed in this chapter, we can show that the physical characteristics are

$$\left(\frac{dy}{dx}\right)_{I,\,II} = \frac{B \pm \sqrt{B^2 - 4AC}}{2A} \tag{12-41}$$

and the characteristic equation is

$$\left(\frac{d\phi_x}{d\phi_y}\right)_{I,\,II} = -\frac{A}{C}\left(\frac{dy}{dx}\right)_{I,\,II}. \tag{12-42}$$

In this case, we note that these physical characteristic curves are nonlinear (since they depend on the current values of ϕ_x and ϕ_y) so they cannot be calculated at the beginning of the solution process, as in the wave propagation in the tensile bar. We must therefore perform the solution process on a point-to-point basis. At each step in the analysis process, we must calculate the values of the dependent variables ϕ_x and ϕ_y at each point in the flow field and the values of the physical characteristics that locate the points in the flow region. Figure 12-13 shows a typical grid point 3, where the values of the variables are required. We assume that the dependent variables are known at points 1 and 2. Using Eq. (12-41) to find the location of point 3, we have

$$\frac{y_3 - y_1}{x_3 - x_1} = \left(\frac{dy}{dx}\right)_I, \tag{12-43}$$

$$\frac{y_3 - y_2}{x_3 - x_2} = \left(\frac{dy}{dx}\right)_{II}. \tag{12-44}$$

From Eqs. (12-43) and (12-44), (x_3, y_3) is uniquely determined. Equation (12-42) yields

$$\frac{\phi_{y3} - \phi_{y1}}{\phi_{x3} - \phi_{x1}} = -\frac{A}{C}\left(\frac{dy}{dx}\right)_I, \tag{12-45}$$

$$\frac{\phi_{y3} - \phi_{y2}}{\phi_{x3} - \phi_{x2}} = -\frac{A}{C}\left(\frac{dy}{dx}\right)_{II}. \tag{12-46}$$

The simultaneous solution of the above equations yields ϕ_{y3} and ϕ_{x3} and thereby completely defines the state of the flow field at point 3. The continual application of the above equations will eventually lead to the solution of the fluid potential equation in the entire field.

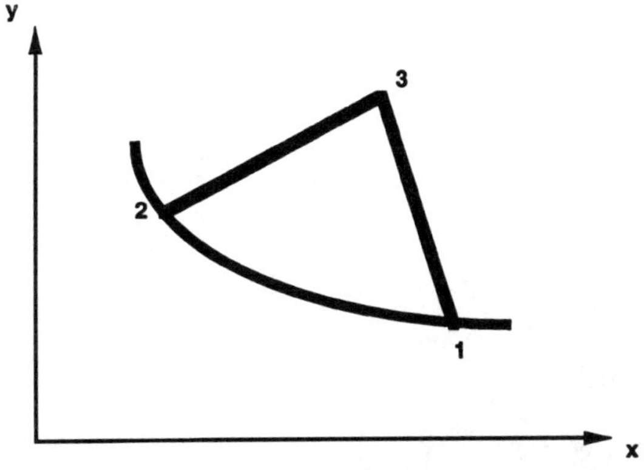

Figure 12-13 Characteristics of a potential flow problem.

12-6 CONCLUSION

This chapter has explored, in an abbreviated manner, one of the most powerful approaches that a researcher can take toward the numerical simulation of hyperbolic differential equations. The methods of Chap. 5 would also be applicable to these differential equations if no discontinuities in the stresses, velocities, or pressures were present in the physical process that is being simulated. However, if engineers limited the scope of their investigations to problems in which the dependent variables (or their derivatives) were continuous, very few significant problems would be solvable.

The method of characteristics therefore provides a solution approach for all kinds of hyperbolic differential equations, regardless of whether a discontinuity in the solution variables exists. The methods presented in this chapter are only a small portion of the actual techniques that may be used when the method of characteristics is used. Readers are encouraged to consult the references and the relevant readings that are listed at the end of this chapter. Additionally, the solution of some of or all the problems that are delineated at the end of this chapter will provide for hands-on experience in using these methods as a computational tool.

REFERENCES

1. Shapiro, A. H. (1954). *The dynamics and Thermodynamics of Compressible Fluid Flow*, Vol. II, Ronald Press, New York.
2. Chou, P. C., Sidhu, H. S., and Zajak L. J. (1964). Attenuation of the Strong Plane Shock Produced in a Solid by Hypervelocity Impact, DIT Report 125-5, Drexel Institute of Technology, Philadelphia, Pennsylvania.
3. Chou, P. C. and Koenig, H. A. (1965). A Unified Approach to Cylindrical and Spherical Elastic Waves by Method of Characteristics, *J. Appl. Mech.* Vol. **xx**, pp. xxx–xxx.

FURTHER READING

Chou, P. C., and Koenig, H. A. (1964). Propagation of Cylindrical and Spherical Elastic Waves by Method of Characteristics, DIT Report 160-5, Drexel Institute of Technology, Philadelphia, Pennsylvania.
Chou, P. C. and Koenig, H. A. (1965). Propagation of Cylindrical and Spherical Elastic Waves by Method of Characteristics, NASA Technical Note NASA TN D-2644, Washington, D.C.
Chou, P. C. and Schaller, R. J. (1966). Propagation of Cylindrical Shear Waves in Nonhomogeneous Elastic Media, AIAA paper 66-444, Washington, D.C.
Schreier, S. (1982). *Compressible Flow*, Wiley, New York.

PROBLEMS

12-1 Consider an elastic circular cylinder, shown in Figure 12-14, whose governing differential equation and constitutive behavior are given by

$$\frac{\partial \sigma_r}{\partial r} + \frac{\sigma_r - \sigma_\theta}{r} = \rho \frac{\partial^2 u}{\partial t^2},$$

$$\frac{\partial u}{\partial r} = \frac{1}{E}(\sigma_r - \upsilon \sigma_\theta),$$

$$\frac{u}{r} = \frac{1}{E}(\sigma_\theta - \upsilon \sigma_r),$$

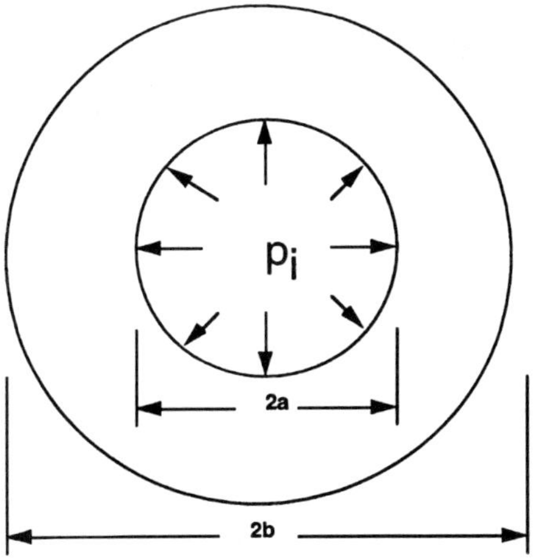

Figure 12-14 Circular cylinder with an internal pressure.

where σ_r is the radial stress, σ_θ is the tangential (hoop) stress, u is the radial displacement, r is the space variable, and t is the temporal variable.

Show that there are three characteristic lines, i.e.,

$$I_-^+ : \quad \frac{dr}{dt} = \pm c = \pm\sqrt{\frac{E}{\rho(1-\nu^2)}},$$

$$II^+ : \quad dr = 0 \qquad \text{(degenerate wave)}.$$

12-2 Show that the characteristic equations for Problem 12.1 are

$$d\sigma_r \mp \rho c\, dv = \{-(\sigma_r - \sigma_\theta) \pm \rho \nu c v\}\frac{dr}{r} \qquad \text{along } I^+ \text{and along } I^-,$$

$$\frac{d\sigma_r}{d\sigma_\theta} = \left\{1 - \frac{E v}{r}\frac{dt}{d\sigma_\theta}\right\}\frac{1}{\nu} \qquad \text{along } II.$$

12-3 Show that the jump conditions for the cylinder in Problem 12-1 are

$$\delta\sigma_r = K r^{-\frac{1}{2}},$$

$$\delta\sigma_\theta = \nu K r^{-\frac{1}{2}},$$

$$\delta v = \mp\frac{1}{\rho c} K r^{-\frac{1}{2}},$$

This shows that discontinuities propagate inversely as the square root of the radial position in the cylinder.

12-4 For the cylinder that is described in Problem 12-1 and shown in Fig. 12-14, the temporal variation of the internal pressure at $r = a$ is shown in Fig. 12-15. The external boundary of the cylinder at $r = b$ remains stress free. Write a computer code, using the method of characteristics, to solve for the stresses σ_r and σ_θ as well as the particle velocity v as functions of r and t. Plot your results. For this problem use a modulus of elasticity $E = 30 \times 10^6$ psi, Poisson ratio $\nu = 0.3$, inner radius $a = 0.5$ in., outer radius $b = 10.0$ in., and density $\rho = 0.278$ lb/in.^{3}. The internal step pressure p_0 has the value of 1000 psi.

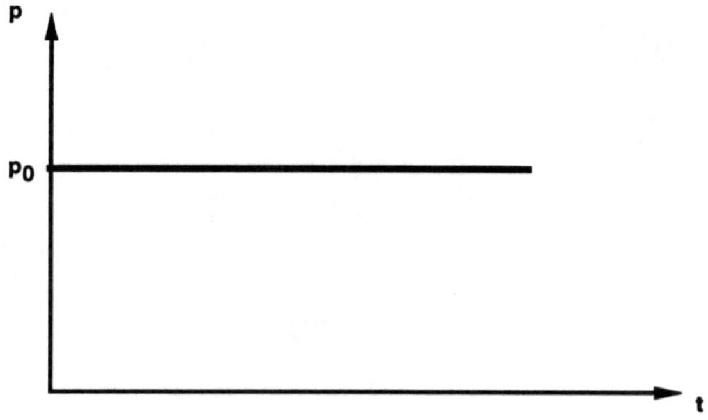

Figure 12-15 Internal pressure in a circular cylinder.

12-5 A circular elastic plate is subjected to normal shear and moment flexural inputs. The plate is shown in Fig. 12-16. The governing equations for this configuration are

$$\frac{\partial M_r}{\partial r} + \frac{1}{r}\left(M_r - M_\theta\right) - Q_r = \frac{\rho h^3}{12}\frac{\partial \phi_t}{\partial t},$$

$$\frac{\partial Q_r}{\partial r} + \frac{1}{r}Q_r = \rho h \frac{\partial w_t}{\partial t},$$

$$\frac{\partial M_r}{\partial t} = D\left\{\frac{\partial \phi_t}{\partial r} + \frac{\nu}{r}\phi_t\right\},$$

$$\frac{\partial M_\theta}{\partial t} = D\left\{\frac{\phi_t}{r} + \nu\frac{\partial \phi_t}{\partial r}\right\},$$

$$\frac{\partial Q_r}{\partial t} = k_2^2 Gh\left(\phi_t + \frac{\partial w_t}{\partial r}\right),$$

Show that five physical characteristics result. They are

$$I_-^+ : \quad \frac{\mathrm{d}r}{\mathrm{d}t} = \pm c_P = \pm\sqrt{\frac{E}{\rho(1-\nu^2)}},$$

$$II_-^+ : \quad \frac{\mathrm{d}r}{\mathrm{d}t} = \pm k_2 c_2 = \pm k_2\sqrt{\frac{G}{\rho}},$$

$$III : \quad \mathrm{d}r = 0.$$

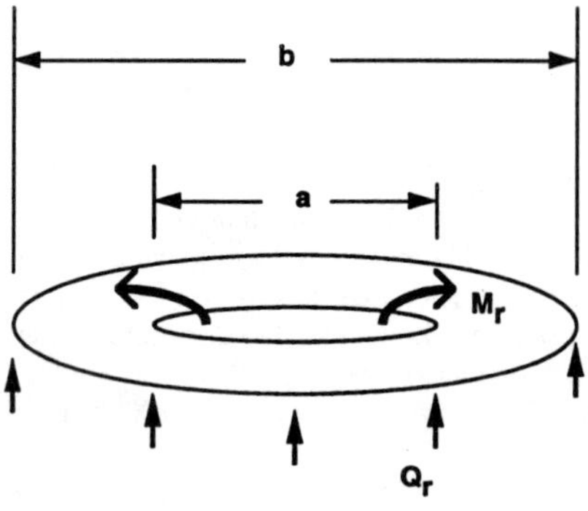

Figure 12-16 Thin circular plate of thickness h.

These results show that, in a thin circular cylinder, there are two wave speeds and a degenerate characteristic. The wave, whose speed is C_p, is called the dilatation wave and carries the impulsive signals from the moment and from the angular velocity. The wave, whose speed is k_2C_2, is called the shear wave and carries the impulsive signals from the transverse shear and from the linear velocity.

12-6 For Problem 12-5, show that the characteristic equations for the variables are

$$dM_r \mp c_P \frac{\rho h^3}{12} d\phi_t = \left(rQ_r - M_r + M_\theta \mp \nu c_P \frac{\rho h^3}{12}\phi_t \right)\frac{dr}{r} \qquad \text{(along } I\text{)},$$

$$dQ_r \mp \rho h k_2 c_2\, dw_t = -(Q_r \mp \rho h k_2 c_2 r \phi_t)\frac{dr}{r} \qquad \text{(along } II\text{)},$$

$$dM_\theta - \nu\, dM_r = \frac{Eh^3}{12}\phi_t \frac{dt}{r} \qquad \text{(along } III\text{)}.$$

12-7 For the thin circular plate, show that the jump conditions can be expressed across I as:

$$\begin{aligned} \delta\phi_r &= Ar^{-\frac{1}{2}}, \\ \delta\phi_t &= \mp c_P A r^{-\frac{1}{2}}, \\ \delta M_r &= DAr^{-\frac{1}{2}}, \\ \delta M_\theta &= \nu DAr^{-\frac{1}{2}}, \\ \delta Q_r &= \delta w_t = \delta w_r = 0, \end{aligned}$$

and across II as

$$\begin{aligned} \delta w_r &= Br^{-\frac{1}{2}}, \\ \delta w_t &= \mp k_2 c_2 B r^{-\frac{1}{2}}, \\ \delta Q_r &= k_2^2 GhBr^{-\frac{1}{2}}, \\ \delta\phi_r &= \delta\phi_t = \delta M_r = \delta M_\theta = 0. \end{aligned}$$

Here, once again, as in the cylinder of Problem 12-1, the discontinuities propagate inversely proportion to the square root of the radial position within the plate.

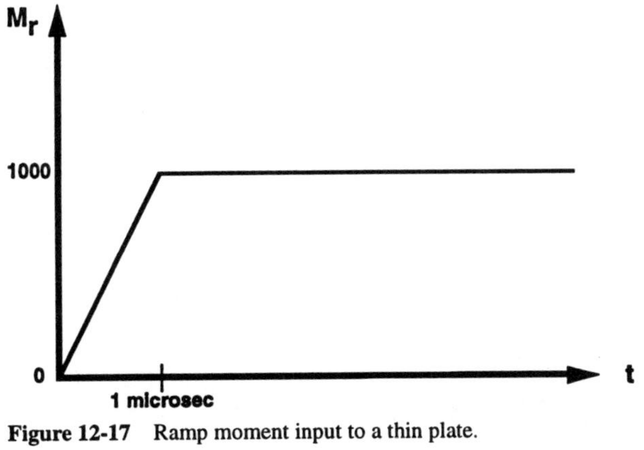

Figure 12-17 Ramp moment input to a thin plate.

12-8 For the plate that is described in Problem 12-5 and shown in Fig. 12-16 the temporal variation of the internal bending moment M_r at $r = a$ is shown in Fig. 12-17. The external boundary of the plate at $r = b$ is fixed (i.e., all the velocities vanish). Write a computer code, using the method of characteristics, to solve for the moments M_r and M_θ, the shear Q_r, the linear velocity w_t, as well as the angular velocity ϕ_t, as functions of r and t. Plot your results. For this problem use a modulus of elasticity $E = 30 \times 10^6$ psi, Poisson ratio $\nu = 0.3$, plate thickness $h = 0.1$ in., inner radius $a = 5.0$ in., outer radius $b = 10.0$ in., and density $\rho = 0.278$ lb/in.^{3}. The plate modulus D is given by

$$D = \frac{Eh^3}{12(1 - \nu^2)}.$$

CHAPTER

THIRTEEN

FINITE-VOLUME METHODS

NOMENCLATURE

ψ	angular rotation
ε_{ψ}	angular strain
ω	angular velocity
$\bar{g}$	gravitational vector
e	internal energy
ε_y	linear strain
ρ	mass density
$\bar{U}$	net vector velocity on the faces of a fluid element
$\bar{n}$	outward normal for the faces of a fluid element
ν	Poissons ratio
ε_r	radial strain
S_r	radial stress
ε	strain
$\bar{S}$	stress tensor in a fluid
ε_θ	tangential strain
S_θ	tangential stress
V_c	volume of a fluid element
A	area
c	speed of wave propagation
dt	temporal increment
dx	finite spatial increment
E	elastic modulus
G	shear modulus
I	moment of inertia

K	thermal conductivity
M	bending moment
m	mass
p	pressure
S	stress
T	temperature
t	time
V	shear force
v	particle velocity
W	work done per unit time
y	deflection

13-1 INTRODUCTION

In Chap. 5, a numerical simulation technique known as the finite-difference method was introduced. This was based on the selection of an approximating function for the dependent variables within a specific physical region and the use of the mathematical concepts of a Taylor series. This technique was shown to be applicable to vectorial representations of the physical processes, i.e., differential equations. In Chap. 10, another concept for the numerical simulation of engineering processes was introduced, i.e., the finite-element method. This was also based on the selection of a function that will approximate the dependent variables within a specific region. In this case, however, the mathematical basis for the technique was an energy formulation known as the potential energy. When this scalar approach and the mathematical concepts from the calculus of variations were used, numerical procedures were established.

In this chapter, the focus of the discussion turns to a more fundamental numerical simulation approach than has been delineated in preceding chapters. This third concept is based on the selection of the region under consideration and the direct application of the laws of physics to obtain numerical values for the variables of importance. There will be no need to select an approximating function before the analysis. Additionally, there is no sophisticated mathematical technique that will be required for setting up the numerical methods. A physical region will be selected and the governing engineering principles will be applied. For these reasons, these new techniques are known as the methods of finite volume.

Before proceeding with the development of these techniques, it is important to point out a misconception that usually accompanies these methods. Since the final formulation of the numerical equations usually appears to be similar to those that might be developed from a finite-difference or a finite-element approach, one believes that the techniques of this chapter are mere extensions of the above two methods. This is not the case. The fundamental approach to the physics in this chapter will be greatly different than previously shown. In the finite-difference methods and in the finite-element methods, the mathematical formulation came first. It was not considered important if these mathematical tools would be applied to physical problems. Here, however, the physical regions are the point of commencement, with the numerical tools' being a product of the physics.

This new method is applied to problems in structural mechanics and to problems in fluid mechanics. In structures, we call the technique direct analysis. When we apply the finite-volume method to fluids, we refer to it as the method of finite volumes. This distinction is made because the fundamental physical approaches to fluid mechanics and to structural mechanics are vastly different, as is explained below.

13-2 DIRECT ANALYSIS

When the method of finite volumes is applied to the area of structural mechanics, it is known as direct analysis. It is given this name since the physical laws are applied directly to the body under consideration. Because of the direct use of the physical laws in this methodology, the mathematical content of the development is minimal. Therefore the presentation of direct analysis is made with several examples of the technique rather than with the development with complex mathematics.

As our first example, consider the one-dimensional tensile bar of Fig. 13-1 whose cross-sectional area is A, length is L, mass density is ρ, and elastic modulus is E. The bar is being restrained from movement on its right-hand side while it is subjected to an impulsive stress (load) on its left-hand side, as shown in Fig. 13-2. For this bar, the speed of elastic wave propagation has been shown to be $c = \sqrt{E/\rho}$. When a section is taken from the beam of length dx (a finite value) and stress is applied on each face of the resulting element i, the free-body diagram of Fig. 13-3 results. We can now apply Newton's law of linear momentum directly to element i. This states that the impulse of the force is equal to the change in the linear momentum.

Applying this statement to Fig. 13-3 yields

$$\mathrm{d}v = \frac{S_{i+1} - S_i}{\rho c}, \tag{13-1}$$

where S_{i+1} is the stress on the right face of element i, S_i, is the stress on the left face of element i, and dv is the velocity change of element i during a time interval $\mathrm{d}t = (\mathrm{d}x/c)$. To follow the temporal variations of stress and velocity accurately, we indicate the updated temporal values by a prime. Since Eq. (13-1) defines the change in the velocity of element i, the updated value for this velocity becomes

$$v_i' = v_i + \mathrm{d}v. \tag{13-2}$$

The elastic strain on the right-hand side of the element may now be calculated as

$$\mathrm{d}\varepsilon_{i+1} = \frac{v_{i+1} - v_i}{c}, \tag{13-3}$$

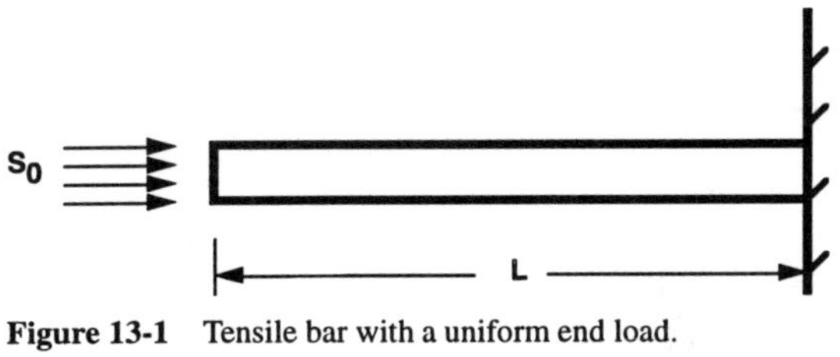

Figure 13-1 Tensile bar with a uniform end load.

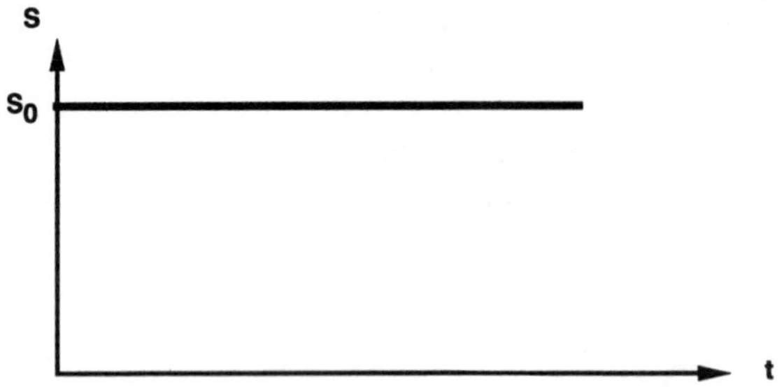

Figure 13-2 Temporal variation of an impulsive end load.

so that the updated strain may be calculated as

$$\varepsilon'_{i+1} = \varepsilon_{i+1} + d\varepsilon_{i+1} \tag{13-4}$$

and the updated stress at the right face of the element is

$$S'_{i+1} = E\varepsilon'_{i+1} \tag{13-5}$$

for an elastic rod.

That is all there is to the direct analysis of a tensile bar. The above equations are cycled from element 1 to element N (the number of elements). When the velocity at the right end of the bar is finally calculated, it must be immediately set to zero to reflect the fixed-end condition. The user now repeats these steps for all time intervals from $t = 0$ to $t = t_{max}$ (which may be arbitrarily chosen) by using the time step dt as the driver.

Appendix 13.1 contains a code called **direct.f** that solves the dynamics of a tensile bar by using the above method of finite volumes.

As a numerical example of the use of direct analysis in the case of the tensile bar, let us select a bar that is 10 in. in length. We then choose the following numerical parameters:

1. Number of elements $N = 10$
2. Modulus of elasticity $E = 30 \times 10^6$ psi
3. Density $\rho = 0.000719\ \text{lb}_m/\text{in.}^3 = 0.28\ \text{lb}_f/\text{in.}^3$

The rod is loaded by a step impulsive load ($\sigma = 1000$ psi) at one end and is restrained at the other end. Figures 13-4–13-6 show the stress histories at the right-hand side of elements 3, 5, and 7. Figures 13-7–13-9 show the particle velocity histories for elements 3, 5, and 7. In all the figures the wave front that is generated at the left end of the rod and propagated within the rod is evident. In the case of the stress, the value of this variable

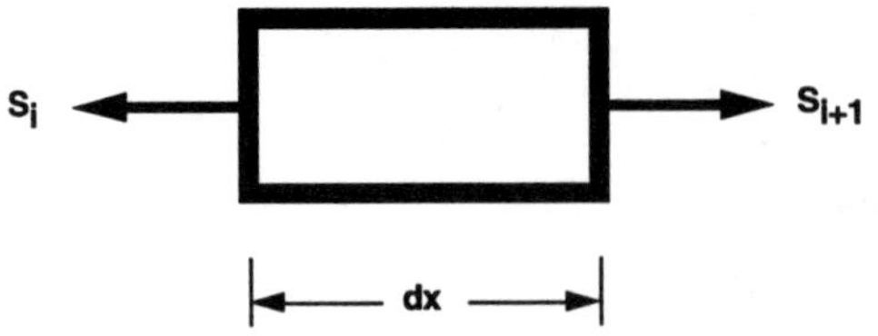

Figure 13-3 Typical element i from a tensile bar.

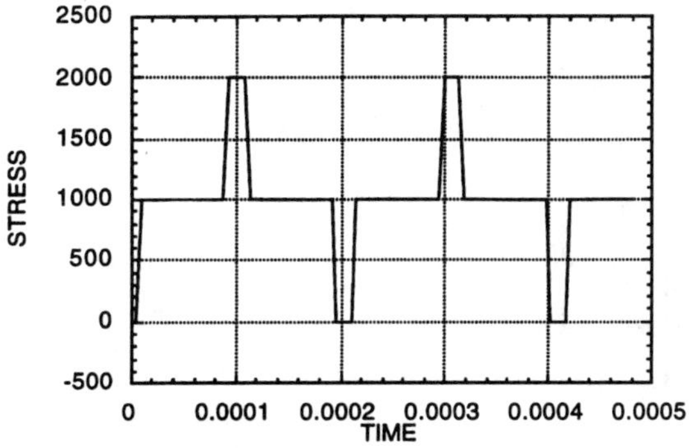

Figure 13-4 Stress history at $x = 3.0$.

remains constant (initially zero) until a wave front arrives at that location to suddenly change its value. Since the right end of the bar is fixed, the stress doubles its value after it hits the fixed end. This higher stress then propagates back into the bar toward the impacted end. Thus the reflected wave doubles the stress value at a particular location on its arrival. The velocity, on the other hand, propagates into the rod with the value that is initially produced by the impulsive stress. When the wave front reaches the fixed end of the bar, the zero-velocity boundary condition is imposed on the bar in the form of the reflected wave front. Since the problem that is being solved is elastic, there is no mechanism by which energy can be dissipated. Thus the wave propagation continues indefinitely toward both ends of the bar and produces a repeatable pattern for the stress and the velocity. The results in Figs. 13-4–13-9 show a typical wave propagation pattern in an elastic medium.

All structural problems are approached in a manner similar to the tensile bar. Therefore the author will now delineate the equations that constitute the direct analysis of several other structural members and will request, in the problems section of this chapter, that the students create an algorithm to obtain numerical results to each of these problems.

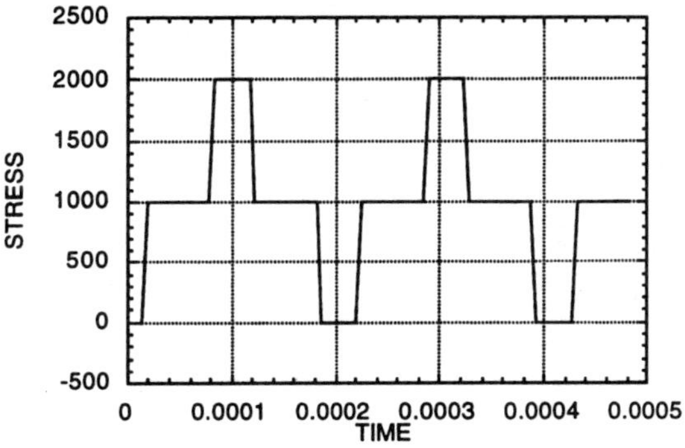

Figure 13-5 Stress history at $x = 5.0$.

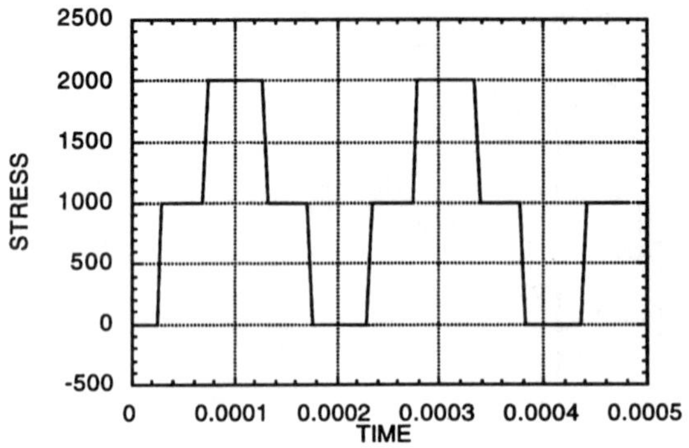

Figure 13-6 Stress history at $x = 7.0$.

13-2.1 Dynamic Analysis of a Spherical Cavity

Consider a spherical cavity whose inner radius is a and whose outer radius is b. The inner radius of the sphere is subjected to an internal pressure loading, as shown in Fig. 13-2. The pressure loading is defined by $p = -S_0$. Figure 13-10 shows a hemispherical free-body diagram with the loads on a typical element i. The direct analysis of this problem is solved by the construction of an algorithm that includes the following equations. The change in the radial velocity of element i is given by

$$\mathrm{d}v_i = \left[\left(S_r^{i+1} A_{i+1} - S_r^i A_i\right) - S_\theta^i (A_{i+1} - A_i)\right] \mathrm{d}t / m_i, \tag{13-6}$$

where S_r^{i+1} is the radial stress on the $i+1$st face of the element and S_r^i is the radial stress on the ith face of the element. A_{i+1} and A_i are the surface areas of the $i+1$st and the ith faces, respectively. S_θ^i is the tangential (hoop) stress on the ith element, and m_i is the mass of the element. The time increment $\mathrm{d}t$ is found from the stability criterion $\mathrm{d}t = \mathrm{d}r/c$, where the wave speed for the sphere is

$$c = \left[\frac{E(1-\nu)}{\{(1+\nu)(1-2\nu)\rho\}}\right]^{\frac{1}{2}}.$$

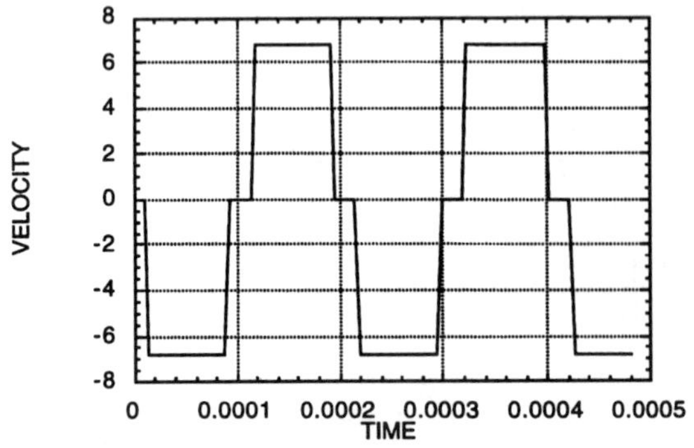

Figure 13-7 Velocity history at $x = 2.5$.

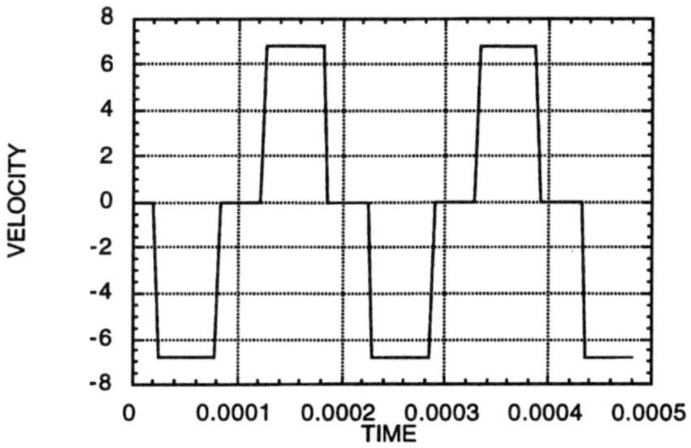

Figure 13-8 Velocity history at $x = 4.5$.

The updated radial velocity of the ith element then becomes

$$v_i' = v_i + \mathrm{d}v. \tag{13-7}$$

The radial strain is updated by

$$\mathrm{d}\varepsilon_r = \frac{(v_{i+1} - v_i)}{c}, \tag{13-8}$$

$$\left\{\varepsilon_r^{i+1}\right\}' = \varepsilon_r^{i+1} + \mathrm{d}\varepsilon_r. \tag{13-9}$$

The radial and the tangential stresses in an elastic sphere are given by

$$S_r^{i+1} = \frac{E\left[(1-\nu)\varepsilon_r^{i+1} + 2\nu\varepsilon_\theta^i\right]}{\nu} \tag{13-10}$$

$$S_\theta^i = \frac{E\left[\nu\varepsilon_r^{i+1} + \varepsilon_\theta^i\right]}{\nu}. \tag{13-11}$$

Finally, the tangential strain is updated by

$$\mathrm{d}\varepsilon_\theta^i = \frac{v_i \mathrm{d}t}{r_i}, \tag{13-12}$$

$$\left\{\varepsilon_\theta^i\right\}' = \varepsilon_\theta^i + \mathrm{d}\varepsilon_\theta^i. \tag{13-13}$$

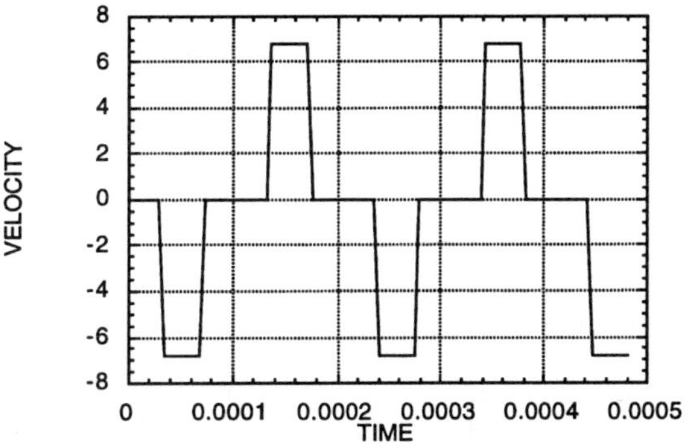

Figure 13-9 Velocity history at $x = 6.5$.

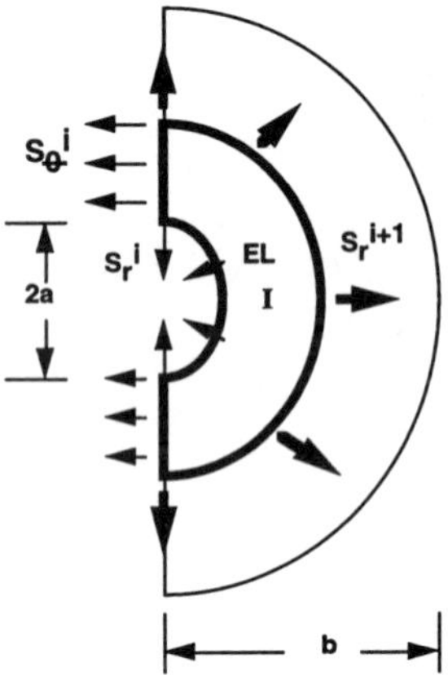

Figure 13-10 Mechanics of a hemispherical element under stress.

Problems 13-3–13-10 ask the reader to use these equations for various boundary conditions and loading conditions.

13-2.2 Dynamic Analysis of a Timoshenko Beam

The assumptions that are necessary to derive the physical laws of a Timoshenko beam may be found in Ref. 1. Consider a straight beam, shown in Fig. 13-11, whose length is L, moment of inertia is I, and area is A, which is subjected to a distributed load q. When the beam is hit by either a shear load or a bending moment, two elastic traveling waves are generated from the point of impact. The dilatation wave speed that produces discontinuities in the bending moment and the angular rotation is given by

$$c = \left\{ \frac{EI_b}{\rho I_i} \right\}^{\frac{1}{2}}, \tag{13-14}$$

where E is the modulus of elasticity, ρ is the density of the beam, I_b is the area moment of inertia that controls the bending, and I_i is the mass moment of inertia that controls the angular momentum. The second wave, called the shear wave, is given by

$$c = \left\{ \frac{A_s G}{\rho A_i} \right\}^{\frac{1}{2}}, \tag{13-15}$$

where G is the shear modulus, A_s is the area that contributes resistance to shear, and A_i is the area that produces resistance to linear momentum. Figure 13-12 shows a typical ith

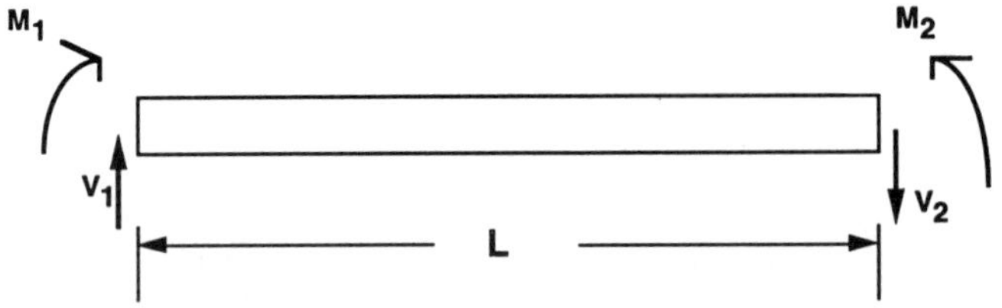

Figure 13-11 Timoshenko beam.

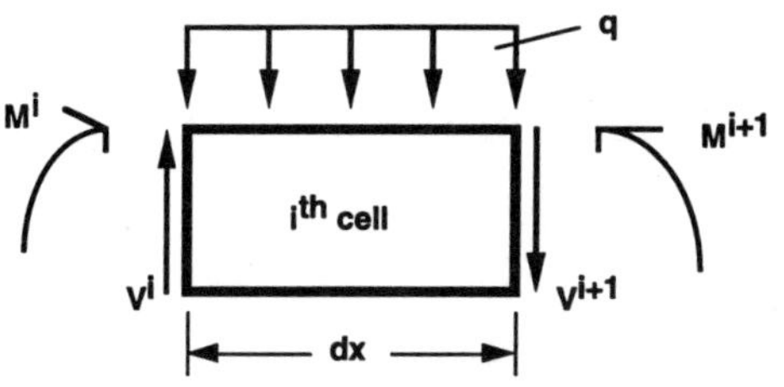

Figure 13-12 Free-body diagram of a typical beam element.

element of the beam of length dx, and Fig. 13-13 shows a deformed element for which the Timoshenko beam assumptions are made. The direct analysis of a beam proceeds as follows.

1. The angular strain is incremented by

$$d\varepsilon_{\psi} = \frac{(\omega^{i+1} - \omega^{i})\,dt}{dx}. \tag{13-16}$$

2. The bending moment increment becomes

$$dM = -EI_b\, d\varepsilon_{\psi}. \tag{13-17}$$

3. The updated bending moment becomes

$$(M^{i+1})' = M^{i+1} + dM. \tag{13-18}$$

4. The angular rotation is updated by

$$\{\psi^{i}\}' = \psi^{i} + \omega^{i}\, dt. \tag{13-19}$$

5. The conservation of angular momentum yields

$$d\omega = \frac{\left[\frac{1}{2}(V^{i} + V^{i+1})\,dx + M^{i} - M^{i+1}\right]dt}{\rho I_i dx}. \tag{13-20}$$

6. We use Eq. (13-20) to update the angular velocity, viz.,

$$\{\omega^{i}\}' = \omega^{i} + d\omega. \tag{13-21}$$

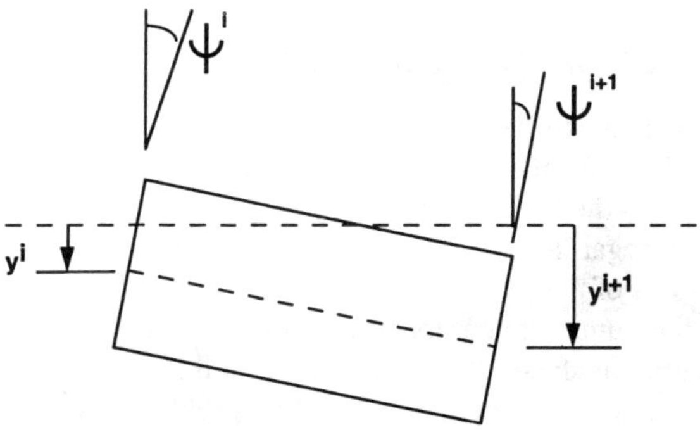

Figure 13-13 Deformation of an element of the beam.

7. The increment in linear strain is

$$d\varepsilon_y = \frac{(v^{i+1} - v^i)\, dt}{dx}. \tag{13-22}$$

8. The increment in the shear force is

$$dV = A_s G\{d\varepsilon_y - \omega^{i+1}\, dt\}. \tag{13-23}$$

9. With Eq. (13-23), the shear force may be updated by

$$(V^{i+1})' = V^{i+1} + dV. \tag{13-24}$$

10. The deflection of the center line of the beam is updated by

$$\{y^i\}' = y^i + v^i\, dt. \tag{13-25}$$

11. The conservation of linear momentum is satisfied by

$$dv = \frac{(V^{i+1} - V^i + q\, dx)\, dt}{\rho A_i\, dx}. \tag{13-26}$$

12. The updated linear velocity is found with Eq. (13-26), viz.,

$$(v^i)' = v^i + dv. \tag{13-27}$$

The successive application of Eqs. (13-16)–(13-27) for each element of the beam will yield the solution for the wave propagation in a Timoshenko beam that is subjected to dynamic loadings.

The above presentations for the tensile bar, the elastic sphere, and the beam are typical of the formulations that may be made for simple structural members. The equations that result from this approach may appear to be derivable from finite-difference methods. In fact, they look similar to the equations that result when the finite differences are applied to differential equations. This would, however, be an erroneous conclusion. Although the equations look similar to the previously obtained results, the latter equations have been derived without the aid of a formal differential equation. The physical laws were directly applied to a free-body section of the structure and the laws were derived forthwith. This method of analysis is therefore known as direct analysis.

13-3 CONTROL VOLUME ANALYSIS IN FLUID FLOW

We now turn our focus to the application of the finite-volume methods in fluid mechanics. Since the main developmental tool in this area of engineering is the control volume, we refer to the numerical techniques herein as the finite-volume methods, rather than direct analysis, as in Section 13-2. However, regardless of the name given to the procedure, these methods are still direct applications of the physical laws.

In fluid mechanics, the level of the numerical computational difficulties increases greatly compared with those in structural analysis. This is because the fluid motion is large, the reference coordinate system is in doubt, and the basic fluid problem tends to be nonlinear. The basic laws that govern fluid flow are

1. balance of mass
2. balance of linear momentum
3. balance of energy

These three laws coupled with an equation of state and a constitutive law form the governing physical principles of fluid flow. These are the laws that form the basis for the numerical computation of the dependent variables of fluids.

The first major decision that must be made before the development of the numerical equations is one of coordinate system reference. Since the movement of a fluid is usually large, one must decide whether to follow the actual fluid (Lagrangian approach) or to select a coordinate system that is fixed in space through which the fluid passes (Eulerian approach). In this chapter, the latter approach has been chosen. Thus we will construct a finite-volume (control volume) that is in a fixed global reference frame and allow the fluid to pass through and interact with this volume.

Before proceeding to the task of developing finite-volume equations for fluid applications, let us define the terms that are used in the derivations. Figure 13-14 shows a control volume (element i) through which a fluid will pass; the net velocities of the fluid that either enters or leaves the faces of the element are given by $\bar{U}$. The volume of the control element is V_c. The outward normals from the faces that accept or reject fluid are $\bar{n}$. The balance of mass may be stated as

> The time rate of change of the mass of a control volume is equal to the net mass that enters across the boundaries of the volume during this time.

Applying this statement, we get the physical equation

$$\frac{\Delta\rho}{\Delta t} V_c = - \sum_{k=1}^{\text{all faces}} \rho^k [\bar{n}^k \cdot \bar{U}^k] A^k \tag{13-28}$$

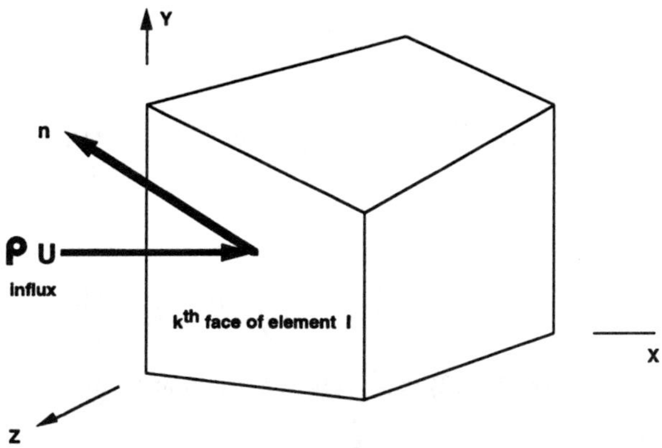

Figure 13-14 Control volume for the balance of mass.

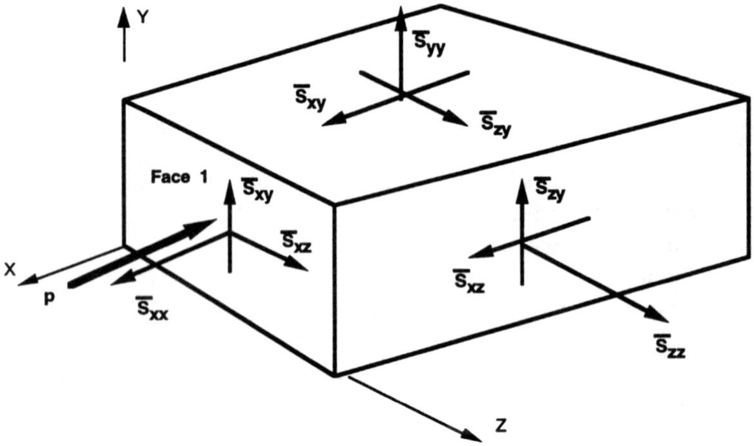

Figure 13-15 Some of the forces acting on a fluid volume.

or

$$\Delta\rho = \left\{ -\sum_{k=1}^{\text{all faces}} \rho^k A^k \sum_{j=1}^{3} n_j^k U_j^k \right\} \frac{\Delta t}{V_c}. \tag{13-29}$$

The new density of element i may be calculated from Eq. (13-29) as

$$\rho_i' = \rho_i + \Delta\rho. \tag{13-30}$$

We can calculate the momentum of the fluid, which is a vector quantity and has three components, by invoking the balance of linear momentum, i.e.,

The net force on the control volume equals the time rate of change of the linear momentum.

Figure 13-15 shows the same control volume as Fig. 13-14, but now shows a representation of the forces acting on it. These forces are the pressure p, the gravitational force g, and the shear stresses $\bar{S}$. Applying the linear momentum theorem to the control volume, we get

$$\bar{F} = \frac{\Delta}{\Delta t}\left\{ \sum_{\text{volume}} \rho\bar{U}\Delta V_c \right\} + \sum_{\text{all faces}} \{\rho(\bar{n}\cdot\bar{U})\}\bar{U}\Delta A_c \tag{13-31}$$

or

$$\bar{F} = \frac{\Delta}{\Delta t}\left\{ \sum_{\text{volume}} (\rho\bar{U})\Delta V_c \right\} + \sum_{\text{all faces}} \rho\bar{U}\left(\sum_{j=1}^{3} \bar{n}_j U_j \right)\Delta A_c. \tag{13-32}$$

It is important to remind the reader that Eq. (13-32) is a vector equation and has three components.

The forces that are acting on the control volume are given by

$$\bar{F} = \rho\bar{g} - \nabla p + \nabla\cdot\bar{S} \tag{13-33}$$

or

$$\sum_{k=1}^{\text{all areas}} F_j^k = \rho g_j - \sum_{k=1}^{\text{opposite faces}} \frac{(p^{k+1} - p^k)}{\Delta x_j} + \sum_{m=1}^{3} \frac{\Delta \bar{S}_{jm}}{\Delta x_m}. \tag{13-34}$$

The stress tensor is related to the velocities by use of a viscosity law such as the kinematic viscosity relationship:

$$S_{ij} = -p\delta_{ij} + 2\mu U_{i,j}. \tag{13-35}$$

Equation (13-35) is given in subscript notation. The equation that would express the balance of angular momentum is automatically satisfied by the symmetry of the stress tensor, as it is in the theory of elasticity.

The balance of energy states that

The time rate of change of the energy (the power) in the fluid is equal to the work done on the fluid.

The same forces that were acting on the control volume for the momentum balance are used to calculate the net work done on the fluid. In addition, energy is supplied to the fluid from thermal sources by conduction.

The kinetic power (kinetic energy per unit time) is given by

$$W_{\text{kinetic}} = \frac{1}{2}\frac{\Delta}{\Delta t}\left\{ \sum^{\text{control volume}} \rho U^2 V_c \right\}, \tag{13-36}$$

where the term U^2 is the magnitude of the velocity vector $\bar{U}$.

The power due to the gravitational force is given by

$$W_g = - \sum^{\text{control volume}} \rho \left\{ \sum_{j=1}^{3} U_j g_j \right\} V_c. \tag{13-37}$$

The pressure and the stress forces produce the following expression for the power:

$$W_f = -p \sum_{k=1}^{\text{all faces}} \sum_{j=1}^{3} -(n_j^k U_j^k) A_c^k + \sum_{k=1}^{\text{all faces}} \sum_{i=1}^{3} \sum_{j=1}^{3} \bar{S}_{ij}^k n_j^k U_i^k A_c^k. \tag{13-38}$$

The power that is transmitted because of heat conduction is

$$W_{\text{heat}} = K \sum_{m=1}^{\text{all faces}} \sum_{j=1}^{3} n_j^m \left[\frac{\{T_{\text{face}1} - T_{\text{face}\,2}\}}{\Delta x_j} \right]^m A_c^m. \tag{13-39}$$

If we define the internal energy change in the control volume per unit time as e, then the balance of energy across the element becomes

$$W_{\text{heat}} - (W_{\text{kinetic}} + W_g + W_f) = \frac{\Delta}{\Delta t}\{\rho e\} V_c. \tag{13-40}$$

In fluid mechanics there are five independent variables, i.e., the density ρ, the three components of velocity, u, v, and w, and the fluid temperature T. We can calculate these variables in each control volume by simultaneously solving Eqs. (13-29), (13-34), and (13-40). The stresses are determined by Eq. (13-35). In a gas, there is usually an equation of state that relates the pressure and the temperature. These equations that were derived by the direct application of the physical laws to the fixed control volume represent the method of finite volume for fluid mechanics. A computer code has been developed with the above-mentioned relevant equations.[2]

13-4 CONCLUSION

In this chapter, computational tools for both solid and fluid mechanics were developed by a direct application of the laws of physics to a control element. This approach is known as the method of finite volumes. There are advantages and disadvantages to using this approach compared with the techniques that were use in preceding sections (i.e., finite differences or finite elements).

One of the advantages includes the avoidance of any complex mathematical formulation to develop the theory. Another advantage to the finite-volume approach is the direct physical approach that is taken. For example, in fluid mechanics, the control volume may be viewed as a box into which and out of which fluid flows while being subjected to external and body (gravitational) forces. The mass balance merely states that the mass of the box increases by the net amount of mass that enters the box. The momentum equations are just an application of Newton's second law, $F = ma$. Finally, the energy equation shows that the net increase in the energy within the box equals the energy entering the box minus the energy that is dissipated by the work done by all the forces. This physical approach then leads to an easy understanding of the processes that are taking place.

The disadvantages of the finite-volume methods lie in the need to apply these laws to relatively simple shapes. In structural mechanics, beams, plates, cylinders, spheres, and rods are easily amenable to the techniques of this chapter. Complex shapes such as airfoils would create a challenging experience for the analyst. In fluids, however, since the entire fluid region is discretized into arbitrarily shaped control volumes, this difficulty is lessened. A final disadvantage exists when finite-volume methods are used. Because of the stability requirement of any numerical analysis, a relationship between the spatial increment and the temporal increment must be established. In the methods of this section, there could be uncertainty as to the proper selection of the relevant spatial increment. Additionally, the speed of wave propagation is usually used as the stability criterion, viz.,

$$\Delta t = \frac{\Delta x}{c}. \tag{13-41}$$

Since the wave speed is not obtainable within the formulation of the finite-volume method, it must be derived from outside the analysis, say, the method of characteristics. This means that these methods are not self contained, which is a distinct disadvantage of the methods.

REFERENCES

1. Timoshenko, S. (1955). *Vibration Problems in Engineering*, 3rd ed., Van Nostrand, Company, New York, pp. 329–331.

2. Koenig, H. A. (19xx). A finite volume approach to the explicit integration of viscous fluid flow (private communication).

FURTHER READING

Abrahamson, H. N., Plass, H. J., and Ripperger, E. A. (1958). Stress Wave Propagation in Rods and Beams, *Adv. Appl. Mech.*, vol. 5, p. 111.

Boley, B. A. and Chao, C. C. (1955). Some Solutions of the Timoshenko Beam Equations, *J. Appl. Mech.*, vol. 22, pp. 579–586.

Chou, P. C. and Koenig, H. A. (1965). Flexural Waves in Elastic Circular Plates by Method of Characteristics, **D.I.T.** Report 60-6.

Davids, N. and Koenig, H. (1966). Direct Analysis of the Flexural Traveling Waves in Beams and Plates, *Eng. Mech. Bull.*, vol. 1.

Davids, N. and Koenig, H. (1967). Double Stress-Wave Discontinuities in Finite, Shear Corrected Beams and Plates, Tenth Midwestern Mechanics Conf. **Publishers name and location.**

Davids, N., Mehta, P. K., and Johnson, O. T. (1965). Spherical Elastoplastic Waves in Materials, *Behavior of Materials under Dynamic Loading*, ASME, New York.

Dengler, M. A. and Goland, M. (1952). Transverse Impact of Long Beams Including Rotatory Inertia and Shear Effects, *Proc. First U.S. National Congress on Applied Mechanics*, pp. 179–186.

Leonard, R. W. and Budiansky, B. (1954). On Traveling Waves in Beams, NACA Technical Report 1173.

Mehta, P. K. and Davids, N. (1966). A Direct Numerical Analysis Method for Cylindrical and Spherical Elastic Waves, *AIAA J.*, vol. 4, pp. 112–117.

Miklowitz, J. (1960). Flexural Stress Waves in an Infinite Elastic Plate due to a Suddenly Applied Concentrated Transverse Load, *J. Appl. Mech.*, vol. 27, pp. 681–689.

Mindlin, R. D. (1951). Influence of Rotatory Inertia and Shear on Flexural Motions of Isotropic, Elastic Plates, *J. Appl. Mech.*, vol. 18, pp. 31–38.

Timoshenko, S. (1955). *Vibration Problems in Engineering*, 3rd ed., Van Nostrand, New York.

PROBLEMS

13-1 The tensile bar of Fig. 13-1 is now unrestrained at the right end. The impulsive loading of Fig. 13-2 is applied to the left end of the bar. Write a computer program, similar to **direct.f** contained in Appendix 13.1, which solves this problem by direct analysis.

13-2 Solve Problem 13-1 with a load at the left end that is shown in Fig. 13-16.

13-3 The sphere whose mechanics are presented in Fig. 13-10 is loaded at its inner radius by the stress

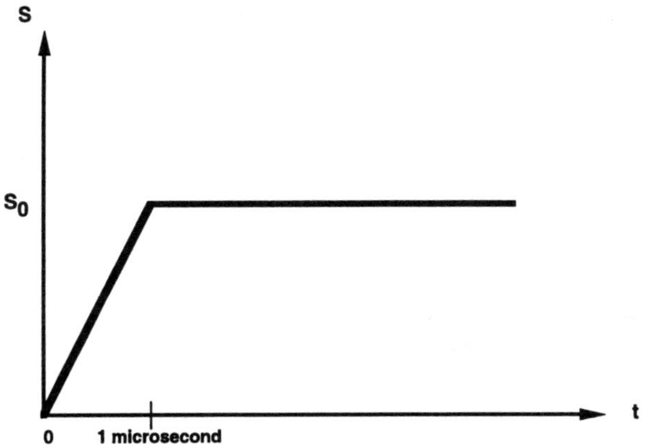

Figure 13-16 Ramp loading of a tensile bar.

history that is shown in Fig. 13-2. Write a computer code that will solve for the unknown variables as functions of space and time by direct analysis. The outer surface of the sphere is fixed.

13-4 The sphere whose mechanics are presented in Fig. 13-10 is loaded at its inner radius by the stress history that is shown in Fig. 13-2. Write a computer code that will solve for the unknown variables as functions of space and time by direct analysis. The outer surface of the sphere is free to expand.

13-5 The sphere whose mechanics are presented in Fig. 13-10 is loaded at its outer radius by the stress history that is shown in Fig. 13-2. Write a computer code that will solve for the unknown variables as functions of space and time by direct analysis. The inner surface of the sphere is free to expand.

13-6 The sphere whose mechanics are presented in Fig. 13-10 is loaded at its outer radius by the stress history that is shown in Fig. 13-2. Write a computer code that will solve for the unknown variables as functions of space and time by direct analysis. The inner surface of the sphere is fixed.

13-7 The sphere whose mechanics are presented in Fig. 13-10 is loaded at its inner radius by the stress history that is shown in Fig. 13-16. Write a computer code that will solve for the unknown variables as functions of space and time by direct analysis. The outer surface of the sphere is fixed.

13-8 The sphere whose mechanics are presented in Fig. 13-10 is loaded at its inner radius by the stress history that is shown in Fig. 13-16. Write a computer code that will solve for the unknown variables as functions of space and time by direct analysis. The outer surface of the sphere is free to expand.

13-9 The sphere whose mechanics are presented in Fig. 13-10 is loaded at its outer radius by the stress history that is shown in Fig. 13-16. Write a computer code that will solve for the unknown variables as functions of space and time by direct analysis. The inner surface of the sphere is free to expand.

13-10 The sphere whose mechanics are presented in Fig. 13-10 is loaded at its outer radius by the stress history that is shown in Fig. 13-16. Write a computer code that will solve for the unknown variables as functions of space and time by direct analysis. The inner surface of the sphere is fixed.

13-11 The beam of Fig. 13-11 is to be analyzed as a cantilever beam with its right end fixed (i.e., $v^{N+1} = 0$ and ω^{N+1} for N elements). The left end of the beam is subjected to a ramp shear loading whose temporal variation is similar to the one shown in Fig. 13-16. Write a computer code, using direct analysis, to solve for the spatial and temporal variations in the moments, shears, deflections, and rotations in the beam.

13-12 The beam of Fig. 13-11 is to be analyzed as a cantilever beam with its right end fixed (i.e., $v^{N+1} = 0$ and ω^{N+1} for N elements). The left end of the beam is subjected to a step shear loading whose temporal variation is similar to the one shown in Fig. 13-2. Write a computer code, using direct analysis, to solve for the spatial and temporal variations in the moments, shears, deflections, and rotations in the beam.

13-13 The beam of Fig. 13-11 is to be analyzed as a cantilever beam with its right end fixed (i.e., $v^{N+1} = 0$ and ω^{N+1} for N elements). The left end of the beam is subjected to a step moment loading whose temporal variation is similar to the one shown in Fig. 13-2. Write a computer code, using direct analysis, to solve for the spatial and temporal variations in the moments, shears, deflections, and rotations in the beam.

13-14 The beam of Fig. 13-11 is to be analyzed as a cantilever beam with its right end fixed (i.e., $v^{N+1} = 0$ and ω^{N+1} for N elements). The left end of the beam is subjected to a ramp moment loading whose temporal variation is similar to the one shown in Fig. 13-16. Write a computer code, using direct analysis, to solve for the spatial and temporal variations in the moments, shears, deflections, and rotations in the beam.

13-15 The beam of Fig. 13-11 is to be analyzed as a simply supported beam (i.e., $v^1 = 0, V^1 = 0, v^{N+1} = 0$, and $V^{N+1} = 0$, for N elements). The center of the beam is subjected to a ramp moment loading whose temporal variation is similar to the one shown in Fig. 13-16. Write a computer code, using direct analysis, to solve for the spatial and temporal variations in the moments, shears, deflections, and rotations in the beam.

13-16 The beam of Fig. 13-11 is to be analyzed as a simply supported beam (i.e., $v^1 = 0, V^1 = 0, v^{N+1} = 0$, and $V^{N+1} = 0$, for N elements). The center of the beam is subjected to a ramp shear loading whose temporal variation is similar to the one shown in Fig. 13-16. Write a computer code, using direct analysis, to solve for the spatial and temporal variations in the moments, shears, deflections, and rotations in the beam.

13-17 The beam of Fig. 13-11 is to be analyzed as a simply supported beam (i.e., $v^1 = 0, V^1 = 0, v^{N+1} = 0$, and $V^{N+1} = 0$, for N elements). The center of the beam is subjected to a step shear loading whose temporal variation is similar to the one shown in Fig. 13-2. Write a computer code, using direct analysis, to solve for the spatial and temporal variations in the moments, shears, deflections, and rotations in the beam.

13-18 The beam of Fig. 13-11 is to be analyzed as a simply supported beam (i.e., $v^1 = 0, V^1 = 0, v^{N+1} = 0$, and $V^{N+1} = 0$, for N elements). The center of the beam is subjected to a step moment loading whose temporal variation is similar to the one shown in Fig. 13-2. Write a computer code, using direct analysis, to solve for the spatial and temporal variations in the moments, shears, deflections, and rotations in the beam.

CHAPTER

FOURTEEN

BIOMECHANICAL APPLICATIONS

14-1 INTRODUCTION

Mechanical engineers tend to be very selective as to the areas of their investigations. They tend to study the basic areas of structures, heat transfer, and fluid flow. Their applications within these areas tend to follow the lines of traditional problems that engineers and physicists are most likely to encounter in industry. Structural engineers, for example, tend to study deformation, materials, and fracture mechanics. There are many unsolved problems in these areas that would enjoy the applications of the methods within this textbook.

This chapter, however, is written to expose readers to the synergistic relationship that engineering enjoys with nontraditional areas of research, namely, biomechanics. It has been this author's experience that the tools of computational analysis that are contained in this book are extremely useful in solving problems in related areas. Therefore the purpose of this chapter is to introduce these biological areas and to demonstrate how the engineer may use the tools that have already been delineated in areas that can benefit greatly from their application.

In this chapter, then, no additional computational techniques are introduced. If any new theories or equations are shown, their main purpose will be to show students how these traditional formulations may be used to solve a nontraditional application. Engineers should turn their attention to these areas in which their expertise will be overwhelmingly received.

14-2 ANALYSIS OF ORTHODONTIC APPLIANCES

The goal of an orthodontist is to provide therapy to a patient that will transform the patient's teeth into an ideal arch. The malocclusions that exist in the oral cavity are due

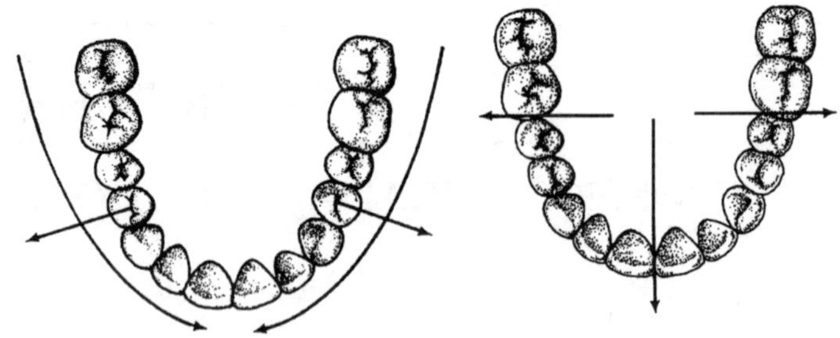

Figure 14-1 Ideal Arch.

to many environmental and genetic factors. Such conditions are an improperly sized mandible, adolescent thumb sucking, etc. Some reasons for orthodontic therapy are functional and some are cosmetic, as may be required by one's economic status. The orthodontist must first determine how the teeth must be realigned in order to produce this ideal arch. The orthodontist must then attempt to design appliances (commonly known as braces) that will move the malaligned teeth to their proper locations.

Since the design and the analysis of orthodontic appliances do not require any knowledge of biokinetics and follow more traditional engineering theories, we study this phase of orthodontics before we study tooth displacement prediction. In this section, therefore, the theory and the computational mechanics that are required for designing orthodontic appliances are discussed. In Section 14-3 the computational tools that are used to predict the movement of one or many teeth are introduced.

Orthodontic appliances may be viewed, in engineering terms, as beams that possess curvature and twist in three-space. Because of the thin geometric design of these orthodontic appliances, the assumption of beam theory is reasonable. However, in the application of these devices during actual clinical practice, the appliances undergo large deformations. Thus a small-deflection beam theory would be inaccurate if it were applied to their design. The objective of any formulation here is to calculate the moments and the forces that are produced in an specific appliance (and therefore transmitted to the teeth) when the wires are restrained in the oral cavity during orthodontic therapy.

In Section 14-3 an explanation is offered for the requirement that these devices deliver light forces to the teeth. In traditional engineering analysis, the force system on a structure may be of any magnitude that allows the stresses to remain below the yield point. This requirement is also placed on appliance design. The reason is that the springback in the appliance must be equal to the desired tooth movement. If this is not met, excessive readjustments would have to be made by the clinician.

Burstone and Koenig[1] initiated the computational approach for the prediction of forces that are delivered by orthodontic wires. Figure 14-1 shows the sign conventions that are used in this analysis, which are modeled so that a clinician can interpret the results. Terms such as posterior, anterior, lateral, medial, mesial, and distal replace the traditional clockwise, counterclockwise, right, and left. Figure 14-2 shows the types of moments and forces that may be required on the teeth to produce proper tooth movement. Koenig and Burstone[2] generalized the theory to large displacements, compared their results with

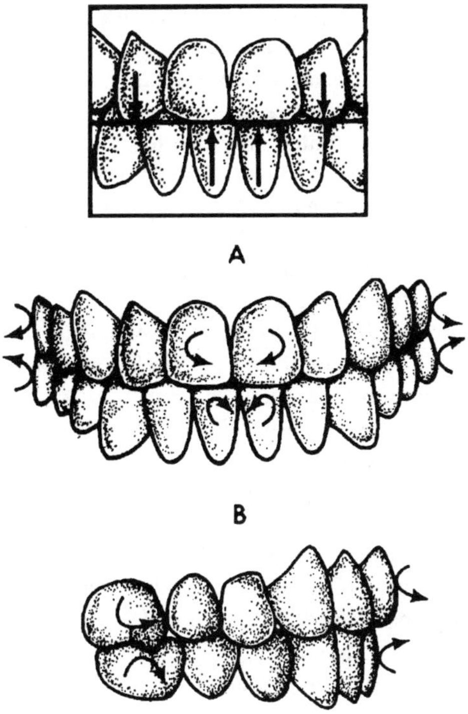

Figure 14-2 Force systems from orthodontic appliances.

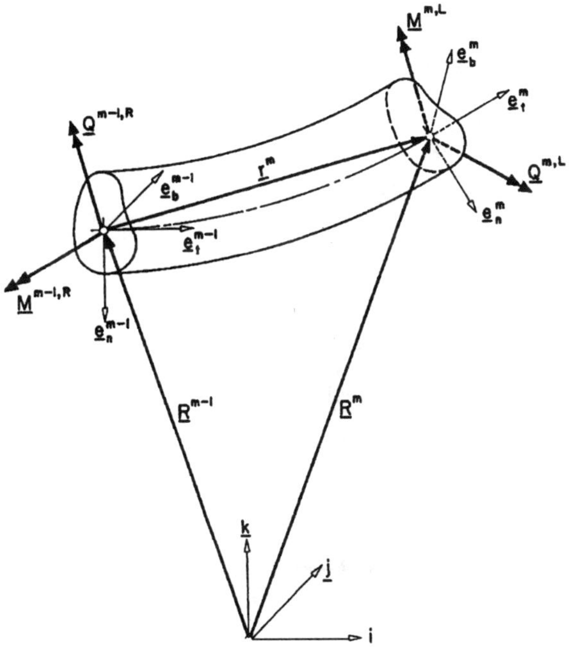

Figure 14-3 Free-body diagram of an orthodontic appliance segment.

the small-deflection theory, and formulated an iterative approach to the resulting matrix set by using the techniques that are discussed in Chap. 2 of this textbook. DeFranco et al.[3] generalized Koenig and Burstone's theory for an arbitrary coordinate system and presented their equations in terms of the geometric curvatures and twists of the wires. With this theory and the resulting equations that are presented in Appendix A of Ref. 3, a robust computer algorithm was developed that could calculate the force system from an existing appliance and form the basis for a synthesis of a new design. In other words, the clinician could predict the behavior of an existing design or design a new appliance that would satisfy the biological requirements of the patient. Figure 14-3 shows the free-body diagram that is required for developing the above analysis.

Goldberg et al.[4] introduced plastic material behavior into the analysis. This was required because of the increasingly common use of nontraditional materials for clinical applications. New materials such as nickel titanium, betatitanium, and chromium cobalt nickel alloys were being introduced into the profession. Koenig and Bolle[5] produced the most ambitious analysis and computer algorithm to date for the purposes of analyzing appliances.

The theoretical development of the above approaches is documented in the references cited above. The theory is well beyond the scope of a basic textbook in computational mechanics. This section was written, as explained above, to introduce readers to areas of computational analysis that are usually ignored by engineers. Reader are invited to use the equations in the above references to create computational algorithms that could solve appliance design problems or any application for which arbitrarily twisted beams exist.

14-3 PREDICTION OF TOOTH MOVEMENT

The movement of teeth in vivo is not a simple deformation process, like the deformations within engineering materials. The geometry that surrounds a human tooth consists of

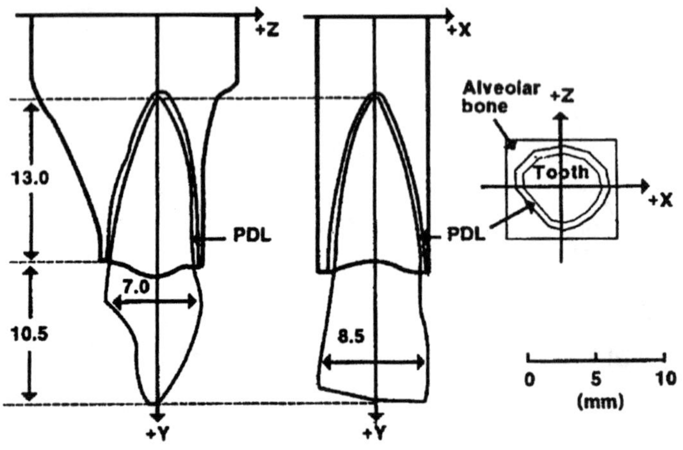

Figure 14-4 Schematic representation of a tooth set.

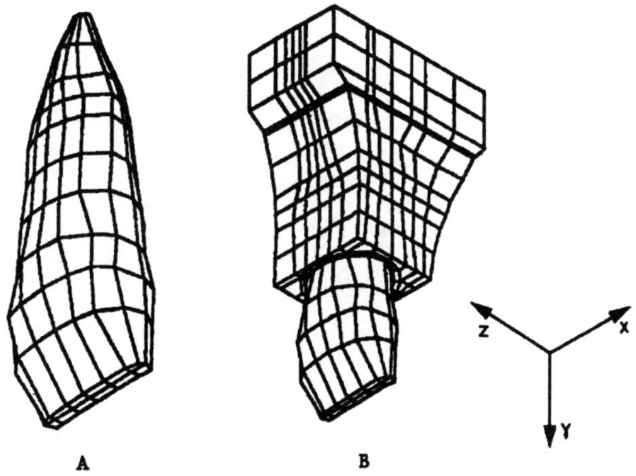

Figure 14-5 A tooth, PDL, and bone set.

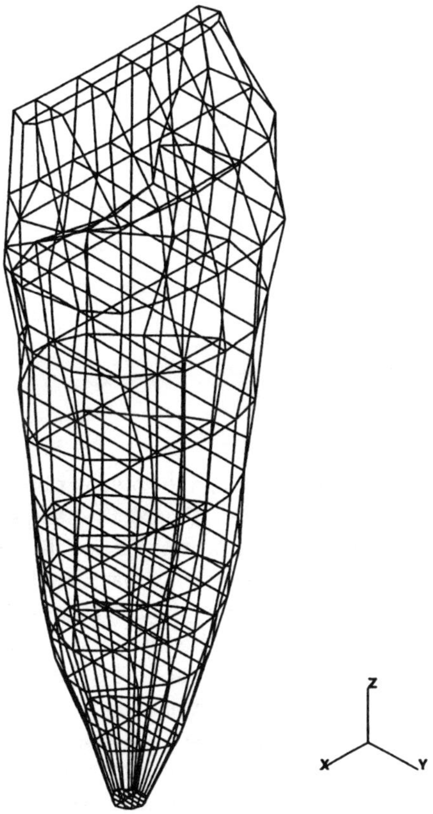

Figure 14-6 Finite-element method model of the tooth.

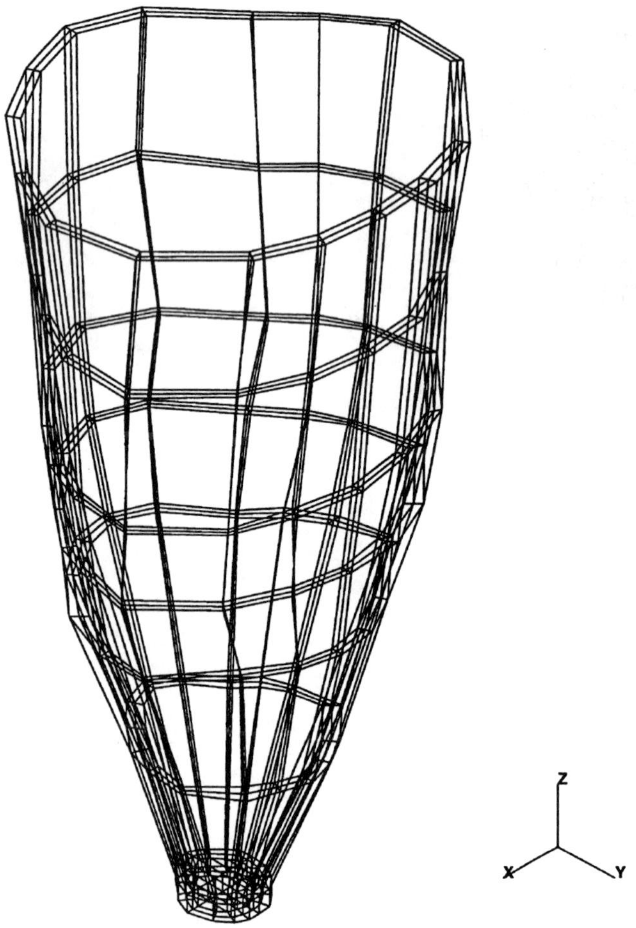

Figure 14-7 Finite-element method model of the PDL.

the tooth, a membrane known as the periodontal ligament (PDL), and the alveolar bone that encapsulates the ligament. We call this geometry a tooth set. Figure 14-4 shows a schematic representation of the geometry that models an upper central incisor. If the deformation process were similar to the typical behavior of an engineering material, then when the forces and moments from orthodontic appliances were applied to this tooth set, the maximum displacement of the tooth would be limited to the thickness of the PDL plus a small deformation in the alveolar bone. This is considerably less than the movement that actually takes place in orthodontic therapy. The question then becomes, "How does a clinician achieve large tooth movements?" The answer is found in the study of the biokinetics. When a force system is applied to the PDL, a growth process commences. The alveolar bone (also known as the alveolar process) is sensitive to the stress distribution that is transferred from the PDL to the bone. Biological processes of

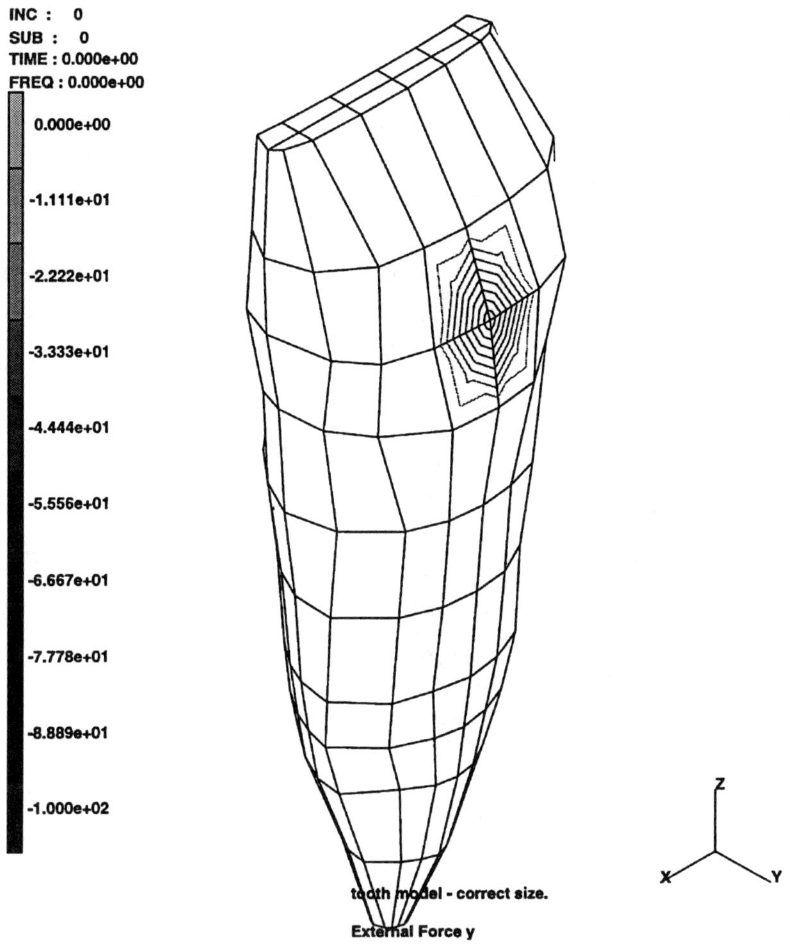

Figure 14-8 Tooth loaded by a concentrated force.

resorption and apposition occur in the bone. Basically, resorption is the phenomenon by which bone is lost and apposition is the process by which bone is created. These two biokinetic processes do not occur in standard engineering materials. Thus the process by which a tooth is moved from one position to another and to another orientation is a growth phenomenon. Bone is either created or lost, and the socket that houses the tooth appears to float from the initial position to its final destination.

If one understands the above biokinetic processes of resorption and apposition, one can understand the need for a light force system to be applied to the teeth by the appliances. If the force system is too heavy, then there will be damage to the PDL. The process is called necrosis. This means that there is death of tissue that occurs in the PDL. When necrosis is present, the PDL will not allow biokinetic processes to occur and all tooth movement stops. Additionally, necrosis produces excessive resorption of the

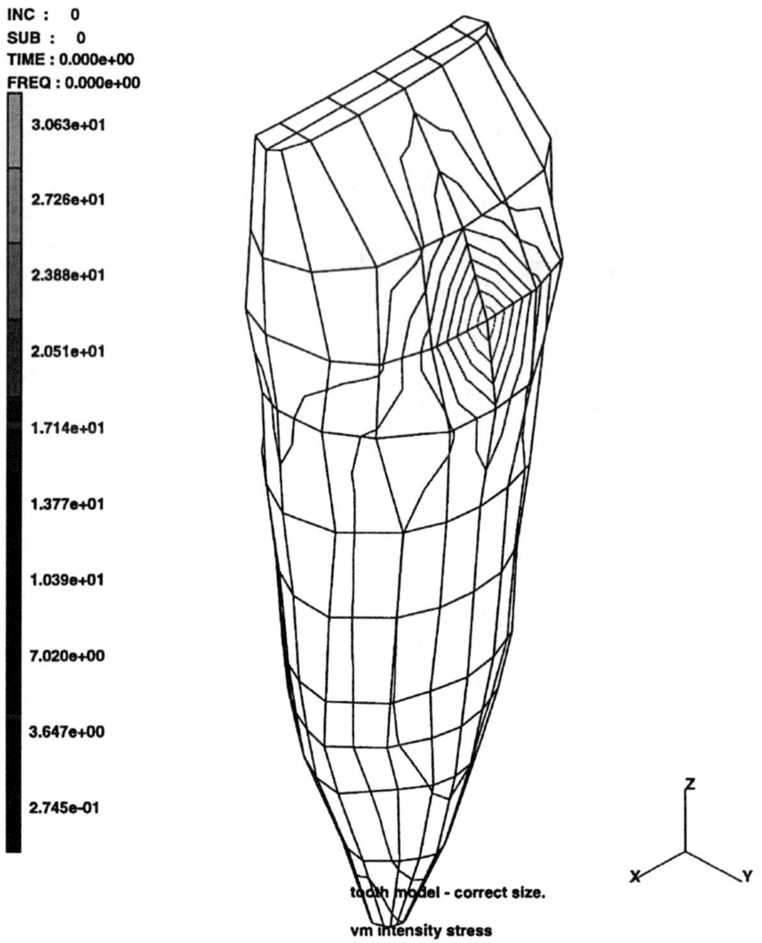

Figure 14-9 Von Mises stresses in the tooth.

bone, which stabilizes the tooth. Therefore, in later years when the joints within a human body become less secure, loose teeth may occur. This could be directly attributable to the loss of alveolar bone due to excessive forces that were delivered by orthodontic therapy.

The application of appliances that exert a force system that is too light will not produce proper tooth movement since these forces will be unable to overcome the normal forces in the oral cavity, such as mastication.

It is therefore the job of the orthodontist to determine the proper level of the appliance force system that will produce ideal forces on the tooth. It is interesting to note that there is considerable disagreement within the orthodontic community as to what constitutes an ideal force system.

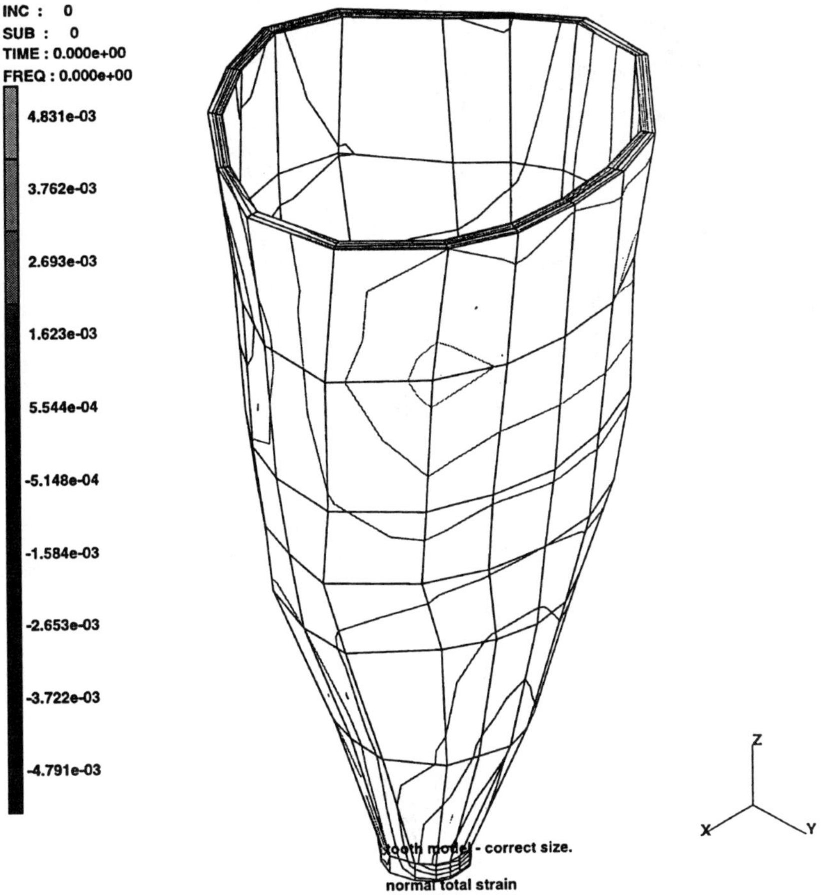

Figure 14-10 Normal strain in the PDL.

In order to determine the stress patterns in the PDL and on the alveolar process, an analysis must be simulated by computational methods on the tooth set. The finite-element methods are useful to accomplish this task. Figures 14-5–14-7 show a tooth set consisting of a tooth, a PDL, and a bone segment that has been discretized by the finite-element method theory in Chap. 10. Solid brick elements were chosen for the tooth and the bone, while thin membrane elements were chosen to simulate the ligament. Tanne et al.[6,7] have shown how this approximation can be used to predict tooth movement due to calculated stress patterns in the PDL. They also showed how the center of rotation for the tooth movement is related to the force system that is delivered by the orthodontic appliances. The results of this research have contributed greatly to the understanding of tooth movement and to the design of orthodontic appliances. Appendix 14.1 contains the data set, **TOOTH.DAT**, that models the above simulation.

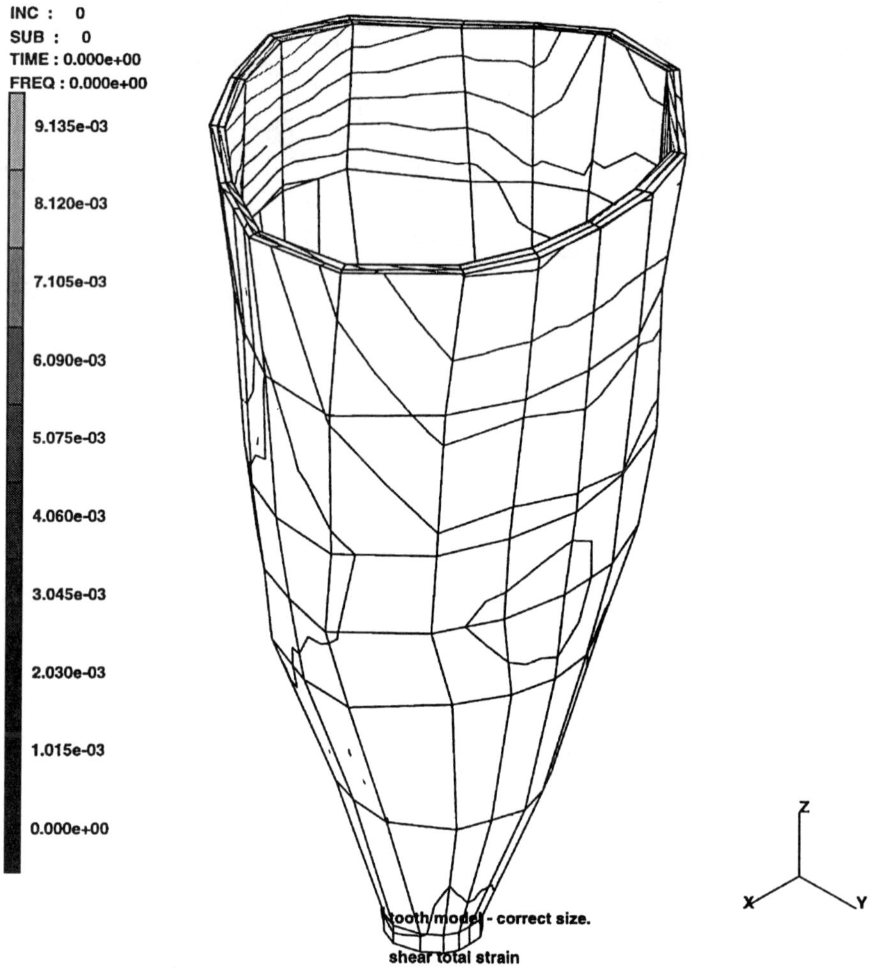

Figure 14-11 Shear strain in the PDL.

Figure 14-8 shows a tooth that is loaded by a concentrated external force in the y direction. This force could be the result of an orthodontic appliance or the result of another force in the oral cavity. The stress distribution on the surface of the tooth that results from the applied load is shown in Fig. 14-9. Figures 14-10–14-12 show the strain and the stress distributions within the PDL. These are the stress patterns that result in the biokinetic responses of resorption and apposition.

Studies of the stress distributions in the PDL allow the orthodontist to predict the nature of the tooth movement when the patient is undergoing therapy. The studies also enhance the ability of the clinician to design special appliances to achieve specific tooth movements.

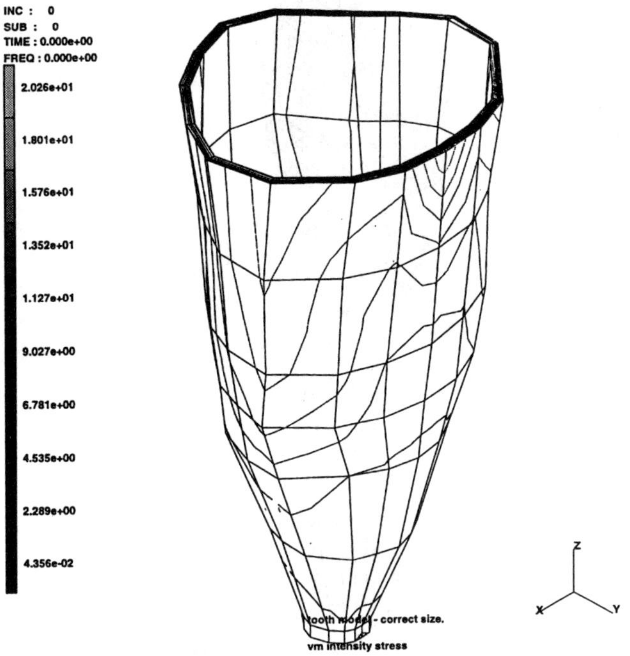

Figure 14-12 Von Mises stress in the PDL.

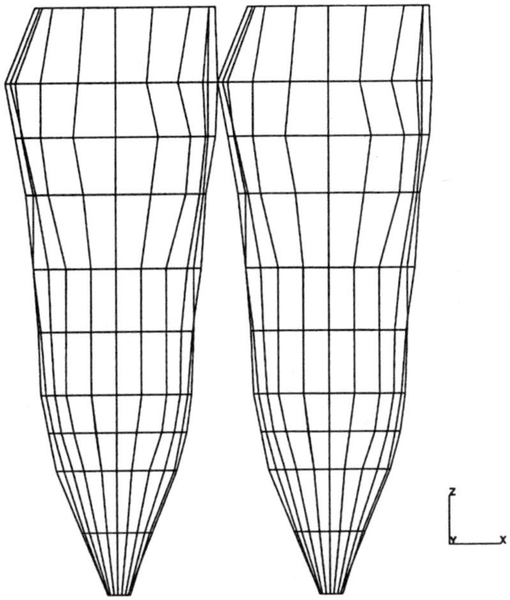

Figure 14-13 Geometry of a two-tooth set.

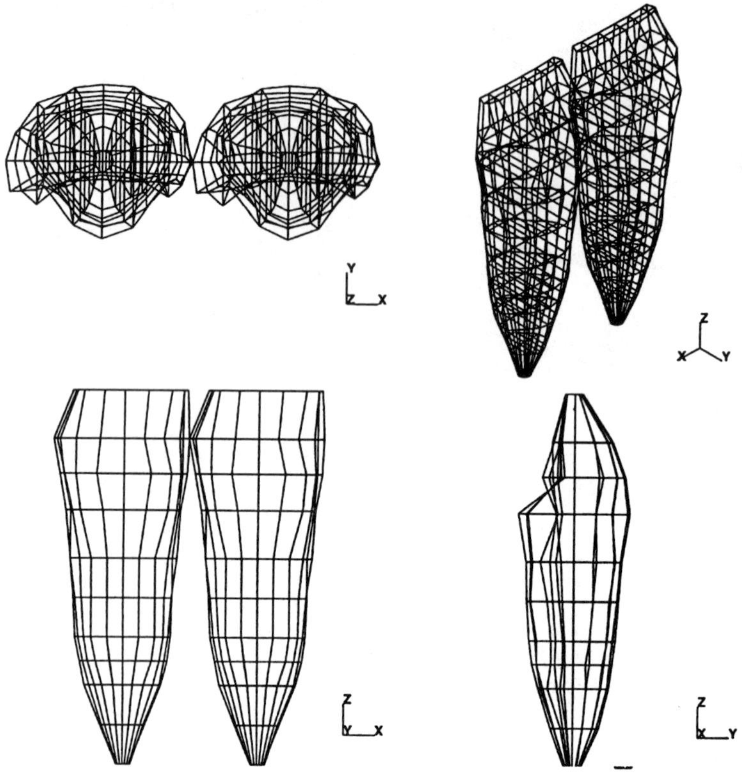

Figure 14-14 Finite-element method model of a two-tooth set.

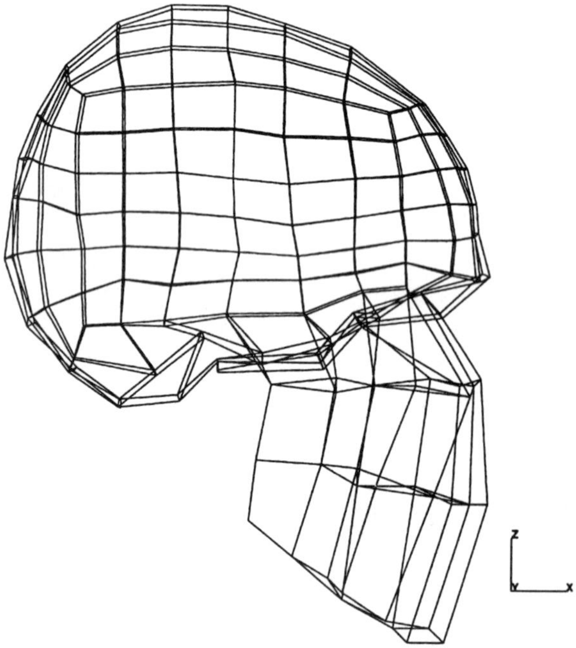

Figure 15-15 Cranium.

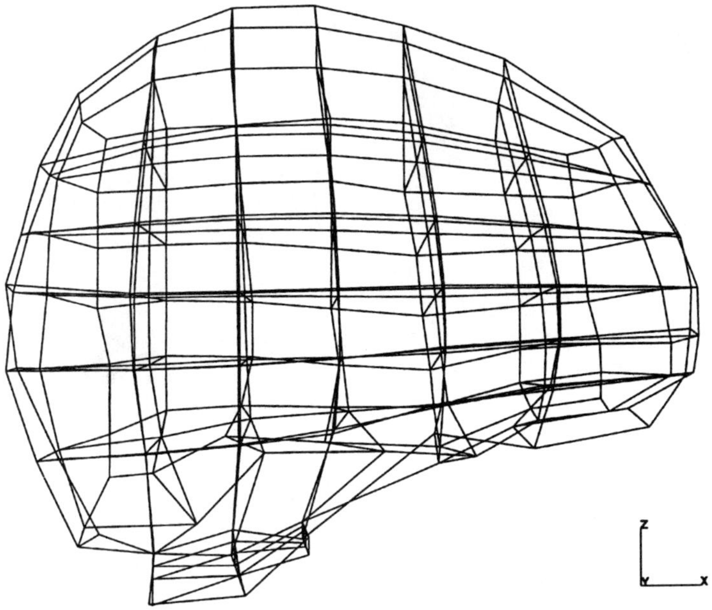

Figure 14-16 Brain.

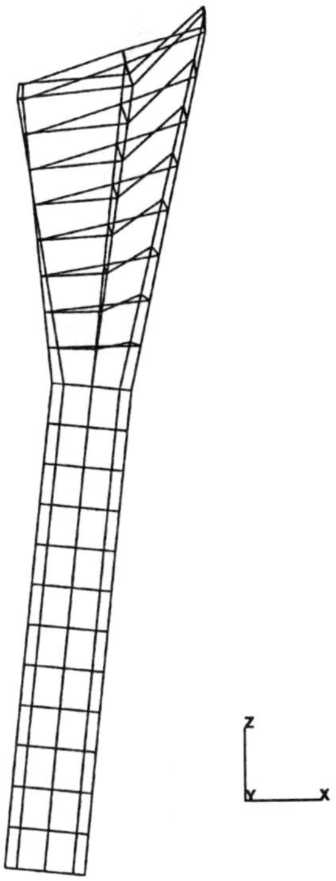

Figure 14-17 Spinal cord.

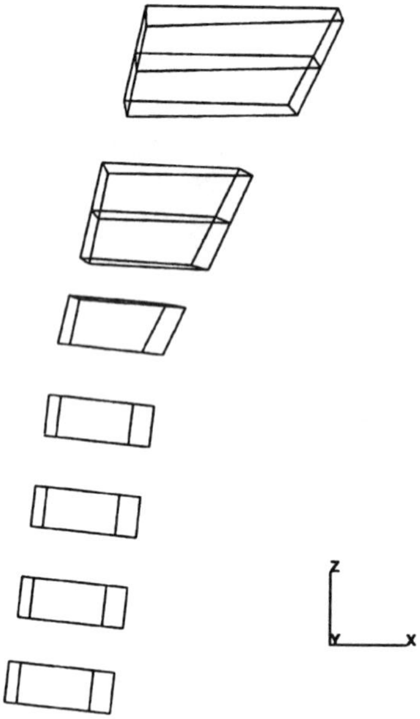

Figure 14-18 Disks.

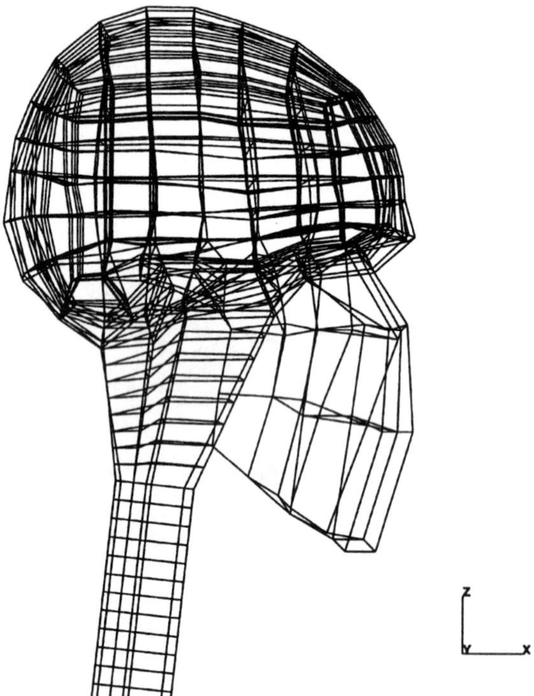

Figure 14-19 Entire head model.

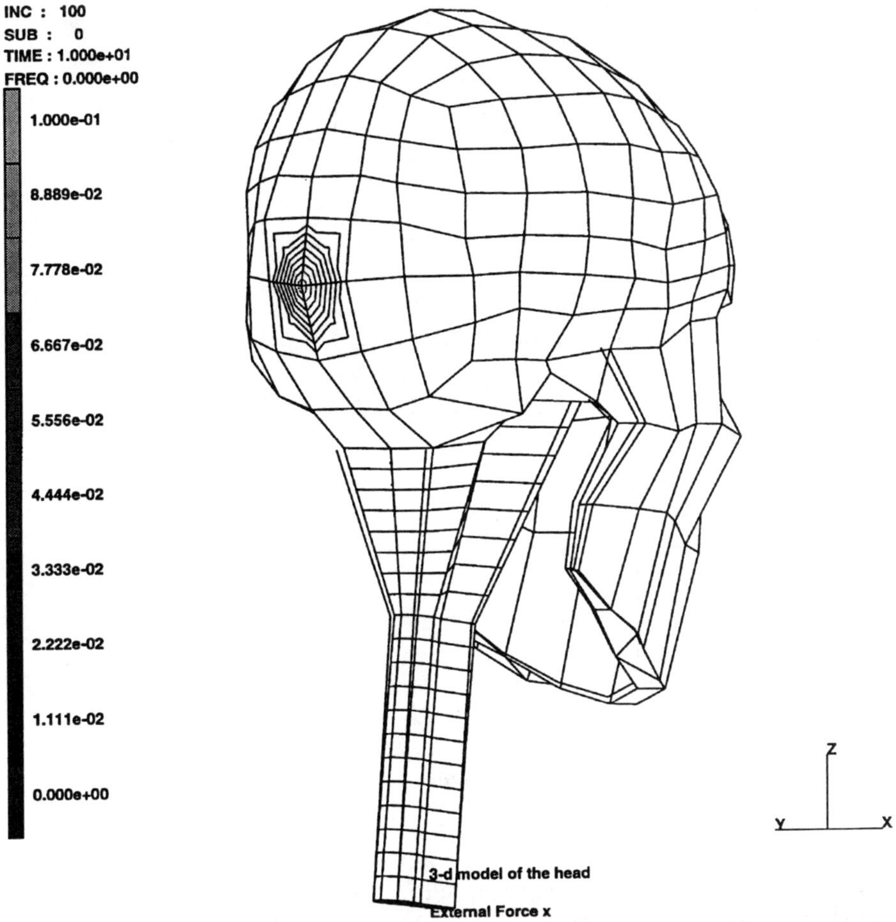

Figure 14-20 External load to the rear of the cranium.

The study of the movement of one tooth is very useful in understanding the biomechanics of the tooth movement process. However, the clinician must treat the entire oral cavity. Therefore the above work was expanded to a computational model that encompasses more than one tooth so that the interaction of the teeth during therapy may be studied. Figures 14-13 and 14-14 show a two-tooth set that has been constructed with the finite-element method. Forces may be applied to this model to study the interaction of the teeth to determine how one tooth interferes with the movement of the other.

The above-mentioned works show how a computational tool (finite-element method) that was devised to study standard engineer problems may be used, with a little imagination, to solve problems in unrelated applications. Section 14-4 expands on this synergism.

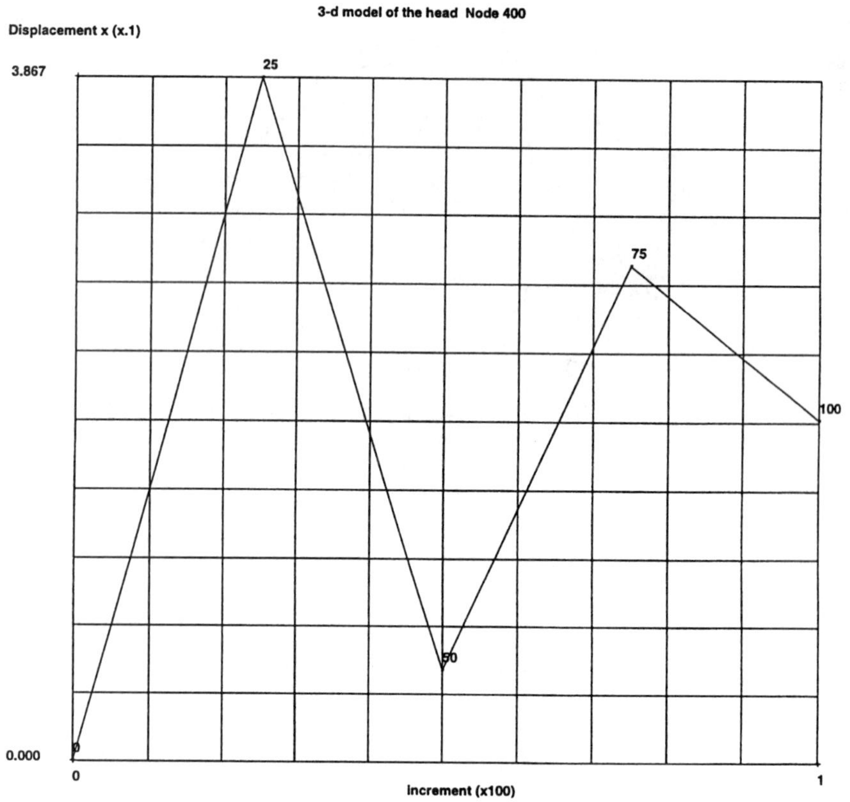

Figure 14-21 Displacement history of the cranium.

14-4 ORTHOPEDIC APPLICATIONS OF THE FINITE-ELEMENT METHOD

There are many areas in orthopedics that can benefit from a computational simulation and analysis. These include artificial joint replacement, head trauma, and the design of protective athletic equipment. This section is devoted solely to the study and the simulation of a human head. An important engineering design application is the design and construction of protective headgear for such applications as football, hockey, and bicycle racing. Also, the study of possible head trauma due to an automobile collision would aid the designers of automotive headrests. In all these cases it would be desirable to study the effect of certain forces on the behavior of the head. A study of the stresses in the cranium, the brain, and in the neck would be useful in designing protection during the above athletic applications.

Figure 14-15 shows a finite-element model of the cranium. Figure 14-16 shows a brain finite-element method model whose material properties simulate a gelatinous substance. Figure 14-17 is an finite-element method model of the spinal cord; Fig. 14-18

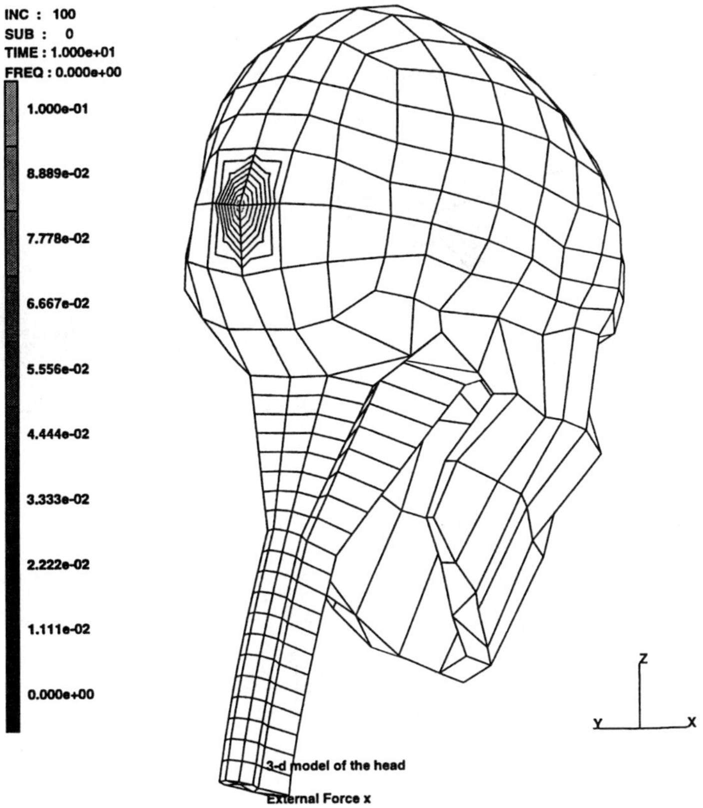

Figure 14-22 Displaced head set at increment 100.

models the disks of the spine. These human components are assembled into a computational model of the head area, which is shown in Fig. 14-19. The purpose of the analysis is to determine the maximum stress levels and locations within this model due to a trauma that is applied to the head in the form of a suddenly applied force system. The goal of the work would be to reduce the following detrimental effects:

1. Skull fracture
2. Brain trauma or tearing
3. Neck injury

A solid-element model of the orthopedic set, which is considered to be restrained at the base of the neck, is struck by a concentrated load. Figure 14-20 shows the restraint and the load. In this case, a nondynamic loading history was applied to the rear of the cranium to demonstrate the ability of the finite-element method to predict the levels of stresses and deformations that can be achieved. In the practical application, however,

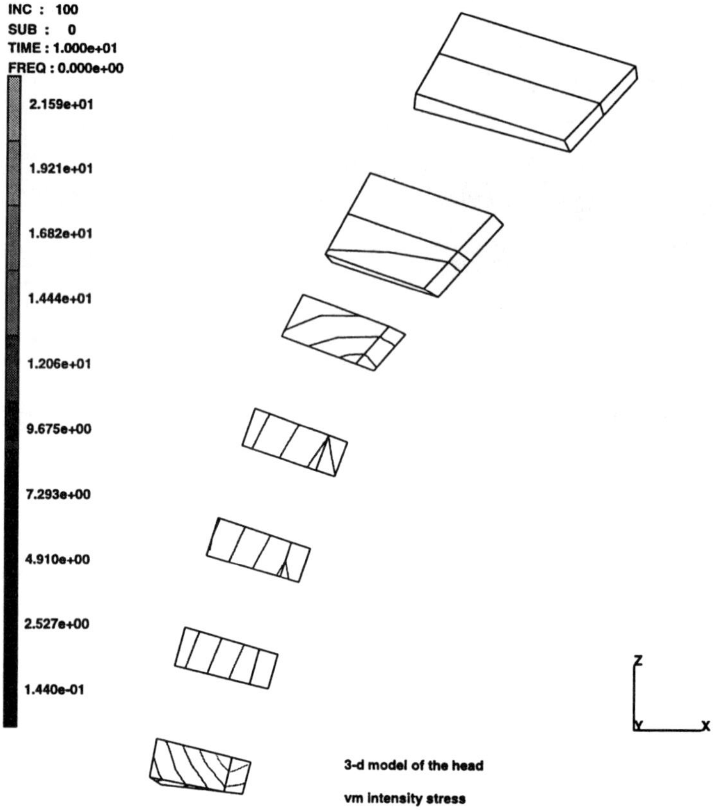

Figure 14-23 Stress distribution in the spinal disks.

dynamic stress analysis effects (such as inertia) would be required with the techniques of Chap. 11. The finite-element method data set, called **HEAD.DAT**, is shown in the Appendix 14.2. Figures 14-21–14-24 show the results of this analysis.

A study of these results clearly shows how the neck injury, commonly known as whiplash, may occur because of a blow to the cranium. Figure 14-22 shows how the head is deflected because of the sudden blow. This configuration can simulate the manner in which the head and the neck are driven during an automobile crash. Figure 14-23 shows the stress levels in the spinal disks that could result in permanent injury. Figure 14-24 shows a history of the stress distribution in the head at a specific point.

14-5 CONCLUSION

There is no need to offer readers any further evidence to substantiate the synergistic relationship that engineering enjoys with nontraditional fields. The areas of orthodontics and orthopedics are but a few of the areas that would benefit greatly from the application of computational methods to the solution of clinical problems. Medical applications such as dental implants, prosthetic hip design, and the flow of blood in vessels are a few

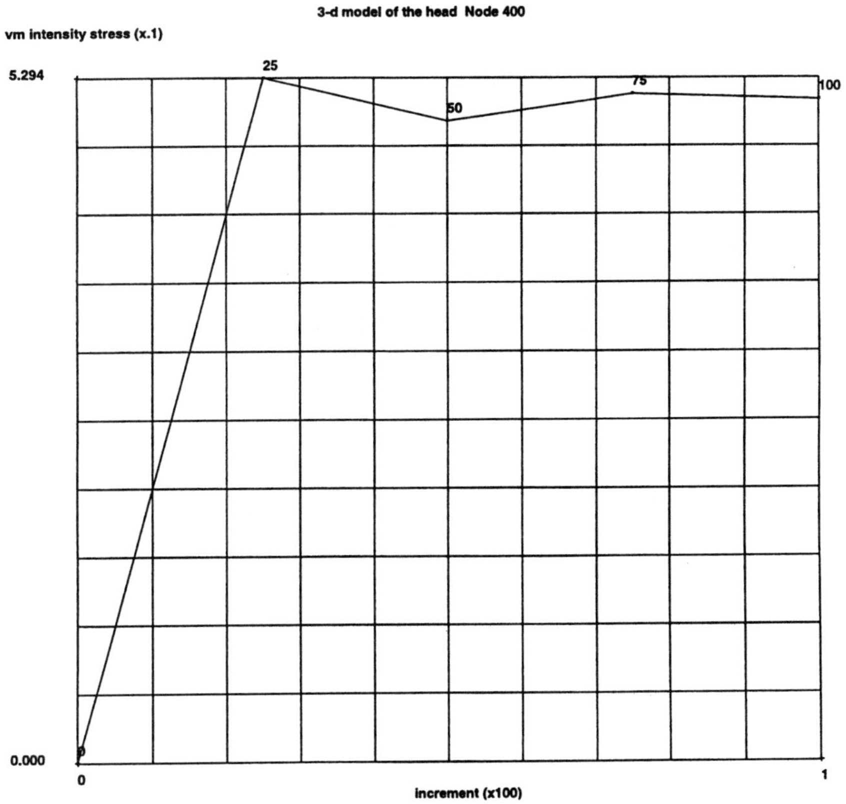

Figure 14-24 Stress history for the impacted head.

additional areas of potential computational application. The author leaves other areas to the imagination of the readers.

REFERENCES

1. Burstone, C. and Koenig, H. A. (1974). Force Systems from an Ideal Arch, *Am. J. Orthod.*, vol. 65, pp. 270–289.
2. Koenig, H. A. and Burstone, C. J. (19**xx**). Force Systems from an Ideal Arch—Large Deflection Considerations, *Angle Orthodontist*, vol. 59, pp. 11–16.
3. DeFranco, J. C., Koenig, H. A., and Burstone, C. J. (1976). Three-Dimensional Large Displacement Analysis of Orthodontic Appliances, *J. Biomech.*, vol. 9, pp. 793–801.
4. Goldberg, A. J., Burstone, C. J., and Koenig, H. A. (1983). Plastic Deformation of Orthodontic Wires, *J. Dent. Res.*, vol. 62, pp. 1016–1020.
5. Koenig, H. A. and Bolle, N. A. (1993). Non-linear Formulation for Elastic Rods in Three-Space, *Int. J. Non-Linear Mech.*, vol. 28, pp. 329–335.
6. Tanne, K., Koenig, H. A., and Burstone, C. J. (1988). Moment to Force Ratios and the Center of Rotation, *Am. J. Orthod. Dentofac. Orthopt.*, vol. 94, pp. 426–431.
7. Tanne, K., Koenig, H. A., and Burstone, C. J. (1989). Effect of Moment to Force Ratios on Stress Patterns and Levels in the PDL, *J. Osaka Univ. Dent. Sch.*, vol. 29, pp. 9–16.

INDEX